Science, Technology and Medicine in Modern History

General Editor: John V. Pickstone, Centre for the History of Science, Technology and Medicine, University of Manchester, England (www.man.ac.uk/CHSTM)

One purpose of historical writing is to illuminate the present. At the start of the third millennium, science, technology and medicine are enormously important, yet their development is little studied.

The reasons for this failure are as obvious as they are regrettable. Education in many countries, not least in Britain, draws deep divisions between the sciences and the humanities. Men and women who have been trained in science have too often been trained away from history, or from any sustained reflection on how societies work. Those educated in historical or social studies have usually learned so little of science that they remain thereafter suspicious, overawed, or both.

Such a diagnosis is by no means novel, nor is it particularly original to suggest that good historical studies of science may be peculiarly important for understanding our present. Indeed this series could be seen as extending research undertaken over the last half-century. But much of that work has treated science, technology and medicine separately; this series aims to draw them together, partly because the three activities have become ever more intertwined. This breadth of focus and the stress on the relationships of knowledge and practice are particularly appropriate in a series which will concentrate on modern history and on industrial societies. Furthermore, while much of the existing historical scholarship is on American topics, this series aims to be international, encouraging studies on European material. The intention is to present science, technology and medicine as aspects of modern culture, analyzing their economic, social and political aspects, but not neglecting the expert content which tends to distance them from other aspects of history. The books will investigate the uses and consequences of technical knowledge, and how it was shaped within particular economic, social and political structures.

Such analyses should contribute to discussions of present dilemmas and to assessments of policy. "Science" no longer appears to us as a triumphant agent of Enlightenment, breaking the shackles of tradition, enabling command over nature. But neither is it to be seen as merely oppressive and dangerous. Judgement requires information and careful analysis, just as intelligent policy-making requires a community of discourse between men and women trained in technical specialties and those who are not.

This series is intended to supply analysis and to stimulate debate. Opinions will vary between authors; we claim only that the books are based on searching historical study of topics which are important, not least because they cut across conventional academic boundaries. They should appeal not just to historians, nor just to scientists, engineers and doctors, but to all who share the view that science, technology and medicine are far too important to be left out of history.

Titles include:

Julie Anderson, Francis Neary and John V. Pickstone
SURGEONS, MANUFACTURERS AND PATIENTS: A Transatlantic History of Total Hip Replacement

Linda Bryder
WOMEN'S BODIES AND MEDICAL SCIENCE: An Inquiry into Cervical Cancer

Roger Cooter
SURGERY AND SOCIETY IN PEACE AND WAR: Orthopaedics and the Organization of Modern Medicine, 1880–1948

Jean-Paul Gaudillière and Ilana Löwy (editors)
THE INVISIBLE INDUSTRIALIST: Manufacture and the Construction of Scientific Knowledge

Jean-Paul Gaudillière and Volker Hess (editors)
WAYS OF REGULATING DRUGS IN THE 19TH AND 20TH CENTURIES

Sarah G. Mars
THE POLITICS OF ADDICTION: Medical Conflict and Drug Dependence in England since the 1960s

Thomas Schlich
SURGERY, SCIENCE AND INDUSTRY: A Revolution in Fracture Care, 1950s–1990s

Carsten Timmermann and Elizabeth Toon (editors)
PATIENTS AND PATHWAYS: Cancer Treatments in Historical and Sociological Perspective

Duncan Wilson
TISSUE CULTURE IN SCIENCE AND SOCIETY: The Public Life of a Biological Technique in Twentieth Century Britain

Also by Jean-Paul Gaudillière

INVENTER LA BIOMÉDECINE: La France, l'Amérique Et La Production Des Savoirs Du Vivant Après 1945 (*2002*)

LA MÉDECINE ET LES SCIENCES (*2006*)

Also by Volker Hess

DER WOHLTEMPERIERTE MENSCH. Wissenschaft und Alltag des Fiebermessens (1850–1900) (*2000*)

DIE CHARITÉ: Geschichte(n) eines Krankenhauses (*edited with Johanna Bleker, 2010*)

KULTUREN DES WAHNSINNS: Schwellenräume Einer Urbanen Moderne (*edited with Heinz-Peter Schmiedebach, 2011*)

Science, Technology and Medicine in Modern History
Series Standing Order ISBN 978-0-333-71492-8 hardcover
Series Standing Order ISBN 978-0-333-80340-0 paperback
(*outside North America only*)

You can receive future titles in this series as they are published by placing a standing order. Please contact your bookseller or, in case of difficulty, write to us at the address below with your name and address, the title of the series and one of the ISBNs quoted above.

Customer Services Department, Macmillan Distribution Ltd, Houndmills, Basingstoke, Hampshire RG21 6XS, England

Ways of Regulating Drugs in the 19th and 20th Centuries

Edited by

Jean-Paul Gaudillière
Director, CERMES3, CNRS, Paris

and

Volker Hess
Chair of the Institute for the History of Medicine,
Charité University Medicine Berlin

This volume is part of the activities of the
European Science Foundation Research Networking Programme DRUGS.

palgrave
macmillan

First published 2013 by
PALGRAVE MACMILLAN

Palgrave Macmillan in the UK is an imprint of Macmillan Publishers Limited,
registered in England, company number 785998, of Houndmills, Basingstoke,
Hampshire RG21 6XS.

Palgrave Macmillan in the US is a division of St Martin's Press LLC,
175 Fifth Avenue, New York, NY 10010.

Palgrave Macmillan is the global academic imprint of the above companies
and has companies and representatives throughout the world.

Palgrave® and Macmillan® are registered trademarks in the United States,
the United Kingdom, Europe and other countries

ISBN: 978–0–230–30196–2

This book is printed on paper suitable for recycling and made from fully
managed and sustained forest sources. Logging, pulping and manufacturing
processes are expected to conform to the environmental regulations of the
country of origin.

A catalogue record for this book is available from the British Library.

A catalog record for this book is available from the Library of Congress.

10 9 8 7 6 5 4 3 2 1
22 21 20 19 18 17 16 15 14 13

Printed and bound in Great Britain by
CPI Antony Rowe, Chippenham and Eastbourne

10 069918 7X

In memory of Harry Marks

Contents

List of Tables

List of Figures

Notes on Contributors

Christian Bonah is Professor of History of Medical and Health Sciences at the University of Strasbourg. He has worked on comparative history of medical education, the history of medicaments, as well as the history of human experimentation. Recent work includes research on risk perception and management in drug scandals and courtroom trials as well as studies on medical film. Selected publications include "Packaging BCG: Standardizing an Anti-Tuberculosis Vaccine in Interwar Europe," *Science in Context* 21(2), 2008, pp. 279–310, and coedited volumes on *Harmonizing Drugs: Standards in 20th Century Pharmaceutical History* (2009); *Meat, Medicine and Human Health in the Twentieth Century* (2010); and *Communicating Good Health: Health Education Films, Medicine, and the Cultures of Risk in Twentieth Century* (forthcoming).

Alberto Cambrosio is Professor in the Department of Social Studies of Medicine at McGill University. His most recent project focuses on the development of cancer genomics at the laboratory–clinical–biotech interface. He has coauthored, with Peter Keating, *Cancer on Trial: Oncology as a New Style of Practice* (2012); *Biomedical Platforms: Realigning the Normal and the Pathological in Late-Twentieth-Century Medicine* (2003; Ludwik Fleck Prize of Society for Social Studies of Science, 2005); and *Exquisite Specificity: The Monoclonal Antibody Revolution* (1995).

Maurice Cassier is a sociologist and a senior researcher for the National Centre for Scientific Research (CNRS). He is working in the Medicine, Science, Health and Society Research Centre (CERMES) in Paris. His work is focused on the appropriation of science, life, and drugs. His most recent publications include "Producing, Controlling and Stabilizing Pasteur's Anthrax Vaccine: Creating a New Industry and a Health Market," *Science in Context* 21(2), 2008, pp. 253–78; "New Enclosures and Creation of New Common Rights in the Genome and in Software," *Contemporary History*, 2006, pp. 255–71; "Flu Epidemics, Knowledge Sharing and Intellectual Property," in *Influenza and Public Health: Learning from Past Pandemics*, edited by Tamara Giles-Vernick and Susan Craddock, 2010, pp. 219–38.

Jean-Paul Gaudillière is a senior researcher at the Institut National de la Santé et de la Recherche Médicale, codirector of the Centre de Recherches Médecine Sciences Santé et Société. His research interests have first focused on the molecularization of biology during the 20th century and later on the reconfiguration of medical research after the Second World War. He is the author of several edited volumes on these issues and of *Inventer la Biomédecine* (2002). He is working on the history of biological drugs before the advent of gene-based biotechnology, with strong interests in the dynamics of knowledge production, clinical work, and market construction. He has recently coordinated *Drug Trajectories* (Studies in History and Philosophy of the Biological and Biomedical Sciences, 2005) and *How Pharmaceuticals Became Patentable?* (History and Technology, 2008).

Jeremy A. Greene is Assistant Professor of the History of Science at Harvard University, Instructor in Medicine at the Division of Pharmacoepidemiology and Pharmacoeconomics of Harvard Medical School, and Associate Physician at Brigham and Women's Hospital. His research interests focus on the role of therapeutics in the production, circulation, and consumption of knowledge, including one current project on the history of pharmaceuticals in global health and another on the development of generic drugs. His first book, *Prescribing By Numbers: Drugs and the Definition of Disease* (2007), traced the development of chronic disease categories as markets for risk-reducing pharmaceuticals; he has recently coedited (with Elizabeth Watkins) a collection titled *Prescribed: Writing, Filling, Using, and Abusing the Prescription in Modern America* (2012).

Volker Hess is Professor for the History of Medicine at the Charité Berlin and Head of the Institute for the History of Medicine. He worked on the introduction of the psychochemicals in the GDR and the marketing of modern pharmaceuticals. He is now chairing a research group funded by the DFG on the "cultures of madness" at the turn of the 19th century (FOR 1120). He published a book on the social and cultural history of science and medicine, "Der wohltemperierte Mensch" (2000), and coedited (with Christoph Gradmann) a collection of essays on vaccination in the 20th century (Science in Context 2008). He has been awarded an ERC European Advanced Investigator Grant in 2011 for focusing on the practices of medical note taking.

Axel C. Hüntelmann studied economy, history, and politics. After doctoral and postdoctoral positions in Berlin, Heidelberg, Frankfurt, and Bielefeld, he is now an academic assistant at the Institute for the History Theory and Ethics of Medicine at the Gutenberg University in Mainz. His work focuses on public health institutions between 1870 and the 1930s, laboratory animals, and serum and vaccine regulation in the German Empire and France. He has published articles, monographs, and anthologies on these topics: *Paul Ehrlich. Leben, Forschung, Ökonomien, Netzwerke* (2011); *Hygiene im Namen des Staates. Das Reichsgesundheitsamt 1876–1933* (2008); "Seriality and Standardization in the Production of Salvarsan," *History of Science* 48, 2010, pp. 435–460; "Dynamics of Wertbestimmung." *Science in Context* 21, 2008, pp. 229–52. Currently he is working on the cultural history of growth in 1770–1970.

Peter Keating is Professor of History at the Université du Québec à Montréal (UQAM) and a member of the Centre interuniversitaire de recherche sur la science et la technologie (CIRST). He is presently working on the development of genomic medicine in the field of cancer. His most recent book (with Alberto Cambrosio) is *Cancer on Trial: Oncology as a New Style of Practice* (2012).

Ilana Löwy is a senior researcher at Institut National de la Santé et de la Recherche Scientifique (INSERM), Paris. She is interested in the history of bacteriology and immunology, tropical medicine, history of oncology, and the intersection between gender studies and biomedicine. She had published, among other things, *Between Bench and Bedside: Science, Healing and Interleukin-2 in a Cancer Ward* (1996), *Virus, Moustiques Et Modernité: Science, Politique Et La Fièvre Jaune Au Brésil* (2001), and *Preventive Strikes: Women, Precancer and Prophylactic Surgery* (2009). Her most recent book is *A Woman's Disease: A History of Cervical Cancer* (2011). She is now studying the history of prenatal diagnosis.

Harry M. Marks was Associate Professor at the Institute of the History of Medicine, Johns Hopkins University. His research on the history of clinical medicine in 20th-century US inspired more than a generation of scholars in the US and in Europe. His studies on the uses of clinical and epidemiological data in contexts ranging from 18th-century inoculation controversies in France to drug regulation and social epidemiology in 20th-century US set standards. Harry Marks left us in January 2011 after a long battle with cancer – and we miss him. The contribution in this volume may be the recent publication highlighting his analytical brilliance and authority in the historiography of 20th-century medicine, clinical trials, and public health.

Donna A. Messner is Research Director at the Center for Medical Technology Policy in Baltimore where her research focuses, in part, on changing expectations for clinical evidence related to patient-centered outcomes research, efforts towards alignment of payer and regulator policies for clinical evidence, and impacts of these developments on pharmaceutical and device development. She was previously a fellow in the Penn Center for the Integration of Genetic Health Care Technologies at the University of Pennsylvania and a Gordon Cain Fellow in Technology, Policy, and Entrepreneurship at the Chemical Heritage Foundation in Philadelphia. Among other topics, she has published on drug and device regulation and reimbursement policies, the role of stakeholders in designing clinical research, regulation of direct-to-consumer genetic testing, and methods for clinical comparative effectiveness research.

Andrei Mogoutov is the founder-director of Aguidel Consulting, Paris and affiliated with IFRIS, Institut Francilien Recherche Innovation Société (Marne la Vallée, France). He has a PhD from Paris VI University (1994) in mathematical models of stochastic systems and disordered media. Since then, he has specialized on methodology and data analysis for the social sciences, which combine research activities, teaching, consulting, and software development. He is also developer of the ReseauLu software package for analysis of heterogeneous data.

Toine Pieters is Professor of the History of Pharmacy at the Department of Pharmacoepidemiology and Clinical Pharmacy and a senior fellow at the Descartes Centre for the History and Philosophy of the Sciences and the Humanities at Utrecht University (The Netherlands), and a senior researcher in the medical humanities at the VU Amsterdam Medical Centre. He has published extensively on the history of the production, distribution, and consumption of drugs and the interwovenness with economy, science, and the public sphere (among other things, "Standardizing Psychotropic Drugs and Drug Practices in the Twentieth Century: Paradox of Order and Disorder" in *Studies in History and Philosophy of the Biological and Biomedical Sciences,* 2011). His broader interests include the history of cancer and genetics, and the history of performance-enhancing drugs in sports. He is a research coordinator of multiple projects in the field of e-humanities.

Viviane Quirke is RCUK Academic Fellow and Senior Lecturer in Modern History and History of Medicine, in the Department of History, Philosophy and Religion, Faculty of Humanities and Social Sciences, at Oxford Brookes University. Her current research areas are the history of pharmaceutical R&D, focusing on the history of drug treatments for chronic diseases and the impact of drug safety regulation; the history of company–hospital relations and the development of clusters of innovation in Britain, France, and the United States; and the history of cancer chemotherapy, studied from the perspective both of the researchers and of the sufferers. Her forthcoming book is titled *Drugs, Disease and Industry: The Evolution of Pharmaceutical R&D at Imperial Chemical Industries.*

Stephen Snelders is Research Fellow at the Descartes Centre for the History and Philosophy of the Sciences and the Humanities at Utrecht University (The Netherlands), and Assistant Professor at the University Medical Centre in Utrecht. He has published extensively on the history of drugs in society and medicine, and is currently working on the history of medicine in the tropics, including the production, distribution, and use of drugs.

General Introduction

Jean-Paul Gaudillière and Volker Hess

On November 27, 1961, the pharmaceutical manufacturer Grünenthal withdrew its wildly successful sleeping pill – Contergan – from the market. At the time, Contergan was the most frequently used soporific in West Germany, raking in unprecedented profits for the manufacturer. Just the day before that, Germany's largest weekly newspaper, *Die Welt am Sonntag*, had published a short two-column piece with the headline "Tablets causing birth defects: Alarming suspicion of doctor on widely used drug." The suspicion had not appeared out of thin air. Nerve damage associated with this drug had been reported on over the previous months. In the United States, approval of the drug had been stalled for months as the manufacturer had not yet been able to provide the evidence regarding possible teratogenicity that the Food and Drug Administration (FDA) had requested. The first public speculation regarding the alleged connection between Contergan and the rise of defects in newborns' extremities appeared in the September 1961 edition of *Medizinischen Welt* – a suspicion that in the following weeks would be statistically substantiated by the pediatrician and geneticist Widukind Lenz. By the time the drug was recalled, however, in West Germany alone more than 4,000 children had been born severely deformed. Some 2,800 of these victims are still alive today (Kirk, 1999).

In the middle of December 1961, the public prosecutor's office began an inquiry into possible negligence on the part of the manufacturer. It would take more than five years, however – March 1967 – before any charges would be brought. The trial against a total of seven Grünenthal representatives began one year later. The charges were involuntary manslaughter, injury caused by negligence, injury caused by malicious intent, as well as breach of the German Pharmaceutical Law. More than 400 affected parents were involved as joint plaintiffs. In December 1970, after 283 days of deliberations, the proceedings were closed due to "minor liability" (*geringer Schuld*) on the part of the accused (Wenzel, 1968–71). Grünenthal offered up a sum of 100 million DM for damages, which would be managed by the *Hilfswerk für behinderte Kinder* foundation, on condition that no further claims would be made on the company. In 1978, the Reform of the Pharmaceutical Law (*Gesetz zur Neuordnung des Arzneimittelrechts*) came into effect in West Germany, strengthening dramatically the requirements for the manufacture and approval of drugs for the market. Despite the national government having provided it with ample support, the foundation's funds dried up in 1997, and the state alone is now carrying the burden of annuity payments to the victims.

This story, which has come to be known as "The Thalidomide Disaster" or "The Contergan scandal," (Friedrich, 2005) is instructive for the purpose of this volume for a number of reasons.

First, it demonstrates the failure of traditional forms of state regulation in the age of industrial research, manufacture, and marketing of pharmaceutical drugs. In 1957, when Contergan was first put on the German market, the manufacturer was only required to register the drug. It was up to the company when and how (if at all) it would test its new product. The German Pharmaceutical Law that was first introduced in 1962 would also have been unable to prevent the scandal. In the United States and in East Germany on the other hand, where drug-approval procedures were much more restrictive, there were scarcely any victims.

Second, the Contergan scandal made it clear just how complex the medical market had become. In the course of the Contergan trial, some 29 experts appeared to present their – at times very different – views; about 350 witnesses were questioned; and no less than 20 attorneys on the side of the defendants were involved. Due to public demonstrations, the proceedings were removed to the town casino of a former mine. Radio, TV, and newspaper reporters followed the events warily, and they had good reason to: at the high point of the trial, the leading defense attorney was appointed Minister of the Interior in the local federal state (*Bundesland*). Representatives of the affected parents were at loggerheads, and the leading media reporter was exposed as a Grünenthal employee. On one side of the bargaining table, there were not only the accused representatives of the pharmaceutical company, but also its owners, its top managers, and the doctors and scientists of its "cosmetic–pharmacological laboratories." On the other side, there were not just the affected parents, but also doctors, scientists, lawmakers, and representatives of the state health administration.

Third, this case made it clear that drug regulation is by no means a task that can be left to the state alone. Apart from the question of causality, the central question that remained was that of the means by which (if any) the pharmaceutical company would guarantee the quality of its product. Representatives of the medical profession also appear to have felt obliged to step in. In 1963, the drug commission of the *Deutschen Ärzteschaft* (German Medical Association) put a "Registration form for undesired drug effects" into circulation. For the German public, the figure of an NGO appeared as player for the first time in the form of the *Bundesverbandes der Eltern körpergeschädigter Kinder* – *Conterganhilfswerk* (national organization for parents of injured children – Contergan help organization). The organization had regional and local as well as national chapters, helping to make the cause known and to mobilize public attention. The Contergan scandal was consequently not simply the first worldwide case of a drug having a dramatic and unexpected effect. It also presented a typology of the various forms in which the production, marketing, and application of drugs is regulated today. Using John Pickering's concept of "ways of knowing," we can ideally typically mark out four ways of regulating.

State or administrative intervention is the traditional form of drug regulation. In the case of Contergan, however, the health administration had no legal means at its disposal. The summer before the recall, the *Bundesgesundheitsamt* (National Health Authority) had already recommended that the sale of Contergan be restricted to prescriptions. At that point, this was only a recommendation, one which the manufacturer did not heed until after reports of unforeseen neurological damage had built up massively. Even before November 26, the Düsseldorf Internal Ministry had asked the manufacturer to remove all thalidomide preparations from the market as quickly as possible. The company had declined, however, pointing to the absence of any binding legal regulations and even threatened with counter demands for compensation, should it be further pressed. It was not until the first amendment to the Pharmaceutical Law had been enacted in 1964 that there was any definition of requirements a manufacturer must fulfill in order to have a drug registered. And not until 1976, due to public pressure induced by the Contergan scandal, did West Germany introduce a formal drug-approval procedure, whereby the manufacturer was required to demonstrate the quality, effectiveness, and innocuousness of its product (Daemmrich, 2004). With the introduction of a reporting procedure for undesired drug effects, a classic instrument of state drug regulation was finally completed.

The Contergan scandal put the drug industry into the pillory, in that it was held accountable for the quality of its product. It was expected (albeit tacitly) that the manufacturer would make use of animal and human experiments at its disposal in order to ensure the toxicological and teratological innocuousness of the substance in question. In the case of Contergan though, it was precisely these quality controls that the manufacturer had used to pitch its product: laboratory tests had produced practically no dosage that could be considered lethal. And it was the economic effects of a ban (loss of jobs, stifling of innovation, government intrusion into the realm of free enterprise, etc.) that the manufacturers and manufacturers' organizations held up as arguments against the critical inquiries of doctors, journalists, and state health authorities. Their suggestion instead was to put a warning on the package regarding the use of the drug during pregnancy.

The Contergan scandal also shed light on the professional way of regulating. It was in precisely this discussion of the side effects of Thalidomide that it became clear just how troublesome, protracted, and open to influence this traditional path of the professional way of regulating truly was. Now, however, doctors' organizations became more active. The *Arzneimittelkommission der Deutschen Ärzteschaft* (Drug Commission of the German Medical Fraternity) actually went back to a pre-World War I German Society for Internal Medicine initiative. The commission saw its main task as the publication of an index of drug specialties, judged as positive, negative, or dubious. After World War II, the commission was reconstituted, but it was not until after the Contergan scandal that any sort of systematic inquiry procedure was put into place. The procedure consisted of a central register in which doctors could record observed drug incompatibilities and unforeseen occurrences (Schröder, 2003). With recourse to the corpus of expertise of all specialists involved, the reports could be evaluated, and recommendations and regulatory suggestions regarding indication and dosage could be made. Within the last decade, the working out of guidelines for regulation and therapeutic use has also come to play an important role.

Fourth, the role of the public became visible in the Contergan scandal. First of all, the epidemiologic proportions of the damage made visible the extent to which drugs had become an everyday consumer good. Also, this was widely reported in the media, which put both the manufacturer and the health authorities under pressure. In addition to the passive market public and the critical voice of a media public, the Contergan case ushered in a new, postwar-specific form of public (Schwerin, 2009). A lobby and some pressure groups collected and presented the observations and reports that provided the disaster with statistical evidence. These societies for the affected, self-help groups, and consumer-protection organizations formed a part of the public sphere that differed dramatically from Habermas's general and diffuse public sphere. It intervened actively – and still intervenes – in questions of drug regulation, in the form of grassroots movements, self-help groups, and NGOs.

Last but not the least, the Contergan case showed the increasing role of the court in modern drug regulation. As litigation did not lead to adjudgment of the company, one may suppose that it is not necessary to distinguish the court as a specific, separate way of regulating. The three-year lawsuit, however, introduced new aspects into the regulatory process. Responsibility and causality were shaped by the courts in a very different way from other arenas, for instance, the laboratory. The court may have offered other regulating actors a stage for performing their particular aims and claims of truth. The court performance, however, operated according to its own rule, free of professional or administrative control. It followed its own logic rooted in the codes of procedure as well as in performative negotiations. For this reason, the court and jurisprudence have to be included in the various ways in which drugs are regulated.

Studies of drug regulation are far from a blank territory in the history of science, medicine, and technology. Among industrial goods, pharmaceuticals are certainly those for which the concept of regulation has been most often applied, even though standards and the question of normalization have often been included in historical studies of techno-scientific domains as diverse as the automobile industry, the electrical and electronic industry, or chemical production. One reason for the pervasiveness of the idea of regulation in the case of pharmaceuticals is certainly that therapeutic agents are not like other goods. Their sale and purchase constitute a very distinct market, the control and oversight of which is to protect "public health" – to avoid poisoning as well as the forming of monopolies that might threaten reasonable conditions of access to them – and has been considered a duty of the state since the early nineteenth century, when the pharmaceutical craft was being transformed into a profession.

As a consequence, the historiography of drugs has given regulatory issues a rather institutional meaning. In this perspective, the words "drug regulation" are usually employed with a narrow understanding, which focuses on actions taken by the government or other political bodies to control the activities of drug makers (Marks, 1990; Abraham and Reed; 2002; Timmermans and Berg, 2003; Daemmrich, 2004). Studies then focus on the administrative features that have been used to tame the market or safeguard production, that is, registration, permits, control of composition, mandatory trials of efficacy (Marks, 1990; Keating and Camprosio, 2007; Valier, 2011). Industrial drug making, as it developed within large capitalistic corporations, has in the twentieth century been the focus of analysis for highly visible conflicts between firms, physicians, and public authorities who were caught in a web of market forces and public-health defense.

Drug regulation is of course not just a tale of the progress of better control. Recent studies have shown that the influence of the FDA – an American agency usually considered the reference regulatory body – over the actual practices of invention, production, sales, prescriptions, and clinical use is much more limited that one might consider on the basis of its legal mandate, and this remains a matter of tension (Carpenter, 2010; Tobell, 2008; Marks in this volume). Moreover, as comparisons with other countries easily show, the case of the FDA is far from being a general model. The example of the Thalidomide scandal suggests that this form of state administration has been associated with or superimposed on other forms of collective management, of "regulation," which have targeted not only sales and marketing but the entire trajectory of drugs, that is, research and development, mass production and preparation, circulation and prescription, as well as application in the clinical framework. Stemming from a workshop on these themes, the essays assembled in this volume share the perspective that the historiography of science, technology, and medicine needs a broader approach toward regulation.

Regulation is usually associated with the question of controlling/taming the market – with strong discourse about the "capture" of regulatory bodies by firms and corporations.

But the market is not a given actor or a stable entity in the same way as the interests of economic actors are historically constructed. Various actors, not only companies, are involved in the "making" of markets. Their development operates at several levels, among which is (narrowly defined) administrative regulation. It is for instance remarkable that the role of the health-insurance system and its decisions regarding pricing, reimbursement, recommended indications, and so on have been granted little attention in the historiography in spite of the fact that they are obviously shaping the markets and uses of therapeutic agents, and that such shaping reveals huge differences in time and place, for instance, along national lines.

Our contention is that various levels of comparison must be taken to explore the relations between research, therapeutic intervention, and marketing. Given the emphasis placed on administrative and legal tools of regulation, it is not surprising that comparisons of national frameworks have – up to the present – been given the highest priority. In agreement with a strong focus on the FDA, studies of "drug politics" have for instance contrasted the adversarial and highly political regulatory culture prevailing in the United States with the rather consensual collaboration between pharmaceutical firms and physicians' professional bodies dominating the German scene (Daemmrich, 2004). The essays gathered in this book follow this pattern but add cross-national perspectives and comparisons between periods, cultures, institutions, and therapeutic agents.

Contemporary studies of science and technology have often made the point that the production and uses of knowledge are interactive, undetermined, and complex processes. The concept of "regulatory science" is usually employed to describe the investments of regulatory bodies, starting with state agencies, in the development of measurement techniques, methodological tools, and decision-making protocols for the standardization, authorization, or control of technological goods. Looking at the regulation of chemicals in the postwar era, science, technology, and society analysts have accordingly used the concept in a broader sense, including the entire spectrum of expert activities conducted in collaboration among scientists, state officials, and politicians; this includes laboratory studies for regulatory purposes, testimonies within political fora such as congressional hearings, as well as court rulings of techno-scientific cases. Within this perspective, "regulatory science" refers to the production of knowledge for administrative, political, or judicial action, but it also implies the idea that this regulatory perspective "feeds back" to science, leading to the development of specific forms of knowledge. Toxicology and the environmental regulation of chemicals are among the most frequent examples of this dual relationship (Jasanoff, 1990; Jasanoff, 1992; Halffman, 2003; Reinhardt, 2008). This must also be granted when therapeutic agents, their invention, mass uses and surveillance come to the fore. In the last two centuries, drugs have become central elements in complex health systems. This complexity needs to be approached from the *longue durée*. It is the belated product of a long-term transformation, which began in the second half of the nineteenth century when the industrialization of drug making coincided with both the rising influence of experimental sciences and laboratory practices in medicine, and the emergence of hospitals as the place where insurance-based health care accessible to the working class would be routinely provided (Cooter and Pickstone, 2000; Jorland *et al.*, 2005). Although state and professional forms of regulation can be traced back to the early nineteenth century, this conjunction deeply affected, that is, *diversified* regulatory practices, setting the pace for new forms of control focusing on standards, homogeneous protocols, or statistical efficacy (Gradmann and Hess, 2008).

The idea of various "ways of regulating drugs" play a central role in the studies gathered in this volume. This typology is not an attempt to accommodate the complexities of regulation within a fixed set of structures and boundaries. It is rather a heuristic model, focusing on the dynamics of social action, useful for bringing some order into the multiplicity of the regulatory practices mentioned above. The concept of ways of regulating seeks to bring to light the internal logic of specific combinations of practices and procedures, describing the various rationalities underpinning the management of therapeutic agents. The approach therefore links the distinct social worlds involved in regulation, the forms of evidence and expertise they mobilized, and the means of intervention they choose or establish (see Table 0.1).

Accordingly, ways of regulating should not be taken as permanent and self-contained social entities whose nature determined either the trajectory of isolated therapeutic agents or the collective fate of drugs in a given time and place. Like Pickstone's *Ways of Knowing* (2000), ways of regulating are sociohistorical products: they have not existed in all eternity; they appeared at some point in history and none have ever disappeared, but their arrangements and articulation have been utterly variable. Ways of regulating are therefore categories or frames used in thinking about, choosing between, and organizing practices that are not "given," but constructed, in a given situation, each representing a "grammar of action" that works in combination rather than in isolation.

In contrast to ways of knowing, ways of regulating are not aligned along a chronological order; they operate simultaneously and in interaction but according to specific hierarchies, which have changed with time and place. Administrative interventions regarding the making and the uses of drugs are for instance as old as the modern state, but their impact and relations with other regulatory actors were radically transformed in the twentieth century, revealing the complexity of pharmaceutical markets. To think about ways of regulating means to think about interrelations and specific patterns of action. Here, one should consider the analogy with the concept of "boundary work," which also targets both separation and interplay (Gieryn, 1983).

The dynamics of a peculiar form of regulation should therefore be discussed at different levels and should take into account the following questions: Which values guide the regulation process? Which problems or adverse practices are targeted? Who are the most important actors? What are the forms of evidence accepted in decision making? Which regulatory tools are mobilized to oversee and control the fate of drugs? Given these questions, our proposal is that drug regulation can be characterized by the interaction of five ways of regulating.

The *professional way of regulating* will be dealt with in three chapters that cover its roots in the early nineteenth century to the contemporary system of drug monitoring introduced in the 1970s. In continental Europe, regulation evolved out of a state delegation, which granted both the physicians' authorization and pharmacists' monopoly over the sale of therapeutic agents.[1] Professional regulation thus went hand in hand with the administrative – both were intertwined in multiple ways. The first approach to this field highlights two models of how professional regulation was linked to an early form of drug approval in the first half of the nineteenth century. The contribution of *Volker Hess* compares the Prussian model with the French one. In Prussia, state authorities recruited professionals to evaluate and to test new drugs. Experts, however, had only to advise the medical officers who made the decision of acceptance or refusal. In contrast, the French state handed over the approval process to the Royal Academy of Medicine, which represented the corporative organization of French medicine. Thus both the Prussian and the French model of regulating drugs were based on professional expertise. Each, however, performed professional regulation in different manners: In France, decision making was executed as independent acts. The corporative organization acted far from state power (even if the state formally delegated its power to the Academy). In contrast, the Prussian model seemed to domesticate professional power in the form of the Scientific Department, which advised but did not decide on the official approval. This argument is developed in the chapter on the institutionalization of sera control, which also deals with the interrelation between professional and administrative aspects.

The second example of professional regulation is found in the chapter by *Harry M. Marks*, which demonstrates how professional governance in the United States led to "clinical autonomy" being valued as an absolute reference. Following the handling of adverse drug reactions in the United States from the 1940s to the end of the twentieth century, Marks shows how the American Medical Association (AMA) remained a major actor in the field while, in contrast to what is usually assumed, the FDA did not gain much power to change (or influence) medical practice. It was the AMA that first managed the registration of adverse drugs and later established a postmarketing surveillance system. These efforts supplemented the FDA's policy by a number of registries run by specialty societies.

The chapter by *Alberto Cambrosio* and *Peter Keating* highlights a third aspect of the professional regulation. Analyzing the clinical trials of anticancer drugs, they conceptualized regulation as the process of "progressive informational enrichment." From the perspective of professional regulation, their contribution focuses on how the protocols for approval of cancer drugs were translated into clinical research strategies and vice versa. The chapter analyzes the interplay between clinical work and standardizing protocols. The authors look at the outcome of interactions between professional bodies and drug companies in terms of therapeutic regimens stabilized by the clinical networks involved in the testing of new cancer drugs and by the same token redefining the boundaries of diseases.

The second way, that is, *the industrial way of regulating*, is investigated in three chapters, all of which look at both the internal dynamics of the activities in industrial firms (research and beyond) and the interplay between companies and physicians, keeping the more classical discussion of lobbying and relations with regulatory authorities in the background.[2] *Jean-Paul Gaudillière's* chapter focuses on the industrialization of preparations that were part of the classical pharmacopoeia in the interwar period following the case of the companies Madaus and Dause. The aim was to explore the transition from a professional way of regulating to an industrial one, associated with the development of mass-produced and widely marketed specialties. The transition is analyzed at the level of research practices, standardization, quality control, and clinical evaluation. *Jeremy Greene's* contribution explores interrelations between consumerism and administrative regulation, which were used by the company to keep the disputed drug in the market. The strategies of the FDA as well as the counterstrategies of the industry resulted in a situation that left patients concerned and uncertain.

Finally, *Viviane Quirke's* study of the Imperial Chemical Industries' new organization of research in the 1960s and 1970s provides an enlightening example of the administrative way of regulating, that is, of how changing administrative practices (in this case the 1962 FDA reform leading to premarketing assessment of both toxicity and efficacy) impacted industrial practices, facilitating the adoption of screening as a regulatory strategy based on large-scale and standardized testing of chemicals, in this case beta-blockers. Her study shows that "industrial regulation" is not only about "quality control" or scientific marketing but includes safety control. The Imperial Chemical Industries enlarged biological–pharmacological testing, which was adopted as a response to the thalidomide affair and viewed as a requirement for entering the US market. Industrial regulation thus appears as a cost/benefit balancing with "safety" exclusively defined on the basis of in-house laboratory testing without much clinical reporting.

Much has been written on the transformation of *the administrative way of regulating* during the postwar years, especially in the public debates and therapeutic crisis, which led to the reinforcement of state regulatory bodies, to the generalization of premarketing evaluation, and to the transformation of controlled clinical trials into a legal means of proof. Four papers in the collection look at twentieth-century administrative regulation and discuss its limitations and boundaries.

Volker Hess offers a long-term perspective, discussing the interplay between the professional and administrative ways of regulating and the nineteenth-century configuration of their frontiers. The case of Prussia with its centralized and strong administration shows how the overall delegation of expertise and decision-making power granted to professional experts prompted a system of state surveillance based on officially appointed committees.

Axel Hüntelmann's chapter focuses on the development and control of Salvarsan in Germany. It shows how the model of expertise and testing by a "public-industrial" venture led by Paul Ehrlich was adapted from the experience of antidiphtheria serum control and resulted in a new form of regulatory science, which linked chemical and bacteriological screening.

Reevaluating the role of the FDA, the late *Harry Marks'* chapter – which could unfortunately not be revised – makes a strong plea for the idea of loose administrative regulation. In spite of all the gatekeeping powers granted to the US drug agency, in order to control a free health market, this body has remained a weak power. According to Marks, the agency has been unable to modify medical judgment and clinical practices, for instance, for off-label uses, as it did not intend to infringe in any way on the autonomy of medical practitioners. The chapter convincingly suggests that many paradoxes associated with the historical trajectory of the FDA's regulatory tools (like its emphasis on labels and information) have their roots in this situation.

The case study presented by *Donna Messner* focuses on the historical circumstances of the mid-1980s, when the first antiretroviral drug (azidothymidine (AZT)) received marketing authorization without qualifying for the FDA's standard evaluation procedures. This case provided the blueprint for introducing so-called fast-track drug approvals, a bypass mechanism to accelerate the marketing of drugs against life-threatening diseases. This was followed by another change of authorization criteria, namely, the acceptance of biological surrogate markers. She analyzes this turn in drug regulation in view of previous and ongoing efforts to reform the FDA and of the 1980s political climate favoring deregulation. The case thus highlights the way in which the administrative way of regulating responded to changes in the industrial way.

Analyzing the development of generics for the treatment of AIDS and Chronic Myelogenous Leukemia (CML) in Brazil and India, *Maurice Cassier* shows that patents not only are the means of the industrial way of regulating, but may also become tools of administrative regulation. This was the case in Europe before the generalization of strong intellectual property rights on drugs but is today becoming increasingly important in developing countries. The Brazilian trajectory of AIDS tritherapy is accordingly rooted in mandatory licensing and copying of drugs in the name of public health and free access. This strategy was supported by a new alliance blurring the boundaries between the Brazilian administrative and industrial sectors.

Finally, *Alberto Cambrosio* and *Peter Keating* look at the very same tools from the perspective of biomedical knowledge and clinical research. They reveal that the machinery of cancer trials, which expanded in the 1960s and 1970s, at first within the National Cancer Institute (NCI), did not target isolated and stable substances as their regulatory understanding would have it, but instead consisted in implementing and modifying complex regimens of treatment, which became the real targets of investigation.

The invention of the public sphere is not a recent development. Even though the policymakers of the nineteenth century imagined a public sphere (which became real in the administrative efforts), the public did not become a powerful concept until the second half of the twentieth century, when the figures of the consumer and the user, and later of collectively organized patients, became highly visible in the drug arenas.[3]

The public way of regulating is approached in four chapters dealing with questions of marketing authorizations, patients' expertise and voices, NGO's new roles.

First, *Stephen Snelders* and *Toine Pieters* follow the ups and downs in the evaluation and marketing authorization of Halcyon in the Netherlands, the United Kingdom, and the United States. They underscore the prominent role of the media in generating legitimacy or distrust, prompting regulatory authority and company choices.

Second, *Jeremy Greene* uses the letters written by anxious patients and worried doctors to the FDA after the news scandalizing the adverse effects of Orinase. Mobilized by the press, patients' voices substantiated the consumer and its increasing role in the regulatory game. Both the administrative authorities and the drug companies tried to influence the patient's voice – with different intentions and results.

How the user became active and expert is also the question raised by *Ilana Löwy*'s chapter, which follows the issue to the present. Focusing on the difficulties in persuading women (and doctors) to use tamoxifen to prevent breast cancer at the turn of the twenty-first century, her case study illustrates the rise and role of the social movement in the public way of regulating. What the reader can see here is the increasing role of the user (who is not necessarily a patient). Women activists developed forms of counter expertise that led to a reexamination of the risk/benefit ratio. Balancing the message of powerful and appealing medical promises with a compelling message of caution, iatrogenic risk, and medical hubris, women users have invented a new mode of engagement, although Löwy's chapter cautiously questions their influence.

Maurice Cassier's study sheds light on the emergence of a new type of actor in the field of international drug regulation: NGOs. These are becoming a serious partner for the administrative authorities in India for developing strategies in marketing generics, as demonstrated in his case study.

Courts and their decisions and judicial proceedings in general have played a role in drug regulation throughout all modern times. In the nineteenth century and the first half of the twentieth century, trials involving pharmacists, drug makers, and physicians have been far from rare, playing a role in settling specific cases, shaping concepts of professional responsibility, or advancing changes in the laws governing therapeutic agents, their production, and uses.[4] The chapters of this collection discuss the more recent roles of the court in the *juridical way of regulating*. They support the idea that these have changed significantly over the past decades, partly as a consequence of interplay with other forms of regulation (the public way of regulating, for example), but also due to the dynamics of the juridical regulation.

Christian Bonah's study of the affair originating in the production and marketing of the antibacterial and dermatological drug Stalinon in France in the 1950s and of the associated criminal trial thus provides a nice example of the interplay between the transformation of a public scandal into an exemplary judicial case that contributed to the introduction of another category of health-related risk, that is, iatrogenic risk, and to changes of the law itself, that is, the reform of administrative drug regulation, which occurred in parallel with the trial. More importantly, the Stalinon trial reveals the transition from a situation in which the court operated as a mediating scene for all the other ways of regulating into a situation in which court proceedings constitute a new way of both compensating and regulating.

Maurice Cassier's chapter, finally, approaches the juridical regulation by focusing on the new role of both court cases and appeal procedures specific to the administrative bodies granting patents. Discussing the case of AIDS and the production of antiretroviral generic therapies in Brazil and other countries of the South, he shows how judicial practices are becoming regulatory tools in their own right.

Although the case studies gathered in this volume all take into account the multiplicity of actors and *dispositifs* constituting twentieth-century regulatory systems, they do not simply stress the novelty, the importance, or the benefits of such diversity but recognize the unequal abilities of actors to shape situations and to control the fate of drugs. Power gradients are at the very core of drug regulation and must be considered as such. Defining

Table 0.1 Five ways of regulating drugs in the 20th century

Way of regulating	Professional	Administrative	Industrial	Public	Juridical
Aims, values	Compliance, competency	Public health, efficacy, access	Productivity, profit, quality	Individual choices, quality of life	Responsibility, rights, guilt, compensation
Forms of evidence	Pharmacology, animal models, dosage, indications	Statistical controlled trials	Animal testing, market research, cost–benefit analysis,	Observational epidemiology, risk–benefit analysis	Adversarial debates, exemplary case
Social sphere	Corporation, scientific societies	Agencies, governmental committees	Firms, business associations	NGOs, patients and consumers, medias	Law professionals and legal associations
Regulatory tools	Pharmacopoeia, prescription, guidelines	Marketing permits, intellectual property rights, public statements, labeling	Intellectual property rights, quality control, scientific publicity, package inserts	Postmarketing surveillance, court decisions	Jurisprudence (including penalties), justifiability, court room proceedings

power relationships as only financial would, however, be a caricature. It is a general feature of the construction of economic markets that it cannot proceed without administrative interventions and public-health policies, which build the terrain on which capital can be invested and goods traded. In the case of the mass production of drugs, this interplay between economic and political aspects may be extended to cultural aspects as the construction of therapeutic markets rests on local visions of diseases that not only define their nature but also their hierarchy.

Notes

1. See Hickel, 1973; Weisz, 1995; Bonah, 2005; Gradmann and Hess, 2008; Balz, 2010.
2. See Liebenau, 1987; Wimmer, 1994; Greene, 2007; Rasmussen, 2008; Chauveau, 1999; Sinding, 2002; Quirke and Gaudillière, 2008; Bächi, 2009;Gaudillière, 2010.
3. Epstein, 1996; Dalgalarrondo, 2004; Timmermans and Berg, 2003; Krieger *et al.*, 2005, Epstein, 2007; Watkins, 2007;Gaudillière, 2010.
4. For the thalidomide trial in Germany see above; for other examples see Bonah *et al.*, 2003; Offit, 2005.

Bibliography

Abraham, J. & Reed, T. 2002. "Progress, innovation and regulatory science in drug development: The politics of international standard-settings." *Social Studies of Science*, 32, 337–69.

Bächi, B. 2009. *Vitamin C für alle: Pharmazeutische Produktion, Vermarktung und Gesundheitspolitik (19331953)*. Zürich, Chronos Verlag.

Balz, V. 2010. *Zwischen Wirkung und Erfahrung – eine Geschichte der Psychopharmaka. Neuroleptika in der Bundesrepublik Deutschland, 1950–1980*, Bielefeld, Transcript Verlag.

Bonah, C. 2005. "The 'experimental stable' of the BCG vaccine : safety, efficacy, proof and standards 1921–1933." *Studies in History and Philosophy of Biological An Biomedical Sciences*, 36, 696–721.

Bonah, C. , Lepicard, E. & Roelcke, V. (eds.) 2003. *La médicine expérimentale au tribunal : implications éthiques de quelques procés médicaux du XXe siécle européen*, Paris: Ed. de Archives Contemporaines.

Carpenter, D. P. 2010. *Reputation and Power: Organizational Image and Pharmaceutical Regulation at the fda*. Princeton, Princeton University Press.

Chauveau, S. 1999. *L'invention pharmaceutique : La pharmacie française entre l'Etat et la société au XXe siècle*, Paris, Institut d'édition sanofi-synthélabo.

Cooter, R. & Pickstone, J. V. (eds.) 2000. *Medicine in the Twentieth Century*, Amsterdam: Harwood Academic Publishers.

Daemmrich, A. 2004. *Pharmacopolitics: Drug regulation in the United States and Germany*. Chapel Hill, The University of North Carolina Press.

Dalgalarrondo, S. 2004. *Sida: la course aux molécules*, Paris, Éditions de l'École des hautes études en sciences sociales.

Epstein, S. 1996. *Impure Science: aids, Activism, and the Politics of Knowledge*, Berkeley, University of California Press.

Epstein, S. 2007. *Inclusion: The Politics of Diference in Medical Research*. Chicago, London, Chicago University Press.

Friedrich, C. 2005. "Contergan. Zur Geschichte einer Arzneimittel-Katastrophe." *In:* Zichner, L., Rauschmann, M. & Thomann, K.-D. (eds.) *Die Contergan-Katastrophe – Eine Bilanz nach 40 Jahren*. Darmstadt.

Gaudillière, J.-P. 2010. "Une marchandise scientifique? Savoirs, industrie et régulation du médicament dans l'Allemagne des années trente." *Annales. Histoire, Sciences Sociales*, 65, 89–120.

Gieryn, T. F. 1983. "Boundary-work and the demarcation of science from non-science: Strains and interests in professional ideologies of scientists." *American Sociological Review*, 48, 781–95.

Gradmann, C. & Hess, V. (guest eds) 2008. "Vaccines as Medical, Industrial, and Administrative Objects," *Science in Context*, 21, 2.

Greene, J. A. 2007. *Prescribing by Numbers: Drugs and the Definition of Disease*. Baltimore, John Hopkins University Press.

Halffman, W. 2003. *Boundaries of Regulatory Science. Eco/Toxicology and Aquatic Hazards of Chemicals in the us, England and the Netherlands, 1970–1995*. Ph.D., University of Amsterdam.

Hickel, E. 1973. *Arzneimittel-Standardisierung im 19. Jahrhundert in den Pharmakopoen Deutschlands, Frankreichs, Großbritanniens und der Vereinigten Staaten von Amerika*. Stuttgart, Wissenschaftliche Verlagsgesellschaft mbH.

Jasanoff, S. 1990. *The Fifth Branch: Science Advisors as Policy Makers*. Cambridge, Harvard University Press.

Jasanoff, S. 1992. "Science, politics, and the renegotiation of expertise at EPA." *Osiris 2nd series,* 7, 195–217.

Jorland, G. , Opinel, A. & Weisz, G. (eds.) 2005. *Body Counts: Medical Quantification in Historical & Sociological Perspectives= La Quantification médicale, perspectives historiques et sociologiques; [Proceedings of the Symposium La quantification dans les sciences médicales et de la santé: perspective historique, held at Musée Claude-Bernard, in Saint-Julien-en-Beaujolais, France, Oct 24–26, 2002],* Montréal: McGill-Queen's University Press.

Keating, P. & Camprosio, A. 2007. "Cancer clinical trials: The emergence and development of a new style of practice." *Bulletin of the History of Medicine,* 81, 197–223.

Kirk, B. 1999. *Der Contergan-Fall : eine unvermeidbare Arzneimittelkatastrophe? – Zur Geschichte des Arzneistoffs Thalidomid,* Stuttgart.

Krieger, N. , Löwy, I. , Aronowitz, R. & Bigby, J. 2005. "Hormone replacement therapy, cancer, vontroversies, and women´s health: Historical, epidemiological, biological, clinical, and advocacy perspectives." *Epidemiol Community Health,* 59, 740–48.

Liebenau, J. 1987. *Medical Science and Medical Industry: The Formation of the American Pharmaceutical Industry.* London: MacMillan Press.

Marks, H. M. 1990. *The Progress of Experiment: Science and Therapeutic Reform in the United States,1900–1990.* Cambridge, Cambridge University Press.

Offit, P. A. 2005. *The Cutter Incident: How America's First Polio Vaccine Led to the Growing Vaccine Crisis.* New Haven, London, Yale University Press.

Pickstone, J. 2000. *Ways of Knowing: a New History of Science, Technology, and Medicine,* Chicago, Chicago of University Press.

Quirke, V. & Gaudillière, J.-P. 2008. "The era of biomedicine: Science, medicine, and public health in Britain and France after the Second World War." *Medical History,* 52, 441–52.

Rasmussen, N. 2008. *On Speed: The Many Lives of Amphetamine.* New York, London, New York University Press.

Reinhardt, C. 2008. Boundary values. *In:* Balz, V., Schwerin, A. V., Stoff, H. & Wahrig, B. (eds.) *Precarious Matters: The History of Dangerous and Endangered Substances in the 19th and 20th Centuries = Prekäre Stoffe / [Workshop Precarious Matters – The History of Dangerous and Endangered Substances in the 19th and 20th Centuries, 22/03– 24/03/07, Max Planck Institute for the History of Science].* Berlin: Max-Planck-Institut für Wissenschaftsgeschichte

Schröder, J. M. 2003. *Die Arzneimittelkommission der Deutschen Ärzteschaft von den Anfängen bis zur Gegenwart,* Köln, Deutscher Ärzte-Verlag.

Schwerin, A. V. 2009. Die Contergan-Bombe. Der Arzneimittelskandal und die neue risikoepistemische Ordnung der Massenkonsumgesellschaft. *In:* Balz, V., Eschenbruch, N., Hulverscheidt, M. & Klöppel, U. (eds.) *Arzneistoffe im 20. Jahrhundert.* Bielefeld: Transkipt.

Sinding, C. 2002. "Making the unit of insulin: Standards, clinical work and industry, 1920–1925." *Bulletin of the History of Medicine,* 76, 231–70.

Timmermans, S. & Berg, M. 2003. *The Gold Standard: The Challenge of Evidence-Based Medicine and Standardization in Health Care.* Philadelphia, PA, Temple University Press.

Tobell, D. 2008 "Allied against reform: Pharmaceutical industry–academic physician relations in the United States, 1945–1970." *Bulletin of the History of Medicine,* 82, 878–912.

Valier, H. 2011. *Cancer, Chemotherapy and the Clinic: The Organization of Cancer Research in Britain and the USA in the 20th Century.* London, Palgrave Macmillan.

Watkins, E. S. 2007. *The Estrogen Elixir. a History of Hormone Replacement Therapy in America*. Baltimore, John Hopkins University Press.

Weisz, G. 1995. *The Medical Mandarins: The French Academy of Medicine in the Nineteenth and Early Twentieth Centuries*. New York, Oxford UP.

Wenzel, D. 1968–1971. *Der Contergan-Prozess*, Bensheim-Auerbach [u.a.], Theilacker [u.a.].

Wimmer, W. 1994. *"Wir haben fast immer was Neues": Gesundheitswesen und Innovationen der Pharma-Industrie in Deutschland, 1880–1935*, Berlin, Duncker & Humblot.

1

Secrets, Bureaucracy, and the Public: Drug Regulation in Early 19th-Century Prussia

Volker Hess

Introduction

Drug regulation is usually regarded as an outcome of contemporary history closely connected to the rise of a large pharmaceutical industry, the growing drug market, and the expansion of state authority (Chauveau, 1999; Daemmrich, 2004; Bonah and Menut, 2004; Löwy and Gaudilliere, 2006). It may then seem strange to start a special volume on drug regulation with an analysis of drug approval starting in the early nineteenth century. Indeed, there are many differences between today's regulation and the kind of regulation presented here. Neither the aims, nor the tools, nor the objects seem to be comparable with our current concept of drug surveillance (Abraham and Reed, 2002; Abraham and Davis, 2006). Today, the drug market is dominated by ready-made specialties portioned in galenic preparations. The active ingredient is mostly comprised of a single chemical or biological substance, and more rarely two or three substances. The effects are usually explained in physiological terms (aside from psychological effects, as with a placebo). Regulation is therefore about maximizing effectiveness, reducing toxicity, and avoiding adverse reactions, and involves state authorities, as well as professional groups (physicians and pharmacists), companies, and NGOs (grassroots, self-help, consumer-support and advocacy organizations, patients, and families).

In contrast, in the late modern period, drugs were usually mixtures of different components of herbal or animal origin. They could only be obtained in drugstores and delivered personally by trained pharmacists. The corporately organized profession ensured availability, freshness, purity, and fair trade in drugs on behalf of the state authority, though subjected to its supervision. Regulation served not only to protect people

from quacks, overpriced drugs, and ineffective remedies, but also to maintain pharmacists' traditional privileges (Huhle-Kreutzer, 1989). As such,
regulation was an integral part of the give and take of corporate rights
and duties.

A closer look, however, reveals some analogies between late-modern and
current ways of regulating. One class of drugs in particular, the so-called
secret remedies, were tested and approved in a way that shows similarities
with modern practices. First, secret remedies were treated as a drug product.
Often produced in proto-industrial ways, secret remedies rather anticipated
the later development of large-scale production (Huhle-Kreutzer, 1989). Like
current drugs, secret remedies were also aggressively marketed in newspapers and broadsheets (Ernst, 1975). We should therefore not believe the
enlightened choruses of philosophers, physicians, and officers condemning
secret remedies for their secret and mystical aura. Second, secret remedies were considered intellectual property. In contrast to contemporary
officinal drugs, their formulas were not listed in the *materia medica*. Like
recent drugs, their compositions remained secret – and the state authorities
acknowledged the status of intellectual property (for the German case, see
Wimmer, 1994; for the French side, Chauveau, 1999). Third, secret remedies were subjected to administrative regulation. They became the object
of an official approval procedure, in which they were tested for their active
ingredients, harmlessness, and effectiveness (Kibleur, 1999; Thoms, 2005).
Fourth, already then, the public and publicity played a crucial role. Like
contemporary drug products, secret remedies depended necessarily on
publicity: producers sought public attention for their marketing and tried
to generate among the public assurance on the potency and virtues of the
drug being supplied. In fact, the public sphere made the difference between
a secret remedy and an unknown product. As a consequence, the status
of a secret remedy was in fact constituted by the public. Secret remedies
might then differ from recent pharmaceutical products as to their ingredients and composition, but on the medical market, they were more similar
to current drugs than the official *materia medica*. In short, secret remedies
can be considered as the legitimate precursor of industrially manufactured
products. This also shows the regulation process as precursor of modern
state regulation.

There are consequently reasons for using the heuristic model outlined
in the introduction (Table 0.1). Distinguishing the different ways of regulating drugs in the late-modern drug market will trace back contemporary
configurations to their beginnings, which may help to understand the
main separations made between the state, science, and the *Öffentlichkeit*
(public sphere). The German term encompasses more meanings than
the English one. It includes the idea of the spatial representation of an
enlightened and universal rationality shared by "all men free to use their

own reason." This Habermasian public sphere might be a phantasm, but it drove action – as its imagined counterpart – when philosophers, physicians, and officials made decisions with respect to the *Öffentlichkeit* (Mah, 2000). This perspective from the origins will also help to characterize the contemporary situation of different aims, actors, procedures of evidence, and regulatory tools. To demonstrate this, I will use the figure presenting the ways of regulating (Table 0.1) to evaluate the Prussian procedure of approving secrets remedies. First, I will consider the industrial way of regulating, in which manufacturers, producers, and traders of secret remedies were involved. The following section will discuss the role of the public sphere, in which consumers make public opinion. Finally, I will analyze in greater detail the reorganization of the approval process in the 1830s, in which scientific expertise was mobilized to justify state concessions. Concessions for secret remedies were increasingly granted on the basis of scientific objectivation as a result of pharmaceutical analyses and clinical trials. These brought experts into play, and their role became critical by intermixing public reason with the state's interest.[1] Societies in the process of industrialization found different models for stabilizing, insulating, and protecting scientific expertise. In Prussia, experts challenged the presumed neutrality and impartiality of the state's authority. As a consequence, drug regulation was organized as a bureaucratic routine that tried simultaneously to integrate and to separate the elements of both an administrative and a professional way of regulating. Developments in France will serve as a counterpart: here, the administrative and professional powers were separated in a different way, which was intended to ensure the impartiality of science.

My argument is based on the archival records of more than 270 applications submitted between 1800 and 1865 (Hess *et al.*, 2006).[2] Many of them resulted in chemical and pharmaceutical analyses, and a few involved clinical trials. The files all contain the owner's application, and many of them contain patients' submissions and recommendations, and the related correspondence with and among the authorities.

Producers

In the late modern period, secret remedies were widespread and often banned by the medical authorities. Unlike other European states, Prussia had never forbidden them. Instead, the medical laws of 1693 and 1725 (as cited in Stürzbecher, 1966: 27–34) made their production and sale conditional to official authorization. We do not know how this authorization was obtained in Prussia prior to the early nineteenth century, but from then on, archival records reveal how the medical authorities organized the approval process.[3] An administrative file was set up for each application to sell a drug or to purchase a secret formula. The range of submissions was broad (see Table 1.1), extending from psychological procedures to cure mental diseases to aggressive chemicals for treating cancerous ulcers. Many cases fell into the category of household remedies and traditional medicine, such as Baltic Balsam or various anticold medicines. Only rarely do we find new treatment methods in submissions to medical authorities. The spectrum of the secret-drug owners was also wide: there were medical professionals, especially from the lower ranks, but also laymen, such as chandlers and merchants, retired civil servants, and clergy, all of whom wished to profit from their secret knowledge.

One may ask what requests for licensing homespun remedies or medical inventions has to do with the commercial manufacturing and sale of drugs. As the files show, however, most applications aimed not at initiating a commercial activity, but at legalizing an existing one. Applicants had not been passively sitting on their secret or waiting for a state license. On the contrary, some even asked for an official authorization by attaching printed leaflets and advertisements for their fabulous remedies.[4] Other applicants praised their remedies based on many years of successful thera-peutic use. In other cases, official requests for approval were combined with pleas to license unqualified medical activities in the face of pending punishment.[5] In short, as commercial products, secret remedies shared the economic rationale that has been well analyzed for the French case (Brockliss and Jones, 1997: 238–45 and 628–30; Ramsey, 1982: 218). The mass of files (and applications) indicates an intensification of traditional commerce in Prussia. Particularly around the 1850s, there were more remedies being produced and sold in a proto-industrial way. One applicant remarked that he had already incurred the cost of several hundred dollars to purchase flasks. For this reason, the applicant, a privy councilor named Meyer, argued that if the ministry did not allow the manufacturing of his new cholera remedy, his very existence was threatened by economic loss.[6] Another applicant justified his request by claiming that he had set up repositories in the whole country: although he already had a conces-sion for his eye water, he argued for a renewal so he could fill contracts to produce thousands of flacons for drugstores. He also argued that the ministry would be responsible for his bankruptcy if the application was not given approval.[7] This marketing technique suggests a professional manufacturing system and a professional distribution system – both of them characteristic of a growing proto-industrial organization starting in the 1850s.

From this perspective, we can explore the producer's way of regulating this kind of medicine. We did not find any indication that the produc-tion processes were regulated. Distribution was made official with restric-tions to a specific region. In practice, however, the long arm of the Berlin

Table 1.1   Secret remedies applied for approval in Prussia, 1800–40 (selection)

1800	Clinical trials of the Reich's fever medicament
1804	Clinical trials of the psychical treatment of the qualified physician G. F. Schmidt
1816	Clinical trials with magnetic cures
1818	Clinical trials of the frost ointment of the Pastor Wahler of Kupferzell (spring)
1818	Testing baltic balm sam (Riga), license for J. Fr. Conradt, Russian Imperial Counselor (1818)
1819	Clinical trials of the cancer medicine of the canoness Louise Grasshoff (spring)
1819	Clinical trials of the cancer medicine of the surgeon Jentzsch in Kollodrau (summer)
1819	License for the universal band-aid of the Orphanage of Züllichau
1823	Clinical trials with the frost medicine of the superintendent Kampfhenkel
1824	Clinical trials with the cancer medicine of the former customs officer Hellmund
1824	Clinical trials with Grabe's treatment
1825	Clinical trials with Schramm's drug against articulation tumor
1825	Clinical trials with Hirtel's remedy against apoplex
1826	Clinical trials with Kampfhenkel's Frost medicine
1828	Clinical trials with Peschel's method for fractures and distortions
1829	Clinical trials with Dr. Plagge's method
1832	Clinical trials with Willmann's remedy against Tinea Capitis
1832	Clinical trials with Aqua binelli seu Liquor haemostaticus
1835	Clinical trials with Extractum papaveris somnifere
1835	Clinical trials with Meyer's remedy
1836	Clinical trials with the Kleinertz' secret remedy against febris intercurrentis
1836	Clinical trials with Schilling's remedy against rheumatism

authorities did not reach the local authorities to enforce such restrictions.[8] Once granted, the license opened the door to the market, in which producers would then distribute their products liberally. Thus, the crucial point in the game was the application procedure, which producers tried to influence and organize. Compared with the scientific methods that we have today for objectifying effectiveness and harmlessness, their tools seemed limited. Indeed, neither of these two modern criteria was used in application documents. Instead, we find two very common and constantly recurring arguments. First, applicants stressed the unique and exceptional character of their product. A secret formula was authoritative experience being passed down from generation to generation. A second argument was that the remedy was the fruit of a long and successful practice to which the applicants laid claim despite their lay status. Documents accompanying the application (such as letters of recommendation) attested to how perfectly well the remedy worked, to a group of patients who tolerated it, and to unanticipated therapeutic benefits.[9] Flyers and other advertisements demonstrated the successful use of the remedy, no matter that they were also proof of illicit activities and commercial interests, which applicants did little to hide.

The second argument is more interesting in the sense that the applicants claimed moral qualities for themselves. If we believe them, financial rewards never motivated their application. The only reason stated for making long-kept secrets public was disinterestedness and duty to humanity (Hess, 2006). Alternatively, the "welfare of mankind" or "the suffering" and "needs of the whole of mankind" drove applicants to disclose their property.[10] Sometimes, such claims prompted sarcastic observations from ministerial officials, confident that rejecting the application would not harm the well-being of humanity.[11] Nevertheless, remuneration was sought as compensation through the promulgation of intellectual property. The amount was not set according to the promised benefit of the remedy but as a factor of personal living conditions: to ensure a living wage, to help in a situation of dire financial straits, or to be able to support old and feeble parents. Thus, one applicant wanted "to secure a partial livelihood for herself and her daughter by marketing the tincture that [she] enclose[d] as a sealed sample."[12] A second applicant hoped to improve her retirement pension by revealing her cancer medicine, and a third requested 600 Reichsthaler for his impoverished father.[13],[14] The wordy illustrations of bad conditions and daily troubles did not call on the state's duties to relieve poverty; they were meant to demonstrate the applicants' personal integrity. Every effort was made to distance future production from any suspicion of selfishness and the odor of money making. In other words, the producer aimed to establish credibility. This was what the applicants tried to impart to medical officials. Knowledge and skills played only a small part, and formal qualifications were not mentioned in letters to the ministry. What the producers were emphasizing for regulation was not so much the physical, but rather the moral qualities of their products – and their own. From the perspective of production, moral values represented the decisive criteria for the evaluation and examination of both the products and the producers. This old economy was no longer working, however, as we shall see in the third section of my chapter. The Berlin ministry no longer shared the traditional system of morals and values.

The public and consumers

As the applications show, producers mobilized public opinion in their attempt to have their products approved and regulated. Relying on personal testimony was necessary to demonstrate the moral character of the product and the producer. Letters of thanks, recommendations, and attestations were not only intended to impress the ministerial readers in Berlin in terms of the efficacy of the remedy, they were also understood as an articulation with public opinion. In one case, 41 Berlin residents had signed the application: teachers, merchants, craftsmen, as well as officers and housewives confirmed producer's statements. They also pleaded for approval – in the interest of mankind. They formed, in a manner of speaking, a consumer organization, which attested to the emergence of civil society in the Prussian capital. The group not only tried to get the ministry to conduct fresh trials with the secret ointment of the veterinarian Friedrich Nickau, it also took a political stand, arguing the scientific and empirical merits of the new medicine. Even if the number of participants was small, the group represented the "public sphere" in the understanding of its members and from the ministerial point of view. The public sphere was constituted by the general reason for which each of the petitioners stood.

As previously emphasized, the public sphere was not limited to consumers. The "public" (in the Habermasian meaning) was also incorporated into the administrative way of regulating. Two brief examples can illustrate this point. The first is the case of Reich's fever remedy, which as far back as the late eighteenth century was reputed to be a rapid cure for fever. Reich's decision to unseal the secret of his new scientific method if he received financial remuneration prompted a heated debate on the inventor's moral responsibilities and on the protection of intellectual property in the *Reichsanzeiger*, one of the leading *Intelligenzblatt*. The second case involved the miraculous magnetic cures of a lay practitioner in the 1820s, when in Prussia animal magnetism had just reached its zenith. In both cases, it was public opinion that prompted medical authorities to test the remedies. Yet in each case, the "public" appears to play diametrically opposed roles. In the first case, the widely circulated newspaper served as the vehicle of an enlightened public sphere (Dietz, 1922). In the second case, rumors and gossip generated an informal public sphere that was inaccessible to print media. Furthermore, the particular setting of the respective agents could hardly have been more different: Reich was a respected professor of medicine in Erlangen, while the miracle healer was a one-time stable boy. In addition, the issues at stake were completely different: whereas Reich based the invention of his fever cure on theoretical considerations of up-to-date medical concepts, the illegal practice of the stable boy named Grabe was from the outset burdened by accusations of heresy and quackery. Both cases are situated prior to the administrative reorganization of drug-testing procedures, although the sources do not provide sufficient evidence suggesting a causal relationship here. Neither case is representative, but both are typical, for they clearly illustrate several critical elements that were especially important for the development of procedures that regulated the granting of concessions.

The first of these was the demand for assistance and cure. In the case of Grabe, the need for help was difficult to manage. It resulted in masses of people streaming into the small Saxon village Torgau, where Grabe practiced under the protective umbrella of an academic physician. Patients and relatives, local doctors and surgeons, mesmerized visitors and worried medical officers all comprised an informal public, sustained by an endless supply of reports on Grabe's miraculous cures. The flip side of this public excitement was a covert operation sponsored by the Prussian Ministry of Culture: the military detained Grabe overnight and took him by dark to the Charité Hospital in Berlin to subject his therapeutic capabilities to intense scrutiny. The broad debate and publicity surrounding this miracle healer was in part prompted by political efforts to stamp out heresy. It also represented a response of reformist bureaucrats to the reactionary turn in Prussian politics (Freytag, 1998; Engstrom, 2006). It was not debate in medical and scientific journals, but public crowds that brought the state to the front of the stage. The publicity and the public scandal prompted the bureaucrats to intervene in the wonder cure.

Second, there was the influence of potential criticism (Broman, 2005). The boundless claims attributed to a panacea would hardly have resonated in a contemporary journal such as the imperial *Reichsanzeiger*, addressed as it was to an enlightened, educated public, had they not emanated from a member of the academic community.[15] By virtue of his academic position, Reich, the professor from Erlangen, participated in the universal discourse of enlightened reason. Furthermore, he could effectively mobilize public debate as a means of establishing his moral rectitude and financial probity. Between 1799 and 1802, nearly 30 articles on Reich's cure for fever were published in the *Reichsanzeiger*, not to mention numerous publications in other journals as well (Hess *et al.*, 2006).

It would be a mistake, however, to reduce the role of the public either to the heated exchanges in these journals or to the rapid spread of rumor and gossip, both of which have been the subject of extensive studies.[16] The impact of the public cannot be adequately grasped from evaluating accounts in newspapers and weeklies of the time. The actions taken by state bureaucrats in reply also need to be considered. In the case of Grabe, those actions are obvious: the tenets governing administrative action incorporated the potential reactions of an informal public. Regardless of whether the fears of administrators were justified or not, the mere anticipation of potentially public ramifications had a significant impact on the bureaucrats' conduct. First of all, the measures taken to detain and transport the miracle healer were all part of a covert strategy designed to avoid public attention. Yet once the masses began congregating before the gates of the Charité Hospital, the bureaucrats adopted proactive measures. They began to draw on outpatients to serve as probands, thus opening the clinical trials to participation from the street. To some extent, the evaluation of Grabe thus included wide public participation as a consequence of the very broad and thorough documentation to which his techniques were subjected.[17] Unlike the standard procedures of the time, which involved extensive casuistic accounts of only one or two typical cases, the parameters used to evaluate Grabe were exceptional in the number of probands used, in the discussions about inclusive and exclusive criteria, in the documentation on the course of illness, in the avoidance of subjective influences on Grabe, in the authorization of protocols, and in the uniformity of trial procedures. As in the case of Reich's trials (see below), the methods used to provide empirical evidence anticipated that the public, after the trials had been completed, would share all the information and would evaluate the detailed results. This mode of "anticipated scrutiny" characterizes an imagined public sphere. Its appearance and effects have to be regarded not in the sense of an "actor" on the stage of history, but rather as an indirect influence on the conduct of officials. The rationale of administrative actions addressed this public sphere even if the public may have only been a figment of the bureaucrats' imagination.

This is especially evident in the case of Reich. Prussian bureaucrats became involved just when the public debate had come to a consensus: in various letters to the *Reichsanzeiger*, Reich's insistence on material remuneration was questioned on moral grounds, because Reich – as an academic and a member of the "Republic of Letters" – was obliged to serve the public good. One commentator weighed "Professor Reich's responsibilities to humanity" against his "responsibilities to himself" and society's responsibilities to him (K-y, 1799). Another was moved to respond because of his "sense of the importance of the case for all of humanity and especially for the medical public (-n-, 1799)." A third observer, writing in the name of humanity, expressed his "moral convictions" regarding Reich's responsibilities, and a fourth considered demands for the publication of the fever remedy to be "a matter for philosophy, religion, and society [to resolve]" (Anonymous, 1799). The participants in this public debate who stood for universal and enlightened reason thus tended to stress their high moral convictions.

Consequently, it would be a mistake here again simply to reduce this enlightened public to the specific readers of the *Reichsanzeiger*, that is, to a subgroup of the public (Broman, 1998; Mah, 2000; Broman, 2005). The *Reichsanzeiger* was an important if not the most important *Intelligenzblatt* of the period. Although it served a relatively small readership of only 2,000 to 3,000 subscribers, its contributors still saw themselves as speaking for humanity as a whole. They wrote not as readers of the *Reichsanzeiger*, but as the voices of reason. None of them spoke for themselves, a specific group, or a profession. Instead, as the anonymous voices of reason, the contributors spoke for the "reasonable and right-thinking portions of the public." The use of Reason in the public sphere was not to include any reference to social criteria, which could have been used to identify and limit the perspectives to an individual reader or author (Kant, 1977). Even when disciplinary or medical arguments were introduced into the debate, they constituted only another facet of Reason in which all readers of the *Reichsanzeiger* could partake. This way, Reich's claim to derive economic advantages from his fever cure was negotiated "before the eyes of the entire nation," with no distinction between general, exoteric knowledge, versus specific, esoteric knowledge.

Ultimately, the readers of the *Reichsanzeiger* agreed that Reich should be remunerated by way of a public subscription. This public in the Habermasian sense rejected state intervention arguing that it was an illegitimate and arbitrary abuse of power: "The just claims of a humble and well-meaning man such as Prof. Reich can only be satisfied by an educated public (B., 1799)." Given such broad, universal jurisdictional claims, every action taken by the Prussian medical administration was considered a priori to be prejudiced.

Indeed, the clinical trials for which Reich was summoned to Berlin were viewed suspiciously from the outset. The appointed commission made every effort to ensure impartiality and objectivity. Whereas the stable boy from Torgau had been subjected to rigorous scrutiny, Reich was more or less able to design his clinical trials as he saw fit: he oversaw the selection of probands, the daily therapy, and the documentation of the treatments. The commission also insisted on utmost secrecy once it had been informed by Reich of the composition of the remedy and the theoretical basis for its effectiveness. In its final report, the commission contributed only a general introduction, leaving to Reich to write up the summary of the results of the trials. Moreover, the members of the royal commission did not even assume their responsibilities as state officials: although they unanimously confirmed that Reich was a "learned physician and an upstanding man," hence a member of the critical public, they explicitly refrained from passing judgment on the trial results. They concluded that Reich's remedy had produced "some good, some ambiguous, and some poor results." They also conceded that in some cases the remedy had exhibited fortuitous and rapid effects (Anonymous, 1800: 65) – faint praise indeed, in light of the claim that the remedy could prevent death in feverous patients "within two hours" (Lenz, 1799). The commission did, however, insist that it had "in no way been instructed ... to reach a given conclusion" (Anonymous, 1800: 65). Instead, the state-appointed experts considered "the public [*Öffentlichkeit*] to be the judge" and handed their task back to the public sphere, "convinced that publicity was the sole means of judging and applying the theory and practice of a secret remedy" (Selle *et al.*, 1800: 315).

The experts' restraint was certainly also due to the exhaustive reporting that had come about: in spite of the best efforts to keep the trials secret, details were leaked and published in the most prominent review journal of the period. Clinical reports on the trials appeared in the *Medizinisch-chirurgischen Zeitung*. Indeed, thanks to the new media, the entire medical community participated in the trials directly, even though they were carried out behind the walls of the Charité Hospital.

Similar issues were at stake in the clinical testing of the remedies. In both cases, performing those tests anticipated public responses. The criteria that were laid out were geared toward transparency and public participation, regardless of whether the public was imagined to be a great danger (as was the case for Grabe) or whether it was evoked by the "manufacturer of the remedy" (as was the case for Reich's fever remedy). Unlike the usual casuistic descriptions of the time, these clinical trials differed in that first, they were serial in nature, second, they took into account subjective factors, third, they confirmed the results, and, fourth, they were correctly presented and published (Hess *et al.*, 2006, Hess, 2008). This time 20 to 30 probands were selected whereas previously 1 or 2 typical cases had usually been chosen. This time, criteria of inclusion and exclusion were clearly formulated whereas previously chance had been relied upon. This time, only the course of the illness was documented whereas previously verbose interpretations of individual cases and detailed descriptions of every detail had been common practice. This time, every effort was made to exclude subjective factors whereas previously the personal authority of the observer had given the observation the stamp of validity. Last but not the least, this time, the documentation of the trial procedures had to be countersigned by the probands, whereas previously, the eloquence and reputation of the author had sufficed (see the work on the role of the Royal Society by Shapin, 1994). Consequently, both clinical trials, though they were very different, broke with the contemporary "ideology of observation." Both cases were to become typical for the bureaucratic procedures that evolved over the following three decades for the approval of secret remedies.

Administrative regulation

In the archival files, we found an emerging evaluating and testing routine starting in the 1830s. For earlier decades, there are only a few special cases that provide insight into regulatory practices (see Table 1.1). We may assume that this administrative regime was established and formalized in and with the Prussian reforms because it started during the reorganization of the medical authority (Münch, 1995). Two agencies were involved: the Medical Department, which was integrated into the Ministry of Cultural Affairs in 1817; and the Scientific Deputation, to which the most prominent physicians and scientists of the capital were appointed. Both agencies arose from the Collegium Medicum, which did not represent the medical corporation like in France, Italy, and the German imperial cities, but was an administrative chamber of experts (Hess, 2010).

Usually, applications were sent to the Medical Department, which conducted the regulation procedure. Submitting a remedy was only the first step in the long bureaucratic road (see Figure 1.1). Medical officers were rarely impressed by the fabulous stories of miracle cures, altruism, and charity. To mobilize counter expertise, they sought scientific evidence. They asked applicants to send a sample of the remedy and its formula in a sealed envelope. The promise that the authority would "keep the secret in strictest confidence" was never broken.[18] In many cases, the official assurance of confidentiality is the last document in the archival files, presumably because the applicant decided not to reveal the secret remedy. When the requested sample and formula arrived, the Medical Department was true to its word and passed the sample on for further testing. Then the Scientific Deputation was asked for expertise.

Breaking the seal opened on to the second step. Only the members of the independent commission saw the sample and tested its ingredients.[19] In general, they initiated a so-called technical assessment of the sample.

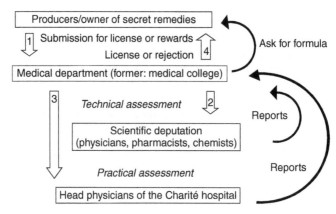

Figure 1.1 The procedure of drug approval (1830s)

If this theoretical assessment seemed to be favorable, the remedy was subjected to practical clinical trials. These trials were conducted neither by the Medical Department nor by the Scientific Deputation, but were commissioned to chief physicians at the Charité Hospital. If the Charité lacked suitable patients, the ministry turned to the public and asked for outpatient volunteers (Hess and Thoms, 2002). When the clinical trials were recorded and reports were sent, the Scientific Deputation discussed the records, commented on the reports, and drafted an expertise summarizing the theoretical considerations and clinical results. On this basis, the medical officers of the ministry drew up the concluding statement, which was to stop or to approve the application. A license usually allowed the remedy to be sold in small quantities or – especially in the early decades of the century – the state to purchase the secret (the early-modern way of regulating drugs).

This bureaucratic organization reveals an interesting division of competences. On one side, the Medical Department represented the political power. The procedure conceded decision making to the ministerial authorities but denied its medical officers the scientific vote of judging and evaluating. On the other side, the Scientific Deputation served as advisory board. The procedure conceded scientific qualification to the commission but denied the scientific elite the power of decision. The physicians in the Charité Hospital also served as experts for the clinical evidence, but were not involved in the official decision-making process. In so doing, the procedure separated the political power and the scientific expertise as distinct domains.

Even if this distinction seems a bit crude, I would like to emphasize it for analytical reasons. Table 1.2 shows why distinguishing between administrative and professional regulation made sense. In assessing different modes of regulation, the two authorities pursued different aims, used different tools, and dealt with different objects. The Scientific Deputation was clearly an institution in which scientific values converged with professional interests. These interests were not threatened by the political power, given that the enforcement of bureaucratic procedures targeted personal rather than professional interests.

The members of the Scientific Department served, in fact, as professionals. The technical and clinical assessment, which aimed primarily at preventing secret remedies (Table 1.2), was particularly demanding: of the 100 examinations conducted in the first half of the century, only 6 resulted in positive reports. Compared with the clinical trials, one-fifth of which returned positive results, the theoretical assessment was considerably more arduous. The experts found very different reasons for rejection. Analyzing their reports, we can distinguish four arguments. First, the experts challenged the supposed worth and therapeutic applicability of the remedy. Some applicants were advised that "this disease could just as easily be cured

Table 1.2   Ways of regulation after reorganizing Prussian drug regulation in the 1830s

	Professional	Administrative
Main actors	Physicians, pharmacists, and chemists in the Scientific Deputation	Physicians, pharmacists, and chemists in the Medical Department
Aims	Preventing secret remedies	Fair trade, public healthinventions, and replacements
Tools	*Materia medica*/pharmacopoeia register techniques	Licensure; rewards
Evidence	Formulas and chemical analysis	Scientific examinations(technical and practical)

by using known remedies."[20] Similarly, the deputation argued that the herbal component of the drug was "less effective than other compounds in the pharmacopoeia."[21] In these cases, the remedies were examined with regard to specific components only if the "mixture was less powerful than the similar, popular one."[22] All of these judgments referred to the *materia medica*, which constituted the authoritative store of medical knowledge on the effects and uses of each substance.

The second argument also derived from the corpus of medical knowledge. However, instead of targeting the substances, it was directed at the applicants themselves. The central issue here was the popular belief in panaceas, from which academic medicine had only recently distanced itself. The reports often seem to express relief at not having to deal with such superstitious relics. In the case of Nickau's ointment, for instance, "it was impossible to approve the submission because a water and ointment cure for all diseases makes no sense. Examining such treatment ~~was rather absurd and~~ [sic] could not be demanded of scientific experts."[23] Such views became more prominent by midcentury. One applicant was deemed incompetent because he claimed to cure all diseases. Another was criticized for having mixed ingredients together arbitrarily. He had "specified the quantities and volumes recklessly – what else can one expect from a potion that did nothing more than enhance his self-delusion."[24]

In the midnineteenth century a third argument came to the fore. In the early 1850s, long before the new pathological anatomy came to be widely used in rejecting applications, a report noted that "a toothache is not a specific disease" (see for details Hess, 1993). Whereas in the 1820s, medical officials had still tested the healing power of the wool shaved from a ram's scrotum when the moon was waning, they now condemned secret remedies for the simple reason that applicants believed that fear caused apoplexy and that a fear herb would cure the illness.[25,26] Officials now found "huge ignorance and inconsiderate treatment" in the lower ranks of the medical professions, or, in another case, even wondered if the applicant could be allowed to practice surgery without retaking his medical exams.[27]

These three arguments speak loudly of the importance of professional interests. The members of the board represented the medical and scientific elite of the Prussian capital. Physicians, pharmacists, and chemists did not hesitate to use the board to advance their own interests. They regarded themselves as the gatekeepers of the medical market, in which they tried hard to restrict the access of secret remedies. The tools they used were drawn from academic knowledge. They mobilized both the technical facilities needed for analyzing the chemistry of compositions and the social skills needed to find information in natural-history networks. In doing so, they reduced the verbose applications to a formalized regime, which also seemed to shift the evaluation away from the realm of a moral economy.

In contrast, the Medical Department had first and foremost to take political and especially economic interests into account. Searching for cheap alternatives beyond scientific opinion, looking for home-grown plants such as poppy or rhubarb, and encouraging new methods of treatment for epidemic diseases were important considerations in the minds of the ministerial authorities.

Thus, the frequent, and decisive, argument for rejection did not mainly refer to professional interests. The main and central point on which the ministry insisted was "innovation." This meant that remedies had to be examined to determine whether "their formulas were new, and their ingredients and mixture unused."[28] Not the drug itself, but the composition was assessed. In principle, the deputation accepted that it was possible to mix the drugs listed in the pharmacopoeia to produce new and powerful remedies. This meant that the remedies had to involve more than a simple modification of the recipes found in the *materia medica*. The applicants must have often been aware that the carefully guarded formula was actually known and had been published long ago.[29] The rationale of the argument followed – in a manner of speaking – the principles of patenting. Like patent officers, the members of the Scientific Deputation consulted long registers for the medication. In the case of Seewald's gout balsam, the arbitrator of the Medical Faculty was convinced that the remedy "represented a distinctly new kind of medicine due to the way in which the inventor had prepared it."[30] He voted to continue the practical assessment. The arbitrator obviously did not have the right credentials. The Deputation decided to deny the submission without further investigation because the ingredients and mixture had already been published in Phoebus's *Arzneiverordnungslehre* (Phoebus, 1831).

Comparing the arguments of both producers and professionals reveals that the two rarely overlapped. Indeed, the clash between both reference systems indicates how regulation began to shift secret remedies into a new framework. While the applicants referred to the traditional regimes of poverty relief and enlightened humanity, the medical officers were mired in a web of competing interests, maxims, and concepts. While the Deputation members could not ignore the applicants' professional background, the medical officers tried to foster the notion of "invention."

This points to the administrative way of regulating secret remedies. The medical officers' field of interest was different than that of the Scientific Deputation. Medical officers watched the markets for secret remedies in France or in the Netherlands in search of new developments.[31] They also sought drugs that could replace expensive imports. Officials pored over newspapers and magazines, carefully watching public debates in the hope of discovering new drugs, remedies, or therapeutic regimes for the Prussian state. Thus the ministry dealt with issues of public health as well as with the task of fiscal accounting.

How did this work? How were the officers and experts able to differentiate among these aims? It is obvious that the bureaucratic division that was organized was not really strict: some physicians and pharmacists were part of the Medical Department as well as of the Scientific Deputation. We can also assume that the medical officers (who also belonged to the medical elite) were not able to ignore their own professional background. Both the experts and the officers were well aware of the web of interests and maxims that arose from their conflicting roles and functions. Even more, the regulation procedure seemed to aim explicitly for this kind of separation: defined as a "purely scientific advisory board," the Deputation was neither subordinated to nor integrated into regular administrative decision-making mechanisms. Consequently, the physicians and scientists were able to distinguish clearly between their roles as state officers and as scientific experts.

The best example of this was Johann Nepomuk Rust, the key figure in the reorganization of the Prussian bureaucracy. Rust was part of the medical establishment. He was chief physician at the Charité Hospital, held an academic chair at the University of Berlin, and served as professor at the academy of military medicine. He influenced the medical landscape of the period as personal physician to the Prussian king, as president of the hospital committee, and as Privy Council to the Ministry of Culture – many jobs and interests difficult to handle. In the 1830s he published a critique of the relationship between politics and science. Summarizing the reform era, Rust emphasized that modern bureaucracy depended on scientific expertise and could not function without it. While scientists and physicians could easily appropriate the formal administrative procedures, civil servants had to work hard to learn the "material of the administrative body" (Rust, 1834–40: 29). Scientists were therefore more suited to administrative tasks than administrative servants were to scientific ones. Rust insisted emphatically on the hazards of sharing administrative power. Empowered scientists would easily become despots "soaring above their colleagues," who, he believed, always liked to enforce their ideas and thinking, which made them ill-suited to hold high positions within the government bureaucracy: "Even the most liberal and open-minded scholar had to draw his own conclusions as soon as judgments and decisions of scientific fact were at issue." That was why the Prussian reforms separated government administration from scientific expertise. When it came to professional qualifications, it was not possible to separate state power and scientific knowledge. The solution was to separate not administrative and scientific capacities, but rather the fields of responsibilities: the Medical Department on the one hand and the Scientific Deputation on the other. In the administrative function, physicians and scientists had "neither special judgments nor the decisive vote" in the field of science. Their duty was simply to open and administer the applicant's case, regardless of whether they were personally convinced by it or not. Scientific knowledge was needed for administrative officers only in order for them to know when and to whom they should direct their queries. As for the members of the Scientific Deputation, they served only as technical advisors. They were the "wahre Kunstrichter" (the real critics) (Broman, 2001), responsible for taking decisions but not for implementing them. The two bodies worked separately, although the council in the ministry took part in the Deputation's work. The Deputation was restricted to presenting the existing scientific knowledge. The clinicians involved also reported their observations in a schematic form, backing them up with quantitative arguments. Neither the expertise of the Deputation nor the reports from the trials were given to the applicants. The officers in the Medical Department drew on the language contained in the scientific votes and ultimately couched their rejections in their own terms. The authoritative decision was thus enacted – as seen from the outside – as the performance of autonomous state authorities. This performance demonstrated that the approval process was not submitted to the professional interests of the medical elite but was driven and sanctioned by the power of the state.

This separation was actually ideological, but this was the ideology that shaped the model for modern German drug regulation (Wimmer, 1994, Daemmrich, 2004). In practice, the self-interest of physicians and scientists was also revealed in the administrative functions, but this is besides the point. What is interesting is how the surveys and statements obtained the status of an independent, impartial, and objective opinion by integrating both scientific expertise and state power. This made the German model of controlling and regulating secret remedies very different from other models.

The French way

France adopted another way of regulating secret remedies. Paris was at the forefront in establishing a scientific review system in the late-modern period (the "French classic"). French administration had already included experts into the decision-making procedure during the *Ancien Régime*. These experts evaluated the remedies based on their theoretical and practical merits and articulated their favorable and negative views to the Ministry of Internal Affairs.

In contrast to the Prussian model, the French *Société Royale de Médecine* served as the scientific board of the administrative body before and after the Revolution, when it was renamed *Académie*. The secretary of the *Société*, and later the *Académie*, was put in charge of correspondence and of the examination procedure, which from the 1820s on was conducted by the so-called *Commission des remèdes secrets* (Kibleur, 1999). This commission was one of four boards that would exist into twentieth century (Weisz, 1995: 87) and throughout would retain jurisdiction over the evaluation process. The commission's reports were debated and ultimately accepted or rejected by the plenum of the *Académie* at its regular meetings. The academy then notified the ministry of its decision, which was then simply confirmed and undersigned by the ministry. As state authorities often delayed official certification, the academy's meetings passed de facto judgment on secret remedies via their *procès-verbaux*, or minutes. Starting in the 1850s, the academy no longer waited for the state administration's validation. It formally confirmed certification by publishing their minutes in the *Bulletin* of the medical corporation.[32]

In principle, the French procedure was very similar to the Prussian one. Integrating theoretical and practical considerations, the corporation of learned physicians was not just responsible for actually testing the secret remedies, as was the case for the Charité Hospital physicians, they were also entirely responsible for deciding whether or not clinical trials were necessary.[33] Moreover, whereas clinicians in Berlin were scrutinized by ministerial officials, who sometimes believed that the number of trials or the manner in which they were conducted was inappropriate, in Paris the physicians were responsible solely to the academy, which also set all the standards for the trial and, if approved by the plenum, even gave the remedy a name, advised on its usage, and fixed its maximum price (Kibleur, 1999: 73). In other words, the academy assumed the role of technical board as well as that of a government agency with executive powers.

The actual evaluation procedure, especially its twofold administrative and scientific nature, has already been described by Pascale Kibleur.[34] As in Berlin, the evaluation involved a theoretical part and a practical part. The theoretical evaluation dealt with the formula, the effectiveness of the specific substances, and the composition of the whole. The practical evaluation tested the therapeutic effectiveness of the secret remedy. The results of these evaluations were presented to the plenum of the *Société* and if approved were adopted into the infrequently published *Codex medicamentarius, sive Pharmacopoeia gallica*.

If the accounts of the *Société* are to be believed, the figureheads of French medicine of the time were strict and rigorous in their assessments. Of the 350 remedies submitted between 1772 and 1778, only a handful were approved, 278 were rejected out of hand, and 61 received only limited concessions (Ramsey, 1982: 218). In subsequent decades, approval rates declined: in the 1820s only three remedies received favorable evaluations, and in the second half of the nineteenth century the *Académie* boasted that of the 30 submissions it received annually not a single one had been approved (Riche, 1903). As Ramsey has shown, however, these numbers were based on a very restrictive interpretation of the term "approval": whereas one report indicates that of 800 remedies only 4 were approved, Ramsey's evaluation of 700 reports indicates that 81 were approved – either formally (32 remedies) or in the form of tolerated limited commercial sale (49 remedies) (Ramsey, 1982: 222).

Thus, regulation in France remained for the most part in the hands of physicians: the Ministry of the Interior passed applications directly on to the academy, under the auspices of which the actual testing was conducted. All assessments, tests, and evaluations fell within the jurisdiction of an autonomous body of academic physicians. The consequences that flowed from these decisions were the sole responsibility of the academy. Whereas in Berlin a sharp distinction was made between the administrative decision to "begin the necessary procedure" (Rust, 1834–40: 32) and the role of the medical experts (*Kunstrichter*), in France the whole process culminating in the official sanction of a medication lay in the hands of the *Académie*. Decisions made in Berlin by the state secretary were made in Paris by a plenum within the corporative body of the medical fraternity. Even the academy's responsibility to report to the ministry was the subject of constant dispute. The academy's plenum deliberated over each and every word of the final report, striving to convince the *public* of their independence. Drug regulation therefore showed different modes, in France and Prussia, of balancing political power and scientific expertise.

Conclusions

In reorganizing medical authority in the early nineteenth century, the Prussian bureaucracy installed a formal procedure for scientific testing and approval of secret remedies. Although this procedure was clearly not as efficient as today's, it, nevertheless, addresses one of the main questions of contemporary drug regulation: How can professionals draw on their skills and knowledge to evaluate drugs without compromising the state's monopoly of power or, conversely, the integrity of scientific evaluations? The administrative way of regulating drugs also involved negotiating the epistemological and sociological status of modern expertise. As demonstrated, the public played a significant role in this process. The public did not actually participate as an active third party intervening in the drug regulation process, nor was it the "historical force" driving the transformation of the Prussian bureaucratic regime. Rather, the public played the role of silent participant, as a public sphere that was ever present in the transformation process. There is no mistaking the extraordinary resonance of the Grabe and Reich cases: in both of these, clinical trial and scientific expertise were performed in a way that anticipated the participation of the public. Similarly, the public was involved in reorganizing the bureaucratic process in the 1830s. Understood in terms of Reason, officials also imagined this public sphere when they tried to separate the domains of science and politics. Admittedly, the public sphere was fiction in terms of real audiences and a real impact of journals and newspapers, but it was truly influential fiction in the sense that officials were seeking to preempt public criticism. The reorganization of drug regulation conceptualized this idea of a public sphere by separating the administrative and the professional ways of regulation (Turner, 2001; Engstrom *et al.*, 2005).

Consequently, regulation was not simply a matter of administrative routines and bureaucratic procedures. Rather, regulation should also be understood as a performative action. This performance could follow different scripts as shown by the comparison of the French and the Prussian ways. The *Académie de Médecine* performed drug regulation as an exercise of scientific knowledge. Minimizing or eliminating the state's influence, the French model legitimated the approval of drugs as an object of the medical sciences. Decisions were justified by the independence of the medical corporation. In contrast, in Prussia, drug regulation conveyed another form of decision-making legitimacy. Regulative performances celebrated – in a manner of speaking – the separation of powers.

These were obviously performative effects because German professionals played the same decisive role in approving new medications and evaluating producers as French professionals did. The important question here, however, is that drug regulation implemented a model reflecting the interplay between the sciences and state power.

There is no space here to delineate this argument systematically, but it seems obvious that the strategy of separating and insulating the different spheres strengthened each of them: both the authority of the administrative way of regulating as well as the scientific basis of the professional one. The concept of public sphere – as addressee and de facto participant of this regulation – played a crucial role. When the Prussian bureaucracy developed a modern model by separating the scientific from the political domain, the policymaker also implemented a model of how to mask professional interests in political decision-making processes.

Notes

1. This differs the modern expert from the early modern concept of expertise (Ash, 2010).
2. The research was funded by the Deutsche Forschungsgemeinschaft (He 2220/3) in collaboration with Eric J. Engstrom and Ulrike Thoms.
3. For the Bavarian case, see Probst, 1992.
4. See, for instance, the debit of the Riga Balsam, which had long since been advertised and sold. Debit for Conrad, December 5, 1818, GStA PK (Geheimes Staatsarchiv, Preußischer Kulturbesitz (Berlin)), 1. HA, Rep. 76 VIII A, Nr. 2166 (Die Zusammensetzung u. den Debit des Rigaischen Balsams Okt. 1818 bis Nov. 1855 betreffend, o.P.).
5. Request of Wilhelmine Clementine Martin, July 9, 1821, GStA PK (see note 4), 1. HA, Rep. 76 VIII A, Nr. 2162 (Acta betr. das Geheimmittel der Klosterfrau Wilhelmine Reimers zu Xanten zur Heilung des Krebsschadens, desgl. des Wundarztes Jentzsch zu Kolloschau und der Mariana Lingern, und des Stiftsfräulein Grasshof, der Wilh. Clementine Martin, des Gottf. Resfel zu Kunzendorf, des Hellmund zu Oldendorf, des Heinrich Staack zu Schlieben, Bd. I (1817–1843), pp. 50–57). Another example is the master saddler Carl Müller, who sought sanction for his therapeutic practice because he "was caught up in the misfortune of being punished for the unauthorized sale of drugs." Immediateingabe Müller, January 20, 1852, GStA PK (see note 4), 1. HA, Rep. 76 VIII A, Nr. 2123 (Acta betreffend die zur Prüfung und Approbation eingereichten Arcana, Heilmittel pp, vol. 15 vom Juni 1852 bis Juni 1852, o. P.).
6. Promemoria of Meyer, August 7, 1835, GStA PK (see note 4), I. HA, Rep. 76 VIII A, Nr. 2116 (Acta die zur Prüfung und Approbation eingereichten Arcana, Heilmittel vol. 8, vom Jan. 1836 bis ult. März 1837 betreffend, o.P.).
7. In other cases, applicants made reference to the costs of purchasing thousands of flacons and of printing labels. Letter of the Prussian ministry at Stroinski, October 28, 1857, GStA PK (see note 4), I. HA, Rep. 76 VIII A, Nr. 2172 (den pensionirten Militair Intendantur-Registrator Stroinski und den Debit des von demselben fabricierten Augenwassers (März 1852 bis Aug. 1866), p. 160).
8. Brenn (Ministry of Commerce) to Altenstein (Ministry of Cultural Affairs), October 6, 1831, GStA PK (see note 4), 1. HA, Rep. 76 VIII A, Nr. 2166. Brenn voiced concerns about a city ordinance that prohibited unlicensed trading.
9. See the request of Nickau (mentioned below), the request of pastor Stübner from Casdorf in 1840, GStA PK (see note 4), 1. HA, Rep. 76 VIII A, Nr. 2117 (Acta betr. die zur Prüfung und Approbation eingereichten Arcana, Heilmittel pp. Vol. 9, (Jan. 1837 bis ult. April 1839. (unfol.))), and the approval given to the Hess's *lozenge* in 1853, GStA PK (see note 4), 1. HA, Rep. 76 VIII A, Nr. 2124 (Acta betreffend die zur Prüfung und Approbation eingereichten Arcana, Heilmittel pp. vol. 16, vom Juli 1852 bis Juni 1854).
10. See, for instance, the request of Louise Grasshoff, January 7, 1819, GStA PK, Nr. 2162 (see note 5), the correspondence of Nickau to the Ministry, July 21, 1847, GStA, (see note 4), 1. HA, Rep. 76 VIII A, Nr. 2122 (die zur Prüfung und Approbation eingereichten Arcana, Heilmittel pp betreffend, vol. 14, Juli 1847 bis Mai 1850), the letter of Bertha Adler to Ladenberg, February 16, 1849 (ebenda), and the Immediateingaben of Davidsohn, February 2, 1846, GStA etc, Nr. 2121 (die zur Prüfung pp vol. 13, 15. May 1846 bis Juni 1847) and Müller, January 26, 1856, GStA etc, Nr. 2125 (die zur Prüfung pp, July 1854 bis Dez. 1857).

11. Recommendation of the Scientific Deputy, March 27, 1846, signed by Klug, Horn, Schmidt. GStA PK (see note 4), I. HA, Rep. 76 VIII A, Nr. 2162.
12. Request of the widow Guiremand, December 7, 1848, GStA PK, Nr. 2122 (see note 10), o.P.
13. Cabinetts-Ordre to State secretary Schuckmann, May 3, 1817, GStA PK, Nr. 2162 (see note 5).
14. Immediatseingabe Seewaldt, November 9, 1836, Nr. 2116 (see note 6).
15. On the *Reichsanzeiger* and popular enlightenment, see Böning, 1987, Böning, 2001 and the papers in Böning, 1992. On enlightenment medicine, see Mann, 1966; Porter, 1995. More than the half of all journals dealt with medicine, agriculture, and home economics (Böning, 1987: 93).
16. For both of these cases, this aspect of the public sphere has been investigated extensively. On Reich, see Hess, 2006 and on Grabe see Freytag, 1998, and Engstrom, 2006.
17. The history of clinical trials remains inadequately researched. As Tröhler, 1988 has shown, in Great Britain, clinical trials dating back to the eighteenth century were employed by medical outsiders (Quakers, Scots, and Unitarians). These groups were unable to base their judgments on personal authority or scientific recognition. The work of Tröhler's students on numerical arguments in contemporary journals shows that most clinical observations were based at most on four cases (Tröhler, 1999).
18. Ministry of Culture to Carl August Maerz, December 2, 1836, GStA PK, Nr. 2116 (see note 6), o.P. Although nowhere explicitly articulated, there existed a standardized procedure at this time. The head of the police department referred to a common procedure involving submission of the formula. See the letter to the military surgeon Paris in Wusterhausen a.d. Dosse, August 30, 1831, Archives of Humboldt University (UAHU), Akten der Charité-Direktion 1725–1945, Nr. 1271 (die Versuche mit Mitteln gegen den Krebs (1824–1852) betreffend, Bl. 33).
19. Ministeriums für Geistliche, Medizinal- und Unterrichtsangelegenheiten to W. Berger, March 11, 1837, GStA PK, Nr. 2116, (see note 6), o.P.
20. Schönlein to Bockel, April 20, 1847, GStA PK, Nr. 2121 (see note 10).
21. Survey of the Scientific Deputation on J. W. Cunz's coagulant remedy (report of Klug, Horn, Schmidt), November 28, 1846, 2121, *Ibid.*
22. *Ibid.*
23. Draft letter of the Ministry of Cultural Affairs to Redslob, October 29, 1847, 2122 (see note 10).
24. Survey of the Scientific Deputation on Hertel's remedy against falling sickness, September 8, 1841, GStA PK (see note 4), Nr. 2119 (Acta betr. die zur Überprüfung und Approbation eingereichten Arcana, Heilmittel pp. von September 1841 bis ult. Dec. 1843. (unfol.)).
25. Report of the Medical Council Dr. Kluge about the therapeutic experiments with Schramm's remedy against arterial rheumatism. UAHU (see note 18), Nr. 1297 (Acta betreffend die Versuche mit Mitteln gegen Flechten; von Jonas 1841), pp. 2–5.
26. Survey of the Scientific Deputation (report of Kluge, Horn, Schmidt), May 20, 1846, 2121 (see note 10).
27. Survey of the Scientific Deputation (report of Kluge, Wagner) on the instructions of the surgeon Krebs at Hornberg in the district Magdeburg, September 7, 1836, Nr. 2116 (see note 6).

28. Ministry of Cultural Affairs to the Scientific Deputation, January 6, 1846, 2120 (see note 6).
29. See, for instance, the survey of the Scientific Deputation (report Kluge), September 7, 1836, Nr. 2116 (see note 6).
30. Survey of Prof. Bartels, August 19, 1836, 2116 (see note 6).
31. Protecting the public against overpriced remedies was only one maxim; allowing only pharmacists to sell medicaments was another.
32. Recommendation of the Minster of Agriculture and Trade, May 3, 1850 (Ramsey, 1994: 60).
33. The academy's drug commission was especially busy in the late eighteenth century. During the *Ancien Régime*, the Royal Commission tested nearly 350 secret remedies from 1772 to 1778 (ca. 50 per annum). Whereas the newly established *Société* had more than 800 applications to test between 1776 and 1790 (including surgical aids, chemical inventions, spirits, and cosmetics, ca. 55 per annum) (Ramsey, 1982: 218 and 222), according to a report of the *Académie* only about 60 cases were assessed between 1825 and 1827 (i.e., ca. 30 per annum). In the second half of the nineteenth century, the number of remedies submitted continued to fall (Ramsey, 1994: 33 and 60f).
34. Kibleur, 1999. Kibleur draws only on French-language literature and doesn't seem to have consulted the work of Ramsey and Weisz.

Bibliography

Anonymous 1799. "Aufforderung an den Prof. Reich in Erlangen; zu R. A." Nr. 177 d.J. *Reichsanzeiger,* No 194 (August 23, 1799), 2218–19.

Anonymous 1800. "Nachricht von den angestellten Versuchen mit den von dem Prof. Reich entdeckten Mitteln, in dem Charité-Lazareth zu Berlin." *Medicinische Ephemeriden von Berlin,* 1, 42–77.

-N- 1799. "Ueber Reich's Mittel gegen fieberhafte Krankheiten." *Reichsanzeiger,* No 250 (October 28, 1799), 2849–53.

Abraham, J. & Davis, C. 2006. "Testing Times: The Emergence of the Practolol Disaster and Its Challenge to British Drug Regulation in the Modern Period." *Social History of Medicine,* 19, 127–47.

Abraham, J. & Reed, T. 2002. "Progress, innovation and regulatory science in drug development: The politics of international standard-settings." *Social Studies of Science,* 32, 337–69.

Ash, E. H. 2010. "Introduction: Expertise and the Early Modern State." *Osiris,* 25, 1–24.

B. 1799. "Über die neu erfundenen Heilmittel des Prof. Reich in Erlangen." *Reichsanzeiger,* No 202 (September 2), 2313–17.

Bonah, C. & Menut, P. 2004. "BCG vaccination around 1930 – Dangerous experiment or established prevention? Practices and debates in France and Germany." *In:* Roelcke, V. & Maio, G. (eds.) *Twentieth Century Ethics of Human Subjects Research : Historical Perspectives on Values, Practices, and Regulations,* Stuttgart: Steiner.

Böning, H. 1987. "Das Intelligenzblatt als Medium praktischer Aufklärung. Ein Beitrag zur Geschichte der gemeinnützig-ökonomischen Presse in Deutschland von 1768 bis 1780." *Internationales Archiv für Sozialgeschichte der Literatur,* 12, 107–33.

Böning, H. (ed.) 1992. *Französische Revolution und deutsche Öffentlichkeit: Wandlungen in Presse und Alltagskultur am Ende des achtzehnten Jahrhunderts,* München: K.G. Saur.

Böning, H. 2001. "Pressewesen und Aufklärung – Intelligenzblätter und Volksaufklärer." *In:* Doering-Manteuffel, S., Mancal, J. & Wüst, W. (eds.) *Pressewesen der Aufklärung.* Periodische Schriften im Alten Reich, Berlin: Akademie Verlag.

Brockliss, L. & Jones, C. 1997. *The Medical World of Early Modern France,* Oxford, NY: Clarendon.

Broman, T. 1998. "The Habermasian public sphere and science in the enlightment." *History of Science,* 36, 123–49.

Broman, T. 2001. "Some preliminary considerations on science and civil society." *Osiris 2nd series,* 17, 1–21.

Broman, T. 2005. "Wie bildet man eine Experten-Sphäre heraus? Medizinische Kritik und Publizistik im 18. Jahrhundert." *In:* Engstrom, E. J., Hess, V. & Thoms, U. (eds.) *Figurationen des Experten. Ambivalenzen der wissenschaftlichen Expertise im ausgehenden 18. und frühen 19. Jahrhundert.* Frankfurt: Lang.

Chauveau, S. 1999. *L'invention pharmaceutique: La pharmacie française entre l'Etat et la société au XXe siècle,* Paris: Institut d'édition sanofi-synthélabo.

Daemmrich, A. 2004. *Pharmacopolitics : Drug Regulation in the United States and Germany,* Chapel Hill etc., The University of North Carolina Press.

Dietz, A. 1922. *Frankfurter Nachrichten und Intelligenzblatt. Festschrift zur Feier ihres zweihundertjährigen Bestehens 1722–1922,* Frankfurt a.M., Druckerei der Frankfurter Nachrichten.

Engstrom, E. J. 2006. "Magnetische Versuche in Berlin, 1789–1835: Zur Entkörperung magnetischer Glaubwürdigkeit." *Medizinhistorisches Journal,* 41, 225–69.

Engstrom, E. J., Hess, V. & Thoms, U. 2005. "Einleitung." *In:* Engstrom, E. J., Hess, V. & Thoms, U. (eds.) *Figurationen des Experten. Ambivalenzen der wissenschaftlichen Expertise im ausgehenden 18. und frühen 19. Jahrhundert.* Frankfurt am Main etc.: Peter Lang.

Ernst, E. 1975. *Das "industrielle" Geheimmittel und seine Werbung. Arzneifertigwaren in der zweiten Hälfte des 19. Jahrhunderts in Deutschland,* Würzburg: jal-verlag.

Freytag, N. 1998. *"Zauber-, Wunder-, Geister- und sonstiger Aberglauben". Preußen und seine Rheinprovinz zwischen Tradition und Moderne (1815–1918),* München.

Hess, V. 1993. *Von der semiotischen zur diagnostischen Medizin. Die Entstehung der klinischen Methode zwischen 1750 und 1850,* Husum: Matthiesen.

Hess, V. 2006. "'Treu und redlich an der Menschheit handeln': Medizinische Innovation zwischen Markt, Öffentlichkeit und Wissenschaft um 1800." *Medizinhistorisches Journal,* 41, 31–49.

Hess, V. 2008. "Precarious matters? Magnetische Heilkraft im klinischen Versuch." *In:* Wahrig, B. (ed.). Preprints, *Max-Planck-Institut für Wissenschaftsgeschichte,* 356.

Hess, V. 2010. "Das Medizinaledikt von 1685. Die Anfänge ärztlicher Standesvertretung zwischen korporativer Autonomie und staatlicher Behörde." *Berliner Ärzte,* 47, 16–19.

Hess, V., Engstrom, E. J. & Thoms, U. 2006. *Expertise und Öffentlichkeit Arzneimittelprüfungen in Preußen (1800–1850).* DFG-Abschlussbericht, Berlin: Manuscript.

Hess, V. & Thoms, U. 2002. "Selektion und Attraktion. Patienten im klinischen Versuch im frühen 19. Jahrhundert." *Archiwum Historii i Filozofii Medycyny,* 65, 197–207.

Huhle-Kreutzer, G. 1989. *Die Entwicklung arzneilicher Produktionsstätten aus Apothekenlaboratorien, dargestellt an ausgewählten Beispielen,* Stuttgart: Deutscher Apothekenverlag.

K-Y 1799. "Über die Aufforderungen an den Prof. Reich, seine Fieber-Mittel bekannt zu machen." *Reichsanzeiger.* No. 235 (October 10), 2681–89.

Kant, I. 1977. *Beantwortung der Frage: Was ist Aufklärung? (1783)*, Frankfurt am Main: Suhrkamp.

Kibleur, P. 1999. "L'evaluation et la validation des remèdes par la Société Royale de Médecine (1778–1793)." *In:* Faure, O. (ed.) *Les thérapeutiques: Savoirs et usages.* Lyon: Merieux.

Lenz, C. L. 1799. "Untrügliches Mittel gegen Fieber, entdeckt von Prof. Reich." *Reichsanzeiger,* 128 (Juni 5), 1485–88.

Löwy, I. & Gaudilliere, J.-P. 2006. "Médicalisation, mouvements féministes et régulation des pratiques médicales: les controverses sur le traitement hormonal de la ménopause." *Nouvelles Questions Féministes,* 25, 48–65.

Mah, H. 2000. "Phantasies of the public sphere: rethinking the Habermas of historians." *Journal of Modern History,* 72, 153–82.

Mann, G. 1966. "Medizin der Aufklärung: Begriff und Abgrenzung." *Medizinhistorisches Journal,* 1, 63–74.

Münch, R. 1995. *Gesundheitswesen im 18. und 19. Jahrhundert. Das Berliner Beispiel,* Berlin: Akademie-Verlag.

Phoebus, P. 1831. *Handbuch der Arzneiverordnungslehre,* Berlin: Hirschwald.

Porter, R. (ed.) 1995. *Medicine in the Enlightenment,* Amsterdam; Atlanta: Rodopi.

Probst, C. 1992. *Fahrende Heiler und Heilmittelhändler: Medizin von Marktplatz und Landstraße,* Rosenheim: Rosenheimer Verlagshaus.

Ramsey, M. 1982. "Traditional medicine and medical enlightenment: The regulation of secret remedies in the ancien régime." *Historical Reflections,* 9, 215–32.

Ramsey, M. 1994. "Academic medicine and medical industrialism: The regulation of secret remedies in France." *In:* La Berge, A. F. & Feingold, M. (eds.) *French Medical Culture in the Nineteenth Century.* Amsterdam: Rodopi.

Riche, A. 1903. "Le rôle administratif de l'Académie de Médecine." *Bulletin de l'Académie de Médecine,* 48, 463.

Rust, J. N. 1834–40. "Die Medicinal-Verfassung Preussens, wie sie war und wie sie ist." *In:* Rust, J. N. (ed.) *Aufsätze und Abhandlungen aus dem Gebiete der Medicin, Chirurgie und Staatsarzneikunde.* 4 ed. Berlin: Enslin.

Selle, Fritze, Richter & Formey 1800. "Nachricht von dem angestellten Versuche mit den von dem Prof. Reich entdeckten Mitteln, in dem Charité-Lazareth zu Berlin." *Medicinisch-chirurgische Zeitung,* 11, 289–315.

Shapin, S. 1994. *A Social History of Truth: Civility and Science in 17th-Century England,* Chicago: University of Chicago Press.

Stürzbecher, M. 1966. *Beiträge zur Berliner Medizingeschichte,* Berlin: Walter de Gruyter & Co.

Thoms, U. 2005. "Arzneimittelaufsicht im frühen 19. Jahrhundert. Konflikte und Konvergenzen zwischen Wissen, Expertise und regulativer Politik." *In:* Engstrom, E. J., Hess, V. & Thoms, U. (eds.) *Figurationen des Experten. Ambivalenzen der wissenschaftlichen Expertise im ausgehenden 18. und frühen 19. Jahrhundert.* Frankfurt am Main etc: Peter Lang.

Tröhler, U. 1988. "'To Improve the Evidence of Medicine': Arithmetic Observation in Clinical Medicine in the Eighteenth and Early Nineteenth Centuries." *History and Philosophy of Life Sciences,* 10 Suppl, 31–40.

Tröhler, U. 1999. "Die wissenschaftliche Begründungen therapeutischer Entscheide – oder 'Evidence-Based Medicines' – im Lauf der Geschichte." *In:* Eich, W., Windeler, J., Bauer, A. W., Haux, R., Herzog, W. & Rüegg, J. C. (eds.) *Wissenschaftlichkeit in der Medizin. Teil III: Von der klinischen Erfahrung zur Evidence-Based Medicine.* Frankfurt am Main: VAS.

Turner, S. 2001. "What is the problem with experts." *Social Studies of Science*, 31, 123–49.

Weisz, G. 1995. *The Medical Mandarins: The French Academy of Medicine in the Nineteenth and Early Twentieth Centuries,* New York: Oxford University Press.

Wimmer, W. 1994. *"Wir haben fast immer was Neues": Gesundheitswesen und Innovationen der Pharma-Industrie in Deutschland, 1880–1935,* Berlin: Duncker & Humblot.

2
Making Salvarsan: Experimental Therapy and the Development and Marketing of Salvarsan at the Crossroads of Science, Clinical Medicine, Industry, and Public Health

Axel C. Hüntelmann

In December 1910 a so-called police physician, Richard Dreuw, launched an attack on Salvarsan, a newly developed and marketed remedy against syphilis, by declaring it a failure. Dreuw wished to expand the discussion on experiences with Salvarsan from the medical–clinical sphere to the medical–social one. "Ehrlich-Hata (the former designation of Salvarsan) is not only a question of medicine, but also a question of culture and the general public; this is why it does not deserve to be measured according only to purely clinical points of view. For this reason alone, social, ethical, and purely human considerations have to be taken into account when discussing the question."[1] Dreuw concluded that there was no ethical aspect that could warrant a physician to deliberately subject his patients to a dangerous treatment like Salvarsan. He also advanced the claim that Salvarsan had not been sufficiently tested for commercial reasons. In fact, on an occasion when Dreuw was lecturing at the Berlin Dermatological Society, the chairman tried to prevent him from speaking three times, but was unable to stop him.[2] Dreuw repeated his allegations over the following years, and although he was supported by others, the allegations remained unheeded. The chemotherapeutic remedy had already been widely discussed in the scientific community and the medical press, and the severe accusations were able neither to delay nor to stop the success of Salvarsan. The therapeutic effect had been proven, validated, and approved, and the related unintended effects had been discussed and negotiated within the public sphere.

Here, I will follow up on the development, production, and marketing of Salvarsan in the Institute for Experimental Therapy and the Georg-Speyer-House in the context of *fin de siècle* history in the German empire. I will not dwell on the successive development stages of the laboratory of the Chemical Department of the Georg-Speyer-House and the toxicological testing that was done at the Biological Department. My emphasis will be instead to discuss the scientific practices and overall institutional context that made Salvarsan an effective and officially approved drug.

There are two reasons why the making of Salvarsan appears to be a good example of the development, production, and regulation of drugs in the twentieth century: first, Salvarsan is considered as the first drug ever to be developed systematically; second, the development of Salvarsan occurred at the intersection of science and the state, industry, and the economy.

I will then discuss the idea of a specific therapeutic remedy of biological or chemical origin developed against an infectious disease at the Institute for Experimental Therapy and at its predecessor, the Institute for Serum Research and Serum Testing. When a therapeutic remedy was developed and the chemical compounds were screened and modified in the Chemical Department of the Georg-Speyer-House, and tested in the Biological Department for their therapeutic impact, such development took place under certain conditions. I will outline these conditions and point out their influence on the development of Salvarsan. Afterward, I will describe the clinical testing of Salvarsan, the discussion within the medical sphere of the therapeutic remedy, the marketing of Salvarsan, and end with the so-called Salvarsan trial of 1914. To summarize the whole process – development, testing, public discussion, marketing, and voluntary control of Salvarsan – I will call the process "the making of Salvarsan." The trial and the judicial decision regarding the (mandatory) use of Salvarsan by public-health officials do not show the failure of the remedy, but rather the success of chemotherapy and the successful making of Salvarsan.

Shifting experimental therapy

Although the early history of the development of chemical drugs can be traced back further, this story begins in the first half of the 1890s with the development of antidiphtheria serum. Antidiphtheria serum was developed by Emil Behring in cooperation with Erich Wernicke and Paul Ehrlich in the aftermath of tuberculin, which had triggered a public-health scandal in Germany. When developing antidiphtheria serum, researchers were very careful: in vitro experiments were followed by in vivo experiments on guinea pigs; these were then scaled up to bigger animals such as sheep, goats, and horses. It was only after the harmlessness of the new serum therapy and the curative value of the sera had been validated in large arrays of animal experiments that clinical trials were conducted on hundreds of patients (Ehrlich, 1894; Ehrlich, *et al.* 1894; Ehrlich & Kossel, 1894; Kossel, 1896; Throm, 1995).[3] The research results were published and widely discussed within the medical scientific community. It was not before August 1894, after four years of research and development, that antidiphtheria serum was introduced into the market and made available in German pharmacies. According to Carola Throm, the development of the antidiphtheria serum had already covered all the aspects of "modern" pharmaceutical research (Throm, 1995, Preface), that is, the sequence of numerous in vitro and in vivo experiments were followed by clinical trials that were critically supervised. As soon as antidiphtheria serum had been marketed, a discussion on its regulation arose.

Antidiphtheria serum, because it fought the disease-causing agent instead of curing the symptoms, was seen as a major therapeutic innovation, and celebrated as a milestone in bacteriology and a revolution in pharmacology (Weatherall, 1990; Schott, 1996; Mochmann & Köhler, 1997; Müller-Jahncke, *et al.* 2005). The new serum therapy did not only offer a curative approach against diphtheria and other fatal infectious diseases, it also promised high profits for manufacturers who would be able to stabilize the production process and produce serum in large quantities in so-called industrial production plants. All the needed detailed information for this purpose was freely available in well-known medical publications, and a health professional trained in bacteriology could reconstruct the experiments and produce the serum. The production of serum in a free market and the prospect of high profits for the serum industry, and finally the novelty of serum therapy were all factors in attracting intense state attention to the new serum therapy with a view to minimizing possible associated public-health risks. A system of state control was thus implemented and institutionalized. By the time the new antidiphtheria serum therapy was publicly accused by a well-known pathologist of having killed his son, the new therapy was already well-established. The regulatory actions and security measures that were set up acted in two directions: on the one hand they were designed to protect the patient as consumer and on the other hand they were also meant to protect the new therapy. Measures in the context of serum regulation required that every production step be recorded in a journal, and every serum batch produced be tested during the production process for its harmlessness and potency and officially approved by a central state-run institute. All of this made it possible to trace the whole production process. It could then be proved that the accused serum phial had been certified as harmless and that after the application of other serum phials originating from the same serum batch all negative effects had been reported on request. The investigation on the death case showed that the security measures of serum regulation had functioned, and this even strengthened serum therapy. In the end, the death case was judged as a fatal coincidence, a fatality, so to speak (Hüntelmann, 2006).

The institutionalized part of serum regulation was to become the foundation of the Institute for Serum Research and Serum Testing in 1896. As reflected in the name of the institute, there were two working tasks and their associated departments. In practice, for the main working task of the institute, institute members were to test the antidiphtheria serum on potency and harmlessness on the one hand and on the other they were to improve and standardize the evaluation methods that were applied to evaluate the therapeutic efficacy of antidiphtheria sera. Later on, these technical issues and practical tasks of serum testing led the members to also become involved in the development and improvement of evaluation methods for other sera and biologicals (Dönitz, 1899; Otto, 1906). On the theoretical side, the director of the institute, Paul Ehrlich, conducted further research on the constitution of diphtheria toxins and antitoxins. At the microbiological level, Ehrlich postulated that vital and pathological processes could be reduced to their (bio)chemical nature: toxins and antitoxins match and neutralize each other in a way comparable with the chemical reaction of acid and alkaline solution (Ehrlich, 1897a). Referring to previous works, he developed a broad theory explaining immunization processes, which would become known as the "side-chain" and later the "receptor" theory. Ehrlich posited that the body cell had specific receptors that were able to bind nutrients and other vital substances that were needed to nourish the cell, building specific side chains. The molecules of the toxic substance also have a specific relationship to body cells – similar to the molecular structure of nutritional substances – and Ehrlich considered that they were able to entrench with the body cells and build side chains. This entrenchment resulted, however, in malfunction and decay of the side chain, finally damaging and threatening the body cell. Ehrlich saw these processes as regenerative. According to Carl Weigert, a cousin of Paul Ehrlich, the failed side chains were reproduced and replaced by the body cell. If the toxic substances also occupied the regenerated side chains, these were also reproduced and replaced – again and again. During this immunization process, the body cell would be trained to reproduce side chains in ever greater dimensions. Side chains that were overproduced and no longer needed were rejected by the cell and injected into the blood circulation. Ehrlich concluded that the abundant side chains were the antitoxins because they had a specific affinity to the toxic substances once again, and were able to bind them and neutralize them (Ehrlich, 1897b; Prüll, 2003).

On the one hand, sera containing organically or biologically produced antitoxins seemed to be an ideal therapeutic remedy. They neutralized the toxins produced by the bacteria perfectly without affecting the infected organism – theoretically. In practice, several side effects appeared. Patients might have idiosyncratic reactions to the natural components of serum. Beyond this, the curative effect and the potency of the serum were difficult to evaluate in a complex procedure. Taking these disadvantages into account, Ehrlich introduced his article on the evaluation of antidiphtheria serum with the hope that once it was possible to produce chemically pure antitoxins and toxins, the complex process of evaluation and the production of "test" or "normal" sera could be reduced to a simple weighing procedure

(Ehrlich, 1897b: 300).[4] As a consequence, Ehrlich aimed for a chemically pure antitoxin, chemically produced and purified, from any natural component. A pure chemical seemed to be more calculable, more standardized, and easier to produce and to handle. Later on, when working on chemotherapeutics, antidiphtheria and other sera served as an ideal example for a chemical "magic bullet," as they targeted and sterilized the disease by causing pathogens, or neutralized the toxin produced by the bacteria (optimally parasitotropic) without affecting the infected organism (minimally organotropic). Although Ehrlich later referred to serum therapy, he was working on an ideal chemical compound that was admittedly linked to and modeled like a biological one, but was even better, more effective, calculable, and pure. Shortly after he had finished his work on the principles of the evaluation of antidiphtheria serum, he started experiments on the components of antitoxins and on the effective processes of immunization, while he also continued his former experiments on the curative effect of dyestuffs, that is, the effect of methylene blue on Plasmodiidae.[5]

During his research on the therapeutic effect of dyestuffs, Paul Ehrlich connected with chemists who were close to him – his nephew Franz Sachs, his cousin Siegmund Gabriel, and his nephew Georg Pinkus, for instance, were working in Emil Fischer's laboratories – clinicians treating patients with malaria in their hospitals, dyestuff companies, and related chemical enterprises such as the Dyestuff Industries at Hoechst, Dyestuff Industries Bayer, and the Baden Aniline and Soda Factory, to name just a few (Hüntelmann, 2011). In this context of shifting experimental therapy, the state regulation of biologicals, at the interface between laboratory, clinical medicine, and industry fostered new research on the therapeutic effects of dyestuffs and arsenic compounds on certain pathogens.

Pure science in the silence of the laboratory and the "development" of arsphenamine

Collaboration between the director of the Institute for Experimental Therapy, clinicians, chemists, and the chemical–pharmaceutical industry became more intense around 1905. Already in 1903, a trained chemist was employed as a member of the institute, and in 1905, several chemically educated assistants followed. These assistants, including Alfred Bertheim, formed the core staff of the Chemical Department of the Georg-Speyer-House, which was founded 1906. The institute's recruitment of chemists had several implications on the structure of the research. At first, contacts with academic chemists outside the institutes diminished and there was less communication between the heads of the institutes and chemists working in university laboratories. Second, the higher staff numbers and their continuity of presence made it possible to focus on a more systematic research schedule. Cooperation between the institute and the chemical industry also changed.

The director of the institute, who had been in contact with several chemical companies at the turn of the century, was cooperating five years later mainly with Dyestuff Industries Hoechst, more particularly with the associated company Leopold Cassella & Co., which at the time was one of the leading dyestuff producers in the world. Soon after the director of Cassella, Arthur Weinberg, founded a pharmaceutical department in 1900, he worked with the Institute for Experimental Therapy and became a close friend of Ehrlich (Bäumler, 1989; Plumpe, 1990). He supported Ehrlich, especially by providing him with chemicals: Weinberg, that is, Cassella, supplied the institute with specific dyestuffs that Ehrlich had asked for, and after promising therapeutic results in vitro and in vivo, he provided the institute with larger quantities of material. When test results were more ambiguous, chemical compounds were modified in the institute or in the Cassella laboratories. Although Ehrlich had had problems finding clinicians when treating malaria patients around 1900, just a few years later he became closely involved with colonial medicine and especially with the sleeping-sickness expedition organized by Robert Koch (Gradmann, 2005; Neill, 2009). In this closer network, members of the Institute of Experimental Therapy and the Georg-Speyer-House screened systematic dyestuffs and arsenic compounds on their therapeutic impact, mainly on trypanosomes and on spirilla. The theoretical idea behind the "screening" referred to the side-chain theory: Ehrlich was searching for the "magic bullet," a chemical substance in which the chemoreceptors perfectly matched the receptors (i.e., nutriceptors) of the spirilla – or even better, a combination of compounds in which the receptors had an affinity with the receptor of the parasite.[6] The idea of an attack on two or more fronts seems to be related to the military idea of a two-front war that was being discussed at that time (Mendelsohn, 1996).

This "screening" of chemical substances points to an indirect cooperation of partners in the scientific community of chemists. Ehrlich read several chemical journals as well as patent literature and corresponded with other physicians and scientists.[7] He drew new ideas from his readings and epistolary discussions. New dyestuffs or arsenic compounds were reported and ordered for the Speyer-House, where they were tested for their therapeutic impact, and new chemical methods and procedures were adapted and tested with known compounds. Inspired by his readings, the director drew blueprints for experiments with references to the original papers. These outlines were forwarded to the assistants as work instructions. For Alfred Bertheim, for example, Ehrlich noted: "In the *Chemischen Neuesten Nachrichten* I read about a patent taken on by Hoechst, and this might be interesting. It is probably possible to oxygenate As=As into AsO by using Bromatin in an alkaline solution. As a test, try No. 418 – the quantity of bromide should already be calculated in the article."[8] These multiple levels of cooperation were to constitute the basis for further developments and the making of new compounds.

The story of the development of arsphenamine is quickly told. After British life-scientists at the Liverpool School of Tropical Medicine had learned of the sterilizing effect of an arsenic compound labeled "Atoxyl" on trypanosomes, further experiments were performed by different scientists – Paul Ehrlich among them. In the course of time and after clinical trials had been conducted in different countries, it turned out that Atoxyl had severe side effects on the optic nerves. Ehrlich then began to remodel Atoxyl: chemical molecules were extracted and replaced by others. By 1906, members of the Chemical Department of the Speyer-House had developed a new compound, acetyla-toxyl, which was more effective and had lesser side effects. In cooperation with Dyestuff Industries Hoechst, further compounds were synthesized and produced in larger quantities (Riethmiller, 2005). Neither acetylatoxyl nor the later synthesized arsenophenylglycine, however, fulfilled the hopes that had been generated after their development, when early in vitro and in vivo experiments had seemed successful. Between 1907 and 1909, larger quantities of both compounds were manufactured at Speyer-House. The chemical substances and intermediate products were provided by Dyestuff Industries Hoechst. The compounds had been sent to clinicians, medical officials, and experts in tropical medicine in the French, British, and German colonies. Ehrlich provided them with arsenic compounds and in return the clinicians sent him medical records, case histories, and reports on the clinical trials. Late in 1909, it became clearer and clearer that acetylatoxyl and arsenophe-nylglycine also had side effects, harming the optic nerves in particular.[9] In 1909, however, the next promising arsenic compound, arsphenamine, had been synthesized. Alfred Bertheim summarized the ongoing research around May 1909. On the basis of the intermediate product nitrophenol-arsonic acid, which was prepared in larger quantities at Cassella & Co. by Ludwig Benda, 3-Nitro-4-Oxy-Phenyl-Arsinacid was combined with sodium hydro-sulfite and heated to between 55 and 60 degrees to reduce the arsenic acid radical to an arseno-group and the nitro-group to an amino-group.[10] After an initial exploratory analysis, the compound was purified from ash residues and sulfates, resulting in a 3.3'-diamino-4.4'-dioxy-arseno-benzol, registered as No. 592. The purified compound was registered later as No. 606 in the "Präparate"-Book.[11]

Subsequently, arsphenamine was transferred to the Biological Department where it was tested in vitro for its sterilizing effect on different pathogens – trypanosomes and different types of spirilla (recurrent fever, chicken spiril-losis, syphilis) – and in vivo for its toxicity on laboratory animals in order to determine the dosis letalis. For the in vitro test, the compound was diluted in water or alcohol, scaled up with saline solution, and mixed with diluted blood that contained spirilla. After an hour, the preparation was analyzed using Dark Field Microscopy.[12] A criterion for measuring the impact of the compound on the spirilla was the mobility of the parasites: their immobility was considered to be a successful result. After these promising results – all the spirilla stopped moving and it was expected that they would be killed – in vivo experiments were conducted in the Biological Department of the Georg-Speyer-House with chickens previously infected with chicken cholera by a Japanese guest researcher, Sahashiro Hata. Just a small quantity of the new compound was sufficient to immobilize or kill the spirilla in the chickens' blood. Further experiments with a series of chickens confirmed these results. After positive results had also been achieved with recurrent fever, the experiments were extended to syphilis (Hata, 1910; Ehrlich, 1910). In the following month, in summer and early autumn, several in vivo experiments were conducted. The laboratory animal of choice was the rabbit, and the experiments achieved amazing results. Even severe ulcers were reduced and healed within a short period of time.

The development of arsphenamine was the result of close cooperation between chemists of the Georg-Speyer-House and the chemical companies Dyestuff Industries Hoechst and Leopold Cassella & Co. Cassella and Dyestuff Industries provided the Georg-Speyer-House for free, but this was not for altruistic reasons or because Ehrlich was a close friend of Arthur Weinberg, the director of Cassella. The director of the Speyer-House and the Institute for Experimental Therapy had multiple and reciprocal relations with the chemical–pharmaceutical industry. Both companies expected high profits from the development of new pharmaceuticals, and by supporting the Georg-Speyer-House they did not have to set up their own research and development departments (Wimmer, 1994). Behind the investment of resources and chemical substances – or financial aid and donations – was the expectation that the scientists at the Speyer-House would be able to develop a new remedy in the long run. Short-term investments were meant to deliver long-term profits. This expectation could be described as relying on trust: trust in personal ability and credibility, as well as institutionalized trust by providing risk capital. This semantic ambiguity underscores the personally and financially entangled relationship between the Institute for Experimental Therapy and the Speyer-House on the one hand and the chemical–pharmaceutical industry on the other, and it became obvious in the application for patents – the legal protection of ideas that might become future products. On June 10, soon after the first biological experiments with the new compound had proved to be promising, Dyestuff Industries Hoechst applied for a patent for arsphenamine (Bäumler, 1979: 216–20).

Ehrlich and his staff followed a schema that had already proved successful during the experiments with antidiphtheria serum and the experiments on the therapeutic effect of dyestuffs. In vitro tests and the first orientating in vivo tests were followed by large-scale animal experiments. At first, arrays of animal tests were to validate the previous experiments. Second, "dosological" investigations were performed on the basis of animal experiments in order to determine the minimum starting dose, the dosis curativa, the dosis tolerata and the dosis letalis. The minimum dose marked the quantity of the compound that was needed to achieve signs of a therapeutic effect, the dosis curativa marked the quantity of arsphenamine that was needed to cure an infected animal, the dosis tolerata marked the maximum quantity of the compound that was tolerated by the organism without any toxic effects, and the dosis letalis marked the quantity of arsphenamine that would simply kill the test animal. The different doses, related to the weight of the animal, were needed to orient further human experiments and to establish the proper dosage (Hata, 1910). At the same time and in close cooperation with the Biological Department, scientists of the Chemical Department conducted experiments to improve the compound and the solubility of arsphenamine, and optimized the production process within the department.

The making of Salvarsan: Clinical trials

After promising therapeutic results in animal experiments, these were transferred to humans. Based on the established dosis tolerata in animals, Paul Ehrlich cooperated closely with the director of the mental asylum of Uchtspringe in conducting experiments designed to establish a dosis tolerata for humans. In September 1909, Ehrlich informed the director of the asylum, Konrad Alt, of the effectiveness of the new compound arsphenamine against spirillosis and asked whether he could test the dosage (Alt, 1910a/1911). The experiments were designed to determine the dosis tolerata for humans, to investigate the physiological processes, and to find an ideal method of application. Alt was also able to conduct initial therapeutic tests because the asylum he ran, like asylum houses in general at that time, dealt with cases of tertiary syphilis such as paralysis or tabes dorsalis. The results were presented to the scientific community in early March 1910 (Alt, 1910b/1911; Ehrlich, 1909, 1910, 1911a and b). Alt reported that the patients treated with arsphenamine showed no severe side effects and that the compound had had a positive effect on the disease, such as the regression of ulcers or a negative Wassermann reaction that indicated the absence of pathogenic spirilla. Alt defined 0.3 grams as the initial therapeutic dose (Alt, 1910a/1911: 74). The physiological progress was investigated by an assistant of Alt (Fischer & Hoppe, 1910/1911). At the same time in Saint Petersburg, Julius Iversen carried out clinical trials on recurrent fever. He, too, came to the conclusion that the new compound was effective against the pathogenic germs (Iversen, 1910 and 1910/1911).

In the spring of 1910, other dermatologists started to test the compound on the basis of Alt, E. Schreiber, J. Hoppe, and Iversen's claims (Alt, 1910b/1911; Schreiber & Hoppe 1910/1911) that arsphenamine had a positive effect on syphilis. Given that the Georg-Speyer-House had no hospital wing like the Institute for Infectious Diseases in Berlin or the Pasteur Institute in Paris, Ehrlich had to contact clinicians to test the therapeutic effect of arsphenamine for him. Ehrlich first asked closely related colleagues like his old friend Albert Neisser in Breslau or Wilhelm Wechselmann in Berlin if they wanted to test the compound (Wechselmann, 1911/1912; Neisser, 1914). From former experiences with atoxyl, arsacetin, and arsenophenylglycine, he knew that transfer from animal to human experimentation was difficult and that side effects or relapses might occur in one patient in a thousand or after a long period of time. Paul Ehrlich therefore planned to test arsphenamine on thousands of patients in the hope of acquiring additional information and assessing the side effects, relapses, indications, and the best form of application (Ehrlich 1911b: 1). The first trials had been conducted by clinicians whom Ehrlich trusted and knew well. The network of cooperation was rapidly enlarged, however, to dermatologists, notably specialists in venereal diseases, who were running an infirmary in a hospital and who were able to conduct clinical trials. For instance, in the summer of 1910, arsphenamine was sent to Bruno Bloch (Basel), Alfred Blaschko (Berlin), A. Broden (Leopoldville), Robert Doerr (Vienna), Piero de Favente (Trieste), a former assistant of the Speyer-House who was participating in an expedition to Africa (Robert Kudicke), J. E. R. McDonagh (London), Arthur Sarbo (Budapest), M. Truffi (Savona), and Wilhelm Weintraud (Wiesbaden), to name only a few.[13] One of these clinicians was Heinrich Loeb, a dermatologist at Mannheim Hospital. Ehrlich and Loeb made contact around March 1910, and Loeb contacted E. Schreiber, who informed him about the clinical trials, the method of application, and the compound. In April 1910, Loeb received phials with arsphenamine for the first time, with the proviso that he should submit regular reports on his experiences. He was asked to select the probands carefully. Patients with severe organic diseases, persons who had already been treated with arsenic compounds, and even patients over 45 were to be excluded from the treatment.[14] In their correspondence, which resembled the previous correspondence Ehrlich had had with clinicians dealing with arsacetin and arsenophenylglycine, Ehrlich discussed the dosage, the form of application, the possible side effects, and the solution of the compound.[15] At intervals, Ehrlich encouraged Loeb to increase the dosage to 0.3 or 0.4 grams because a high initial dose would sterilize the spirilla completely and prevent relapses. Furthermore, he encouraged Loeb to deliver arsphenamine intravenously, which was new to most of the physicians.[16] In June, Loeb presented the results of 130 case studies to the scientific community in a public lecture (Loeb 1910/1911a and b). Similarly, most of the clinicians collaborating with Ehrlich and performing trials published their results in medical journals. At the end of 1910, a bibliography published by Kurt von Stokar included more than 260 publications dealing with arsphenamine-based treatment (Stokar, 1911).[17] Between September 1909, when Alt and Iversen's first trials started, and September 1910, when Ehrlich presented his results at the annual meeting of the German Natural Scientists and Physicians in Konigsberg, Ehrlich had already collected more than 10,000 case studies on arsphenamine (Ehrlich, 1910, Ch. C; 1910/1960, 1911b).

The clinical trials and the use of arsphenamine were discussed at length in the medical scientific community. Besides the presentation of therapeutic success in reference to the case studies, occurring side effects had been a main topic of the publications. After the injection of arsphenamine, sometimes a rash, fever attacks, headaches and muscle pain, and exhaustion, accompanied by a worsening of the symptoms had been reported, but the side effects vanished after a short period of time and health conditions improved (Wechselmann, 1911: 46–55; Stokar, 1911: 14–17). According to Ehrlich, some side effects were caused by wrong application, when arsphenamine was diluted with water and injected in a patient. For instance, when arsphenamine reacted with oxygen, it was transformed into an arsenic oxide combination that caused local corrosion and had noxious effects. Applied carelessly, it could cause painful abscesses and necrotic effects from the injection. Furthermore, if the treating physician did not use distilled water for the dilution, bacteria might also cause abscesses and heavy inflammations from the injection.[18]

The question of side effects and sensibility or the threat of a patient's hypersensibility to arsenic compounds, which could cause an anaphylactic shock, and the issue of relapses or the development of resistances was closely interlinked. Ehrlich favored a high dose of arsphenamine at first treatment, or what he called "therapia magna sterilisans." The high dosage should then eradicate all the parasites from the start in order to prevent any relapses, which again might result in resistances. On the other hand, a high dosage could cause inflammations or provoke a severe anaphylactic reaction. During the clinical trials and during the introduction of arsphenamine into the market in particular, negative effects affected its implementation and induced more cautious physicians to hold on to the traditional mercury cure. The treating physician had to strike a balance between potential negative effects caused by a high dosage on the one hand and the possibility of relapses caused by weak dosage on the other hand – taking into account the patient's constitution and anamnesis.

Negative effects of arsenic compounds on the optic nerves had been discussed, given that the human experiments with atoxyl, arsacetin, and arsenophenylglycine had caused (temporary) blindness in several cases. As Ehrlich and the clinicians involved were very attentive to any rumors on amaurosis, in every presentation Ehrlich gave on the use of arsphenamine, he did not forget to mention that no negative effects on the optic nerves had ever been observed and reported after its use.[19]

Another strain of controversial discussion in the context of "unintended effects" had been that of cases of death after the use of arsphenamine. When in July 1910, a case of death had been discussed in a daily newspaper as related to the use of arsphenamine, Ehrlich became outraged. To Heinrich Loeb, he complained of the agitation that had been initiated against arsphenamine and soon published a refutation.[20] In September 1910, he discussed further cases of death. He distinguished between two different situations in which patients had died and offered two explanations. First, patients in the stage of tertiary syphilis who had been treated with arsphenamine might have died anyway because of the state of advancement of the disease and their bad health status: these had been lost cases (*verlorene Fälle*), stupid idiots (*verblödete Idioten*) with severe brain degenerations with one foot in the grave already. Here, the injection of arsphenamine and death were only a coincidence. Even though there might be a direct relationship between the two events and the injection might have caused death, he argued, they would have died in the near future because of their terminal illness. As part of an experimental setting, their worthless life and their death had been given value for society.[21] Second, a woman ill with syphilis and with a weak health condition had died after an arsphenamine injection. Ehrlich judged her death a misfortune, a fatality that should not have occurred. If a patient, due to an unfortunate accident or due to his or her bad health status, died hours after the injection, one ought not to blame the arsenic compound for this fatality (Ehrlich 1911b: 5). Compared to these fatal cases, thousands of patients had been treated successfully with arsphenamine without any (greater) complications or side effects. The very large numbers of cured cases had to be set against the very few cases of death, which were rare exceptions. Ehrlich reminded that the use of chloroform and the smallpox vaccination also occasionally caused cases of death – nevertheless, no one would want to abandon these fundamental achievements of the life sciences.[22] Interestingly, the same analogy between the successful use of chloroform and the thousands of lives that had been saved by smallpox vaccination that had to be charged up against the few unfortunate cases of death had been constructed when Ernst Langerhans died after an injection of anti-diphtheria serum – as mentioned previously (Hüntelmann, 2006). In the end, Ehrlich summarized, one had to know the contraindications, and the physician had to assess the mortality rate of the disease and the risk of side effects. As a result, Ehrlich defined severe affections of the nervous system and diseases of the heart and the blood vessels as contraindications.[23] In a circular to dermatologists and physicians in October 1910, he declared that the experimental phase had ended, and starting in December 1910, arsphen-amine was launched on the market under the brand name Salvarsan.[24]

Use of and treatment with arsphenamine had been discussed in detail and at length in the medical press from early spring 1910 to the end of that year all over Europe. Whereas the development of arsphenamine took place in the silence of the laboratory involving only few scientists within the Georg-Speyer-House, the making of Salvarsan took place in the context of a wider public. I have reviewed the discussions on the determination of the dosage for humans, investigations on the physiological processes and the ideal method of application, and the therapeutic impact, and the discussions on the unintended effects of arsphenamine. This allows me to contend that arsphenamine was not just launched on the market, but that the therapeutic effect was negotiated within a network of clinicians including the medical scientific community. The discussion on side effects or the critiques, again, had effects on the work in the Chemical Department where scientists were dealing with the improvement of the chemical modification and the solubility. I would describe this whole process of negotiations among chemists, life scientists in the laboratory, clinicians, and the public as the making of Salvarsan.

Production of Salvarsan: The industry

Clinical trials required large quantities of material and by the end of 1909, Ehrlich had instructed Bertheim to prepare a sufficient quantity of material to insure uninterrupted work: "The demand will become hyperbolic."[25] Production of arsphenamine was very difficult: the compound had to be produced, dried, and portioned in a vacuum. Based on the experience with arsenophenylglycine, the head of the Chemical Department rearranged the technical devices to produce large quantities of the compound in the laboratory. Eleven assistants worked for the production process under the supervision of Alfred Bertheim with a view to maintaining the devices, weighing the portions, and then sealing and shipping up to 700 ampoules per day (Bertheim, 1914: 468–69). In the course of 1910, members of the Chemical Department continued their work on arsphenamine in order to improve the compound and to ensure the production process.[26] Since April, all the resources of the Speyer-House had been focused on the production, organization, and testing of arsphenamine. Ehrlich constantly corresponded with Ludwig Benda and Arthur Weinberg of Leopold Cassella & Co., and with engineers working at Dyestuff Industries Hoechst. In July, Dyestuff Industries Hoechst began to build a large-scale production line for arsphenamine. Although the first samples produced in a small test plant had been evaluated as of ideal quality (*hyperideal*), production at an industrial scale generated manifold problems. The longer distances that chemical substances had to travel through the conduit pipes affected the quality of the end product negatively. Hence, in the following months, the production procedure was optimized in cooperation with members of Speyer-House. By November 1910, Dyestuff Industries Hoechst were able to produce arsphenamine, respectively Salvarsan, in large quantities of reliable quality.[27]

The first batches produced in the Chemical Department and then at Dyestuff Industries Hoechst were tested for toxicity in the Biological Department. A certain quantity of arsphenamine was injected into a group of mice. These would die after a certain period of time, but the earlier death of some of the mice was taken as a criterion that the compound was too poisonous. Testing for toxicity was done analogously with the state quality control of antidiphtheria serum at the Institute for Experimental Therapy.

For one, insofar as side effects might discredit the therapy as such, testing was designed to ensure that a compound and more generally the therapeutic approach itself were harmless and effective when administered to patients. Testing for toxicity was designed to ensure that no phial would be filled with noxious material. In response to complaints, Ehrlich could cite the test results and, consequently, there had to have been an error in treatment or abnormal behavior in the patient, such as hypersensitivity. As with serum, testing for toxicity made the compound an approved remedy, regardless of its therapeutic efficacy.[28] After World War I, a test for toxicity was developed in the Georg-Speyer-House on behalf of the League of Nations Hygiene Section and its Permanent Commission on Biological Standardization, and this test became mandatory for all arsenic preparations that were sold on the market (Mazumdar, 2010). International (state) regulation of Salvarsan and biological standardization took place in the context (and along the lines) of the standardization of biologicals such as antidiphtheria serum (Hüntelmann, 2008).[29]

Salvarsan achieved great economic success. In the first year after its marketing, Dyestuff Industry made a profit of 3 million marks.[30] While the long-standing investment in pharmaceuticals began to pay off, public complaints about the high price of Salvarsan were launched in some national conservative newspapers, in which Paul Ehrlich and the Dyestuff Industries Hoechst were characterized as unscrupulous moneymakers. This discussion on the price was only one aspect of a broader campaign against Salvarsan.

Endurance test: The patient and public health

Already in December 1910, Richard Dreuw went public and presented his conclusions on Salvarsan. His opinion was that Salvarsan was not able to sterilize the spirochetes permanently and was only effective against the symptoms of syphilis. He also focused on the side effects, such as infiltrates, painful swelling, necrosis from the injection, and even cases of blindness and death from the medical literature, and argued from his own experience as a police physician (*Polizeiarzt*). In this capacity, he had to examine prostitutes at regular intervals. According to Dreuw, the side effects did not justify the use of Salvarsan, because Salvarsan was unmature and the testing of Salvarsan had been insufficient. His arguments were picked up by a few national conservative German newspapers and brought up during lectures on Salvarsan or discussions on its therapeutic and unintended effects. Over the next few years, the few hundred cases of deaths caused by Salvarsan circulated in the press. Overall, Salvarsan was still very successful and there were only a few doubts, expressed by a minority, on the therapeutic effect of the remedy.[31]

Opponents of Salvarsan, among them only a few physicians and dermatologists, were mainly based in Berlin and Frankfurt. They got more attention after 1913, when Neo-Salvarsan, which was more effective and had better solubility, had already been marketed. Richard Dreuw then contacted members of parliament and the Imperial Health Office. He urged that an official investigation be conducted on the negative effects of Salvarsan, and in March 1914, with regard to Salvarsan and its unintended side effects, more particularly, there were debates in the German parliament on cases of death and whether or not Salvarsan had caused these deaths. In 1913, Karl Wassmann and his brother Heinrich, two journalists in Frankfurt, had written articles in their gazette, *Freigeist*, reporting that prostitutes had been treated against their will with Salvarsan and that several of them had died after this compulsory treatment. A member of the Frankfurt magistrate and the directing physician in the dermatological department of the hospital sued the two brothers for slander. The trial in June 1914 attracted wide public interest. To summarize, the trial was not about the effectiveness of Salvarsan but about whether Salvarsan was an up-to-date remedy, accepted and used by the scientific community. Beyond this lay the implicit question as to whether the testing phase had been extended during the practical use of the drug, or the other way around: Wassmann, Dreuw, and others concluded that patients treated with Salvarsan had been inappropriately used as guinea pigs for the chemical industry.[32]

Once the rumors were propagated, they developed a momentum of their own and practically became fact evidence in the argumentation of the nation's Salvarsan opponents. The arguments and accusations raised by Richard Dreuw since the end of 1910 were repeated and amended with new dubious or fictitious cases and exaggerations. In the spring of 1914, the public-health administration reported on lost confidence, uncertainty among the public, and bewilderment among patients, and on the threats of chemotherapy itself. As a consequence of opponents' agitation, rising bewilderment among the public, and the official petition to the parliament, the Imperial Health Office organized a request to all university clinics and large hospitals in the German empire to report on their experiences and any related side effects related with Salvarsan treatment.[33] Statistics confirmed that Salvarsan was an effective and very potent cure, but had to be handled with care because its arsenic compound might cause severe side effects. Any cases of death had to be correlated to the millions of injections that had been performed. As such, these statistics functioned as a technology of trust (Porter, 1995). The Salvarsan trials in Frankfurt would finally recapture trust in Salvarsan (Hüntelmann, 2006).

From the very beginning of the trial, it was clear that the opponents' accusations had been unfounded. Despite their having been stated repeatedly, they had been disproved in court: reports of cases of blindness emerged as fake because no prostitutes could be found in this case, so those reports were obviously an invention; symptoms like paralysis were part of the disease; the treatment and the death cases were not recognized as being in a causal relation, and so on. In the end, the editor of *Freigeist* was handed a prison sentence.[34]

The Salvarsan trials, as they were called, or the war on Salvarsan, as Ernst Bäumler described the agitation against Paul Ehrlich and Salvarsan, were a matter of interest because the trials, that is, the agitation, focused on resistances against the remedy and on the patient. As a remedy, Salvarsan was very difficult to handle. Taking this into account, Salvarsan was only obtainable in a pharmacy and on prescription. The target group for the Dyestuff Industries Hoechst was the practitioner or the dermatologist, not the patient. Direct advertising for remedies was seen very critically and deemed unethical by physicians (Binder, 2000). Salvarsan promotion by Dyestuff Industries was very low key. Advertising was not needed, however, because due to the numerous publications, Salvarsan was well-known in the scientific community. We have to distinguish between the dermatologist and public-health personnel performing the injection on the one hand and the patient as the recipient on the other hand. For certain diseases, the patient is very visible – for diphtheria, for example, suffering children or concerned mothers were mentioned, imagined, or shown in pictures. Suffering from syphilis, however, was different, and the patient usually wished to remain invisible – especially as in this disease, the associated moral lapse was often its visible part (Sander 1998). Male patients in particular often tried to remain invisible, and although female patients were invisible as patients, prostitutes were the focus of public attention. In the campaigns combating syphilis, they were the first group addressed, and with their agreement or against their will, their treatment with Salvarsan was compulsory.[35]

Human experiments by the chemical industry, as alleged by some opponents, and resistance against compulsory treatment with Salvarsan lead to the biopolitics or public-health aspect. I will summarize the different biopolitical aspects of Salvarsan, which was developed at the intersection of several transformations: the most rapid phase of German industrialization, the constitution of the pharmaceutical industry, the standardization of different spheres of society, and an increase in biopolitical activities to improve public health in the national "struggle for survival." This involved the foundation of public-health institutions, such as the Institute for Experimental Therapy. The primary task of these institutions had been to minimize public-health risks by reinforcing protection of the population from harmful drugs as well as the protection of (therapeutic) concepts from misuse (approval of serum and later, approval of Salvarsan), which again was supposed to protect the population from epidemics and widespread diseases. The concept of chemotherapy corresponded best to the biopolitical concept of fighting a widespread moral disease like syphilis in its specificity and the goal of sterilizing the parasite – in contrast with other concepts such as that of social hygiene, which propagated the fight against diseases like tuberculosis through social or political change or, in the case of syphilis, through moral change.

Despite Dreuw's agitation, despite the severe accusations (if not slander) and the resistance in Frankfurt, and the "Salvarsan trials," the rumors did not really affect the success of Salvarsan as a remedy. The accusations that led to the Salvarsan trials could be interpreted as a certain endurance test to find whether safety measures like clinical trials had been sufficient. The uncertainty among the public and bewilderment among patients lasted for only a short moment. After the trials, confidence in chemotherapy and Salvarsan regained strength. The accusations were not able to threaten the remedy because countless clinical trials had been performed, published, and discussed for the therapeutic as well as the unintended effects of its use. The accusations did not hit the mark and were easy to rebut. The extensive clinical trials, the permanent observation of the usage of Salvarsan even after it had been brought to the market, the voluntary toxicological testing, and the open discussion of side effects proved to be very beneficial and in the end strengthened chemotherapy.

The making of Salvarsan

When Paul Ehrlich was working under Robert Koch in the Berlin Communal Hospital Moabit testing tuberculin on its therapeutic value *after* the tuberculin scandal had swept across the medical community, which was condemning the cure wholesale, and having had to deal with strong criticism, it might have entered his mind that clinical trials were an elementary sine qua non in the development of a new remedy. A few years later, when he was developing antidiphtheria serum with Emil Behring, clinical trials – although in the small number of a few hundred – had been performed before the remedy had been put on the market. Directly after the market launch of antidiphtheria serum, a system of serum regulation had been installed to minimize the public-health risk that might occur after the use of the serum. Paul Ehrlich became head of the newly founded Institute for Serum Research and Serum Testing.

One of the working tasks of Ehrlich had been to further work on the constitution of diphtheria toxin and antitoxin. In this context, he formulated his side-chain theory, which again shifted the program to synthesized artificial chemical "magic bullets" similar to the antitoxin. Salvarsan was developed in this framework of serum regulation and the extensive screening of chemical substances. This regulation aspect, in the sense of risk management, meant to deal not only with unintended effects like side effects but with unintended effects like public resistance, became important in the Salvarsan trials. During the screening of chemical compounds, the investigations of the proper dosage, and the clinical therapeutic testing of arsphenamine, several different groups had been involved: the state, which financed the head (and privy council) of the Institute for Experimental Therapy as well as the Georg-Speyer-House Paul Ehrlich, the city of Frankfurt, the industry, which was supporting Ehrlich with material (and was bound via work contracts), chemists, and within the medical scientific community, clinicians and practitioners, public-health institutions such as hospitals, and patients. They all played a pivotal role in the development and use of Salvarsan, but the making of Salvarsan took place in the process of negotiating the therapeutic effects as well as the unintended effects. Since the development of antidiphtheria serum, the process of development of a new drug covered by in vitro and in vivo series (with upscaling from small to big test animals) of tests and clinical trials, as well as the observation (and regulation) of the use of the drug after its launching on the market had been initially installed and established with Salvarsan, and became a role model for most of the drugs being developed afterward. Thus Salvarsan has to be seen as in line with antidiphtheria serum, especially as it was developed by the same person and in a similar institutional framework. The patient played only a marginal role because the treating physician had the power of interpretation: he defined whether an ulcer had regressed or the Wassermann reaction had been negative. In the treatment of most of the potential patients, prostitutes, and soldiers, these had no power at all: compulsory treatment was widely practiced and led, in fact, to the Salvarsan trials.

In court, Salvarsan was judged to be an up-to-date treatment and seen as a cure against syphilis. This – expected – positive judgment had its origin in the making of the drug within the process of making Salvarsan. The performed clinical trials as an evidence-based system developed a momentum on their own – most of the medical scientific community had already defined (and judged) Salvarsan as a valid cure.

Notes

1. "Es handelt sich bei Ehrlich-Hata nicht nur um eine medizinische, sondern um eine Kultur-Frage und eine Frage der Allgemeinheit, und die deswegen verdient, nicht nur nach rein klinischen Gesichtspunkten gemessen zu werden. Ein Körnchen sozialer, ethischer und rein menschlicher Betrachtung muss daher schon bei der Erörterung dieser Frage mitsprechen." Federal Archive Berlin (Bundesarchiv Berlin, henceforth BAB) R 86/2817.
2. See *Deutscher Tagesanzeiger*, 19.12.1910.
3. See also the report of Dr. Weisser on antidiphtheria serum in October 1894 in BAB R 86/1646.
4. "Wenn es gelungen sein wird, das Antitoxin resp. das Gift in chemisch reiner Form zu gewinnen, wird die Herstellung von Normalprüfungslösungen auf einer einfachen Wägung beruhen und es überflüssig sein, irgendwelche Maßnahmen für die Erhaltung der Konstanz des Massstabes zu treffen." "If we will succeed one day to produce antitoxin resp. toxin in a pure form, the production of a standard solution will then base only on a simple weighing and it will become no longer necessary to adopt measures to keep the standard reference constant" (Ehrlich, 1897b: 300).

5. Paul Ehrlich to BASF, 15.11.1898; and Ehrlich to Rudolf Nietzki, 24.12.1898, Rockefeller Archive Center, 650 Eh 89 Paul Ehrlich Collection (henceforth RAC PEC), Box 4.

6. Paul Ehrlich explained his idea of a magic bullet in several articles. For the side-chain theory, receptor immunology, and the receptor concept, see Parascandola, 1981; Silverstein, 2002; Prüll, 2003; Prüll *et al.*, 2008.

7. The director of Speyer-House wrote several instructions to his assistants to please read and reproduce patent xy, for instance, or during an experiment he referred to patent literature to improve results, as in his note to Robert Kahn: "I wish merely to direct your attention to the latest issue of the *Chemische Neueste Nachrichten*, the Bych Patent, the reduction from acid to aldehyde! Might this also be possible for anthranilic acid?" Note 21.1.1910, RAC Box 33.

8. Paul Ehrlich to Alfred Bertheim, Note 11.2.1910, RAC Box 33, p. 278.

9. See the correspondence between Paul Ehrlich and Felix Mesnil, 1908–1910, Archives Institute Pasteur Paris, SPE.2; correspondence between Paul Ehrlich and German colonial officials in BA Berlin, R 1001/5876–5878, 5889, 5903; correspondence between Paul Ehrlich and Dyestuff Industries in Dokumente aus Höchster Archiven, Vol. 14 (1966) and Vol. 19 (1966) and the letter of Paul Ehrlich to August Wassermann, 4.1.1910, reporting about the failure, RAC PEC Box 26.

10. Instructions from Paul Ehrlich to Ludwig Benda 14.3.1909; notes from Paul Ehrlich to Wilhelm Röhl, 2.4.1909, 18.5.1909, RAC PEC Box 32.

11. For the development of arsphenamine see Ehrlich, 1910: 115–25; idem, 1911; Ehrlich & Bertheim, 1912; Bertheim, 1914: 466–68.

12. For microscopy techniques, see Crary, 1990.

13. A list of cooperating clinicians is given in Ehrlich 1910.

14. See the letters Paul Ehrlich sent to Heinrich Loeb from April 9 and 11, 1910, RAC Box 57, Folder 6.

15. See the mentioned correspondence between Paul Ehrlich and Felix Mesnil between 1908 and early 1910, Archives Institute Pasteur, SPE.2.

16. See the correspondence between Paul Ehrlich and Heinrich Loeb, RAC Box 57, Folder 6.

17. Ehrlich supported his collaborators to publish their results to generate a public debate on his compound, see Ehrlich, 1910/1960: 242. Ehrlich edited some of the publications (Ehrlich, 1911–1917).

18. See, for example, the contributions of Edward C. Hort *et al.*, James McIntosh *et al.*, and S. Samelson on the dangers of saline injections and the prevention of toxic symptoms in Ehrlich, 1911–1917, Vol. III (1913).

19. Until end of 1910, no case of amaurosis had been reported, Ehrlich, 1910: 142. There were complaints about the "fable of blindness" in Ehrlich, 1911b: 2. Wilhelm Wechselmann (1911: 32), a clinician close to Ehrlich, as well as about false rumors on blindness.

20. Paul Ehrlich and Heinrich Loeb, 5.8.1910, RAC PEC Box 57 Folder 6.

21. See Ehrlich, 1910: 140–44; Ehrlich, 1911b: 5; Lecture given at the 82nd Meeting of German Scientists and Physicians in Konigsberg, 20.9.1910, RAC PEC Box 3 Folder 9.

22. Ehrlich estimated the mortality rate of chloroformation at around one case per 2,060. See Ehrlich, 1910: 140–41; Ehrlich, 1910/1960: 243; Ehrlich, 1911b.

23. See Ehrlich, 1910: 140–44; Ehrlich, 1910/1960: 243; Ehrlich, 1911b. Cases of death had been discussed in Ehlers, 1910/1911; Wechselmann, 1913.

24. See Paul Ehrlich, "Permanent Action on Ehrlich's New Remedy 606" in Ehrlich/
    Himmelweit, 1960: 342–43; Circular, October 25, 1910 in the correspond-
    ence to Heinrich Loeb, RAC Box 57, Folder 6. See the clinical tests in detail in
    Hüntelmann, 2009.
25. Paul Ehrlich to Alfred Bertheim, 1.12.1909, RAC Box 33, p. 34.
26. Alfred Bertheim, for instance, received the instruction to improve the solubility.
    Paul Ehrlich to Alfred Bertheim, 18.1.1910, RAC Box 33, p. 207. At the Biological
    Department, research was forced to deal with the problem of hypersensibility
    and relapses that were related to the application.
27. See the report on the introduction of Salvarsan at the Dyestuff Industries Hoechst
    by B. Reuter, Histocom Archive (Hoechst Archive); Bäumler 1979: 237–38, 241.
28. See Hüntelmann, 2009.
29. Also like the international standardization of the Wassermann reaction. See
    Mazumdar, 2003.
30. See the table about the profits with Salvarsan in the personal papers of Ernst
    Bäumler, Berlin Museum of Medical History.
31. See material like newspaper clippings, separata by Dreuw, Wassmann, and other
    opponents concerning Salvarsan in BAB R 86/2812 and 2817; and the chapter
    about the war on Salvarsan in Bäumler, 1979.
32. See material like newspaper clippings, off-prints by Dreuw, Wassmann, and other
    opponents concerning Salvarsan in BAB R 86/2812 and 2817; and the chapter
    about the war on Salvarsan in Bäumler, 1979.
33. One has to mention that clinicians mainly supported the use of Salvarsan.
34. See material like newspaper clippings, separata by Dreuw, Wassmann, and other
    opponents concerning Salvarsan in BAB R 86/2812 and 2817; and the chapter
    about the war on Salvarsan in Bäumler, 1979.
35. Another group had been soldiers, but they do not play a role here and they would
    not have had the possibility of complaining. See Sauerteig, 1996, 1999.

Bibliography

Alt, Konrad 1910a/1911: "Das neueste Ehrlich-Hata-präparat gegen Syphilis."
    Abhandlungen über Salvarsan (Ehrlich-Hata-Präparat gegen Syphilis) ed. Paul Ehrlich,
    Vol. I. Munich: 67–76.
Alt, Konrad 1910b/1911: "Zur Technik der Behandlung mit dem Ehrlich-Hataschen
    Syphilismittel," *Abhandlungen über Salvarsan (Ehrlich-Hata-Präparat gegen Syphilis)*
    ed. Paul Ehrlich, Vol. I. Munich: 17–20.
Bäumler, Ernst 1979: *Paul Ehrlich. Forscher für das Leben*. Frankfurt am Main.
Bäumler, Ernst 1989: *Farben, Formeln, Forscher. Höchst und die Geschichte der industri-
    ellen Chemie in Deutschland*. Munich.
Bertheim, Alfred 1914: Chemie der Arsenverbindung. *Paul Ehrlich. Eine Darstellung
    seines wissenschaftlichen Wirkens. Festschrift zum 60. Geburtstage des Forschers*, by
    Hugo Apolant *et al.*, Jena, 447–76.
Crary, Jonathan 1990: *Techniques of the Observer: On Vision and Modernity in the 19th
    Century*, Cambridge.
Dönitz, Wilhelm 1899: Bericht über die Thätigkeit des Königl. Instituts für
    Serumforschung und Serumprüfung zu Steglitz Juni 1896–September 1899.
    Klinisches Jahrbuch 7 (1899): 359–84.
Ehlers, Dr. Ein Todesfall nach Ehrlich-Hata "606." *Abhandlungen über Salvarsan
    (Ehrlich-Hata-Präparat gegen Syphilis)* ed. Paul Ehrlich, Vol. I. Munich 1911: 274–75.

Ehrlich, Paul 1894 : "Über die Gewinnung, Werthbestimmung und Verwerthung des Diphtherieheilserums." *Hygienische Rundschau* 4: 1140–52.

Ehrlich, Paul 1897a : "Zur Kenntnis der Antitoxinwirkung." *Fortschritte der Medizin* 15: 41–43.

Ehrlich, Paul 1897b : "Die Wertbemessung des Diphterieheilserums und deren theoretische Grundlagen." *Klinisches Jahrbuch* 6: 299–329.

Ehrlich, Paul 1909 : "Chemotherapie von Infektionskrankheiten." *Zeitschrift für ärztliche Fortbildung* 6: 721–33.

Ehrlich, Paul 1910 : Schlußbemerkungen. *Die experimentelle Chemotherapie der Spirillosen (Syphilis, Rückfallfieber, Hühnerspirillose, Frambösie)*, ed. idem and Sahachiro Hata, Berlin: 114–64.

Ehrlich, Paul 1910/1960 : "Die Behandlung der Syphilis mit dem Ehrlichschen Präparat 606." *Collected Papers* ed. Fred Himmelweit, Vol III, New York: 240–46.

Ehrlich, Paul 1911a : *Aus Theorie und Praxis der Chemotherapie*, Leipzig.

Ehrlich, Paul 1911b : "Die Salvarsantherapie. Rückblicke und Ausblicke." *Münchener Medizinische Wochenschrift* 58: 1–10.

Ehrlich, Paul 1911–1917 (ed.): *Abhandlungen über Salvarsan (Ehrlich-Hata-Präparat gegen Syphilis)* 5 Vols. Munich.

Ehrlich, Paul and Fred Himmelweit 1960 (ed.): *Gesammelte Arbeiten.* Vol III, New York.

Ehrlich, Paul and Alfred Bertheim 1912: "Über das salzsaure 3.3'-Diamino-4.4'-dioxy-arsenobenzol und seine nächsten Verwandten." *Berichte der Deutschen Chemischen Gesellschaft* 45: 756–66.

Ehrlich, Paul and Sahachiro Hata 1910: *Die experimentelle Chemotherapie der Spirillosen (Syphilis, Rückfallfieber, Hühnerspirillose, Frambösie).* Berlin.

Ehrlich, Paul and Hermann Kossel 1894: "Über die Anwendung des Diphtherieantitoxins." *Zeitschrift für Hygiene und Infektionskrankheiten* 17: 486–88.

Ehrlich, Paul et al. 1894: "Über Gewinnung und Verwendung des Diphterieheilserums." *Deutsche medizinische Wochenschrift* 20: 353–55.

Fischer, Ph. and J. Hoppe 1910/1911: "Das Verhalten des Ehrlich-Hataschen Präparates im menschlichen Körper." *Abhandlungen über Salvarsan (Ehrlich-Hata-Präparat gegen Syphilis)* ed. Paul Ehrlich, Vol. I. Munich: 45–49.

Gradmann, Christoph 2005, *Krankheit im Labor. Robert Koch und die medizinische Bakteriologie.* Göttingen.

Hata, Sahachiro 1910: "Experimentelle Grundlage der Chemotherapie der Spirillosen," *Die experimentelle Chemotherapie der Spirillosen (Syphilis, Rückfallfieber, Hühnerspirillose, Frambösie)*, ed. Paul Ehrlich and Sahachiro Hata, Berlin.

Hüntelmann, Axel 2008: "The dynamics of *Wertbestimmung.*" *Science in Context* 21: 229–52.

Hüntelmann, Axel 2009: "1910. Transformationen eines Arzneistoffes – vom 606 zum Salvarsan." *Arzneimittel des 20. Jahrhunderts. 13 historische Skizzen von Lebertran bis Contergan*, ed. Nicholas Eschenbruch et al., Bielefeld: 17–51.

Hüntelmann, Axel 2011: *Paul Ehrlich. Leben, Forschung, Ökonomien, Netzwerke*, Göttingen.

Iversen, Julius 1910: "Chemotherapie des Recurrens." *Die experimentelle Chemotherapie der Spirillosen (Syphilis, Rückfallfieber, Hühnerspirillose, Frambösie)*, ed. Paul Ehrlich and Sahachiro Hata, Berlin: 90–108.

Iversen, Julius 1910/1911: "Ueber die Wirkung des neuen Arsenpräparates (606) Ehrlichs bei Rekurrens." *Abhandlungen über Salvarsan (Ehrlich-Hata-Präparat gegen Syphilis)*, ed. Paul Ehrlich, Vol. I. Munich 1911: 343–51.

Loeb, Heinrich 1910a/1911, "Erfahrungen mit Ehrlichs Dioxy-diamido-arsenobenzol (606)." *Abhandlungen über Salvarsan (Ehrlich-Hata-Präparat gegen Syphilis)* ed. Paul Ehrlich, Vol. I. Munich: 101–05.

Loeb, Heinrich 1910b/1911: "Weitere Erfahrungen über '606.'" *Abhandlungen über Salvarsan (Ehrlich-Hata-Präparat gegen Syphilis)* ed. Paul Ehrlich, Vol. I. Munich, 106–12.

Mazumdar, Pauline M. H. 1995: *Species and Specificity: An Interpretation of the History of Immunology.* Cambridge.

Mazumdar, Pauline M. H. 2003: "'In the silence of the laboratory'. The league of nations standardizes syphilis tests." *Social History of Medicine* 16: 437–59.

Mazumdar, Pauline M. H. 2010: "The state, the serum institutes and the league of nations." *Evaluation and Standardizing Therapeutic Agents, 1890–1950,* ed. Christoph Gradmann and Jonathan Simon, Basingstoke.

Mendelsohn, John Andrew 1996: *Cultures of Bacteriology. Formation and Transformation of a Science in France and Germany, 1870–1914.* Diss. Phil. Princeton University, Princeton.

Mochmann, Hanspeter and Werner Köhler 1997, *Meilensteine der Bakteriologie. Von Entdeckungen und Entdeckern aus den Gründerjahren der Medizinischen Mikrobiologie,* 2nd edn, Frankfurt am Main.

Müller-Jahncke, Wolf-Dieter *et al.* 2005, *Arzneimittelgeschichte,* 2nd edn, Stuttgart.

Neill, Deborah 2009: "Paul Ehrlich's colonial connections: Scientific networks and sleeping sickness drug therapy research, 1900–1914." *Social History of Medicine* 22: 61–77.

Neisser, Albert 1914: "Einleitender Überblick. Salvarsan und Syphilis." *Paul Ehrlich. Eine Darstellung seines wissenschaftlichen Wirkens. Festschrift zum 60. Geburtstage des Forschers,* by Hugo Apolant *et al.,* Jena: 515–40.

Otto, Richard 1906: *Die staatliche Prüfung der Heilsera* (Arbeiten aus dem Institut für experimentelle Therapie 2). Jena.

Parascandola, John 1981: "The theoretical basis of Paul Ehrlich's chemotherapy." *Journal of the History of Medicine* 36: 19–43.

Plumpe, Gottfried 1990: *Die IG-Farbenindustrie AG. Wirtschaft, Technik und Politik, 1904–1945.* Berlin.

Prüll, Cay-Rüdiger 2003: "Part of a scientific master plan? Paul Ehrlich and the origins of his receptor concept." *Medical History* 47: 332–56.

Prüll, Cay-Rüdiger *et al.* 2008: *a Short History of the Drug Receptor Concept,* New York.

Riethmiller, Steven 2005: "From Atoxyl to Salvarsan: Searching for the magic bullet." *Chemotherapy* 51: 234–42.

Sander, Gilman L. 1988: *Disease and Representation: Images of Illness from Madness to AIDS,* Ithaca. NY.

Sauerteig, Lutz 1996a: "Salvarsan und der 'ärztliche Polizeistaat'. Syphilistherapie im Streit zwischen Ärzten, pharmazeutischer Industrie, Gesundheitsverwaltung und Naturheilverbänden (1910–1927)." *Medizinkritische Bewegungen im Deutschen Reich (ca. 1870–ca. 1933),* ed. Martin Dinges, Stuttgart: 161–200.

Sauerteig, Lutz 1996b: "Militär, Medizin, Moral. Sexualität im Ersten Weltkrieg." *Die Medizin und der erste Weltkrieg,* ed. Wolfgang U. Eckart and Christoph Gradmann, Pfaffenweiler: 197–226.

Sauerteig, Lutz 1999: *Krankheit, Sexualität, Gesellschaft. Geschlechtskrankheiten und Gesundheitspolitik in Deutschland im 19. und frühen 20. Jahrhundert.* Stuttgart.

Schreiber, E. and J. Hoppe, "Ueber die Behandlung der Syphilis mit dem neuen Ehrlich-Hataschen Arsenpräparat (No. 606)." *Abhandlungen über Salvarsan (Ehrlich-Hata-Präparat gegen Syphilis),* ed. Paul Ehrlich, Vol. I. Munich: 77–83.

Schott, Heinz (ed.) 1996: *Meilensteine der Medizin*. Dortmund.

Silverstein, Arthur M. 2002: *Paul Ehrlich's Receptor Immunology: The Magnificent Obsession*. San Diego.

Stokar, Kurt von 1911: *Die Syphilis-Behandlung mit Salvarsan (Ehrlich Hata 606) nebst einer systematischen Zusammenfassung der bisher veröffentlichten Literatur*. Munich.

Throm, Carola 1995: *Das Diphtherieserum. Ein neues Therapieprinzip, seine Entwicklung und Markteinführung*. Stuttgart.

Weatherall, Miles 1990: *In Search of a Cure: A History of Pharmaceutical Discovery*. Oxford.

Wechselmann, Wilhelm 1911/1912, *Die Behandlung der Syphilis mit Dioxydiamidoarsenobenzol ("Ehrlich-Hata 606")*, 2 vols, Berlin.

Wechselmann, Wilhelm 1913: *Zur Pathogenese der Salvarsantodesfälle*. Berlin.

Wimmer, Wolfgang 1994: *"Wir haben fast immer was Neues". Gesundheitswesen und Innovation der Pharma-Industrie in Deutschland, 1880–1935*. Berlin.

3
Professional and Industrial Drug Regulation in France and Germany: The Trajectories of Plant Extracts

Jean-Paul Gaudillière

Introduction

On September 11, 1941, Marshall Pétain's government of nonoccupied France issued a law reorganizing drug supply in the country. Based on a decade-long discussion on the effects of industrialization on drug production and sales, the reform replaced the previous system – in which new remedies required professional approval from the Académie de Médecine – by establishing an administrative procedure for securing a *visa de spécialité pharmaceutique*. This "visa" was the political recognition of the "great transformation" that had led industrially made preparations to dominate the markets for thera-peutic agents (Chauveau, 2005). The visa was a form of marketing permit. It authorized the introduction of a specialty if, and only if, a committee made up of physicians, pharmacists, representatives of the industry, and health administrators (the *Commission Technique des Spécialités*) gave its green light. According to the law, this technical committee was to assess the compos-ition, the conditions of production, the novelty, and the absence of toxicity of the product. The new law also included a short clause specifying that the state-granted herbalist certificate would no longer be delivered; as a result, the *École d'herboristerie*, which had been associated with the Paris Faculty of Pharmacy, would disappear.[1] Herbalists would thus no longer have an official status, ending a century-long battle with pharmacists over control of medicinal plants. This particular item had been written into the law as part of a trade-off with the pharmaceutical profession, which was otherwise deeply affected by the proindustrial, modernist, and technocratic perspec-tives characterizing the rest of the law.

On February 11, 1943, the national–socialist administration in command of chemical and pharmaceutical production in Germany announced a reinforced control of the drug market in order to cope with the rapidly growing difficulty of finding raw materials and supplying the military.

The 1943 decrees drastically limited the possibility of putting new drugs on the market. The first idea had been to ban all new drugs. Responding to protests from the industry, a complementary decree was finally passed, which introduced a special procedure for obtaining marketing authorizations.[2] New drugs would require the approval of government authorities previous to their being put on the market. Although adopted on the basis of very different concerns, the German *Stop-Verordnung* established a mechanism resembling the French visa in several respects. Producers of pharmaceutical novelties were to obtain a special certificate by documenting the composition, labeling, packaging, dosage, toxicity, and clinical utility of their invention. Documentation was to be provided by the manufacturer, proving the value of the new therapeutic composition per se, as well as in relation to existing therapies. In contrast to its French counterpart, the 1943 decree, however, did not challenge what had been the most important change in the German order of pharmacy since World War I, namely, the 1939 recognition of *Heilpraktiker* as a legitimate profession, in which members were controlled through a state exam and received a license authorizing them to prescribe and prepare a number of therapeutic agents – not only homoeopathic remedies but also medicinal plants and their derivatives – that did not belong to the toxic classes (Haug, 1985; Brinkmann & Franz, 1982).

Adopted under highly peculiar circumstances, these legal changes determined the nature of the postwar drug order in both countries. The French visa system would not be changed until the late 1950s, when the country had to face its first public-health crisis associated with the adverse effects of chemical therapeutic agents.[3] In Germany as well, the 1943 legislation was maintained after the end of the war and, as mentioned in the introduction to this volume, revised only in the early 1960s within the context of the thalidomide/Contergan affair. In both cases, new ways of regulating drugs had been established during the war, signaling an altered power relationship among the industry, the state, and the medical and pharmaceutical professions.

The word "regulation" is often intended to mean an action taken by the government, an administrative body, or any form of state institution to control the marketing of drugs. Closer attention to the diversity of practices involved in the "trajectories" of drugs suggests that a broader definition is worth considering (Gaudillière, 2005). Regulation may then be viewed as a series of *dispositifs*, or purviews, not only targeting commercial practices, but also aiming to define production standards or to set norms for medical uses. Regulation is not exclusively a problem of government control and marketing authorizations; it is also a problem of legitimate patterns of action within the laboratory, the production plant, the doctor's consultation room. As suggested in the introduction to this volume, the past century has seen the emergence of different "ways of regulating," which may be characterized by targeted values and aims, the main actors involved, acceptable forms of evidence, and legitimate means of intervention.[4]

Focusing on the interwar period, this chapter addresses the relationship between the professional and the industrial ways of regulating that surfaced during the first half of the twentieth century, when the majority of therapeutic preparations used in medicine became industrial goods, and started to be mass-produced and – to some extent – standardized. As previously outlined, the professional regulation of drugs takes its roots in the nineteenth-century corporate organization of pharmaceutics. In continental Europe, it evolved out of a state delegation of expertise, which granted pharmacy graduates a monopoly over the sale and preparation of the therapeutic agents included in the national pharmacopoeia. Such regulation was justified as a means to avoid unnecessary competition among pharmacists, as well as to bar the entry into the market of supposedly untrained and unskilled practitioners. Under this regulation, professional governance emanated from corporations, pharmaceutical societies, and their medical counterparts, possibly supplemented by special committees of experts set up by academic journals or public-health authorities. Within this framework, judgment on the value of specific drugs focused on the production of pharmacological knowledge with emphasis on the links among dosage, concentration within the body, and pattern of elimination, and on the balance between toxic and therapeutic responses. The basic idea was that all drugs were potentially poisons if used without care. The routine practice of regulation thus consisted in defining good practices, that is, mandatory protocols for the preparation of each type of drug, appropriate dosage, and lists of recommended indications and contraindications. In addition to the traditional pharmacopoeia, regulatory tools included the many forms of guidelines and recommendations for practice issued by collectives of pharmacists and/or physicians.

During the first decades of the twentieth century, the dynamics of professional regulation were less and less able to cope with the gradual transformation of many workshops into proper factories, with the development of industrial specialties, and with the growing competition between pharmaceutical firms and chemical corporations. Academic pharmacists often perceived ready-made specialties as the reign of secrecy, economic greed, bad quality, unproven efficacy, and marketing lies. Although the leaders in the profession often maintained a more ambiguous attitude toward mass production, they, nonetheless, supported the adoption of measures such as special authorization systems that would limit the introduction of specialties, and if possible place them under the control of the profession. Such attempts were rarely successful until the above-mentioned marketing permits were introduced, albeit with a different purpose, but they triggered important changes within the producing firms.

Standardization and quality control became the slogans of industrial reorganization, which combined two different aims. The first was to maximize the creation of economic value by reducing the consumption of raw materials and increasing manufacturing productivity. The second was to ensure that preparation errors would not result in sales that would generate complaints, a bad reputation, and – worst of all – direct liabilities. After World War I, in parallel to a division-of-labor and task-definition approach inspired by Taylor's rationalization system, large pharmaceutical companies placed increasing emphasis on the need for quality control, all of this relying on a combination of homogeneous processes, sampling procedures, and standardized assays conducted on model organisms. Within this perspective, regulation became an issue of technology and proper management. It addressed the elimination of bad habits and bad products through surveillance, but also the organization of markets. Starting in the 1930s, a growing panoply of interventions to influence prescription practices, accrue clinical knowledge, and define drug uses was developed within the largest firms; these included publications in medical journals, brochures, in-house periodicals, seminars intended for prescribing physicians, product representation, and visits.

In his remarkable book on the transformation of the German pharmaceutical industry between 1880 and 1935, the historian Wolfgang Wimmer underscores the deep conflicts that were prompted by the expansion of large capitalistic drug companies (Wimmer, 1994). In Germany, after the turn of the century, political and administrative debates on drug sales concentrated on the question of whether industrial specialties should require a specific form of control and surveillance, given that the pharmacists buying and selling them were less and less in a position of knowing and checking both the composition and the utility of what they were providing to physicians and to patients. A double-check system, however, prevented any change of the law from occurring until the advent of the above-mentioned *Stop-Verordnung* in 1943. On the one hand, the pharmaceutical profession embodied by the Union of German Pharmacists (*Deutsche Apotheker Verein*) resisted any reform project that would weaken its role in ensuring safety. On the other hand, the large chemical and pharmaceutical companies represented in the industrial union Cepha opposed all schemes for a central and/ or administrative control of the market, insisting on their role as regulator, that is, their ability to guarantee the safety, homogeneity, and utility of their merchandise.

Industrial regulation, its emergence, and the conflicts it brought about, are at the core of this chapter. Its aim is less to illustrate the replacement of one way of regulating with another than to chart the tensions and the articulation processes among various forms of regulation. As suggested by the contrasted fate of herbal medicine in France and Germany, any general vision of regulation must provide room for variation. The trajectory of one type of drugs or one class of compounds cannot be associated with a single way of regulating: most configurations have combined several patterns, even if boundaries have been established and dominant patterns are, as we shall see, not too difficult to identify. Contrasted ways of regulating could characterize the trajectories of drugs in different places, enterprises, or nations. Within the context of this chapter, this diversity will be approached by focusing a single class of therapeutic agents, the transformation of which is particularly revealing of the multiple meanings and effects of the industrialization of drugs in the interwar period, with its scientific, medical, legal, and economic dimensions, namely, plant extracts. These were at the very core of nineteenth-century *materia medica* and remained central to the invention of specialties during the first decades of the twentieth century.

Dausse and its *intraits*: The professional culture of preparation and the production of plant extracts in interwar France.

"We are no longer living in those times when we were forced to ignorance for lack of a proper school to instruct us. If we do not want to perish, we must adapt to the new times and live with our century, which calls for ever more instruction and knowledge to contribute to the welfare of mankind. Who does not advance falls back, and if we do not follow the advancement of professions close to ours, we shall be left behind to the point that everybody will have forgotten us and we shall forever be relegated from the health professions. It is therefore an emergency measure to improve the intellectual level of the profession. It is this goal that the National Federation of Herbalists in France and in the Colonies pursued when it created with its own funding this magnificent National School of Herbal Medicine."[5]

The 1941 defeat of the French herbalists was far from being inevitable and written into the order of twentieth-century pharmacy. During the 1930s, the corporate bodies of herbalists followed a clear and determined path toward their recognition as a legitimate health profession. Originating in the *Bulletin de l'Association amicale des anciens élèves de l'École Nationale d'Herboristerie,* the above statement is just one among many testimonies of the importance reached, at the time, by the question of a state diploma, meaning a diploma that the Faculty of Pharmacy would accept and eventually control. Institutionalization was, however, not the sole strategy to survive in an increasingly national, as opposed to local, drug market. The industrialization of plant extracts was a complementary approach. Just as controversial – since it implied the end of the local collector and maker – it was, nonetheless, able to bring the elite of the corporation to form alliances with a few companies led by licensed pharmacists but engaged in the mass preparation of plant preparations. In the herbalists' literature, Dausse was the firm most mentioned as an envisioned partner.

In Dausse's understanding of pharmaceutical innovation – as well as in its operations – we can see the pervasiveness of professional regulatory practices within the arenas of early twentieth-century industrial drug production.

During the 1920s and 1930s, Dausse was in effect the main company producing plant extracts with a sales catalogue including most of the species for which pharmaceutical preparation was described in the Codex. The expertise of the company was mechanical and "galenic" rather than chemical. Its early history was actually linked to the invention and patenting of a device designed to "displace" the active components of medicinal plants in order to prepare dried – rather than soft – extracts, namely, a machine for mixing up, slowly, the water or the solvents used to concentrate the active principles while keeping the minced and dried parts of the plant under constant temperature and pressure conditions (Dausse, 1836).

As the firm expanded in the early twentieth century, a refined displacement procedure was combined with other ways of handling the plant material, including distillation, lixiviation, solvent separation, or sterilization in large autoclaves. One common feature of all the procedures introduced within the Dausse factory located in Ivry, a southern suburb of Paris, was a strong interest in the relationship between mechanical innovations aiming at an accelerated transformation of plant material and the preparation culture of pharmacists, namely, the search for extracts that would be not only "potent," but above all stable, not too difficult to store, and easy to administrate (Brissemoret, 1908). One innovation Dausse's directors promoted during the interwar period was the so-called stabilization procedure invented by E. Perrot and A. Goris, then professors of pharmacology and *materia medica* at the Paris Faculty of Pharmacy and regular associates of Dausse. In tune with the strong interest of the times in diastases, the basic idea was to avoid spontaneous fermentation, which alters the plant material, by placing the dried organs or the powder resulting from their mincing in boiling alcohol for ten minutes in order to eliminate the enzymes involved in the oxidation of the active principles (Ruffat, 1996).

Dausse's approach to drug preparation was, however, not purely mechanical. An aspect of the industrialization of medicinal plants that attracted the attention of medicinal-plant vendors as well as plant therapists was the development of farming as a means to ensure a more regular supply of raw materials. Although many medicinal plants did not breed very well, Dausse's agricultural production reached 600 tons by the end of World War I. Interests in the replacement of collections in the wild by greenhouse or field growth were not only linked to issues of productivity and mass production. The question of the integrity of a plant as a prerequisite for potency and physiological effect played a critical role in the understanding of what could be a good – and marketable – drug, which was shared by Dausse and the herbalists.

In addition to the (not so common) extracts prepared on the basis of purification procedures listed in the Codex, Dausse was actually promoting the making of other preparations, called plant *intraits* ("intracts" rather than extracts). *Intraits* were mixtures obtained from stabilized fresh plants, later distilled under vacuum, and dried after the elimination of chlorophyll and heavy-fat substances by means of organic solvents.[6] According to the firm's experts, the main value of *intraits*, in addition to long conservation, was their complexity. In contrast to most drug manufacturers, Dausse did not emphasize the purity and chemical nature of its products. Stressing the botanical origins of the preparations was not simply a way to signal their naturalness, it was also a means to link efficacy and complexity. Looking for pure active principles was taken to be very important in the laboratory, a good way to acquire knowledge, but a rather poor therapeutic practice. In contrast to "what many doctors had been trained to believe, purification did not improve or increase the clinical efficiency of drugs." Dausse's managers considered that the effects of fresh medicinal plants were rarely the effect of a single compound, but the consequence of synergies among the various substances they contained. As a result, the preparations of stable specialties faced two problems: (a) that of avoiding the inactivation of fragile active principles induced by many mechanical treatments; and (b) that of avoiding the separation or the elimination of compounds acting in complementary or combined ways. Industrial preparations should not target the simplification of plant material, but its stabilization in order to facilitate storage, circulation, regular supply, and easy dosage.

Even if Dausse's catalogue mentioned pure or crystallized substances, advice to doctors always mentioned the advantages of combinations. Dausse's *intrait* of digitalis was, for instance, documented as having better and less dangerous effects than pure digitaline.

If the present status of science only makes it possible to isolate dubious and unstable active principles, what is the actual meaning of extracts? An extract – if well prepared – is the sum of the active principles of a plant. Unfortunately, a whole generation of physicians has been accustomed to accept as self-evident the idea that isolated active principles are equivalent to the plants just because science has proven able to purify specific substances, sometimes crystallized, and granted them a formula and a structure. This type of chemical medication has replaced the use of extracts... Such practice is perfect when seeking a very determined goal, the exclusive use of active principles, however, impoverish the physician's arsenal, excluding major therapeutic weapons. It is current knowledge that digitaline does not reproduce the action of *Digitalis* extracts, that morphine does not reproduce the action of opium, that quinine does not replace cinchona extracts, that Conicine does not calm pains the way a plaster of hemlock extracts can. (Joanin, 1921–39)

This "holistic" vision of the therapeutic properties of plants was articulated in complex ways with the quantitative pharmacological understanding of drugs. This is well-reflected in the various books on remedies published by the firm. Before World War II, Dausse did not, like many of its competitors, engage in the publication of a journal with new research results and/or clinical reports. Rather, a scientific advertisement for the firm took the form of a Codex-like encyclopedia of its own preparations. The second edition of its *Remèdes galéniques* was a 3,000-page book circulated in isolated chapters throughout the 1930s.

Dausse's masterpiece was organized according to plant species even if it included more general entries such as "biology" or "tests." Each article combined botany, toxicology, and preparation. The entry on *Atropa bella-donna* (Figure 3.1) thus described the form of the plant, its fruit, its botanical location, and its varieties (Dausse, 1836). It then went on to describe the parts employed, underscoring the numerous frauds of supplies in which the leaves of *Belladonna* were mixed with leaves of other species, and signaling the possibility of a botanical diagnosis based on the microscopic examination of the anatomy of the leaves. The book then presented the conditions under which Dausse cultivated the plant on its medicinal farm, emphasizing the agricultural experimentation conducted with A. Goris, who

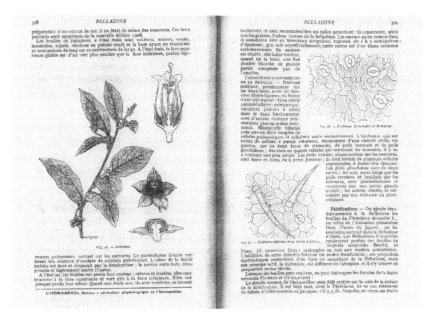

Figure 3.1 Dausse's handbook of plant remedies. Entry for *Atropa Belladona*. Reproduced from *Les remèdes galéniques*, Paris, 1924

correlated various forms of fertilization with the alkaloid concentration of the leaves. The active principles were accordingly taken as the basis for the value of the plant, and a summary was offered of the colorimetric methods used for measuring their global concentration in the leaves. A second section, called *Pharmacodynamie* (pharmaceutical dynamics), presented collected data on toxicity and the effects on animals and humans. The core of the discussion was the question of dosage, thresholds, and sensibility. Taking for granted that the alkaloids of *A. Belladonna* were active at low concentration levels and poisonous when given in too large amounts, the book recalled experiments with animal models showing variable effects, but not a single clinical report was included. At stake was not only the definition of dangerous (if not lethal) doses, but also the inventory of physiological effects, which included changes in eye activity, inhibition of gland secretions, vasoconstriction, and accelerated cardiac rhythm, all of these mimicked by two alkaloids isolated in the plants, thus leaving the reader uncertain as to the supposed synergies involved in the clinical use of the plant. The next section was a Codex-like definition of the right way to handle extracts with the definition of indications (targeting symptoms rather than diseases), legitimated by the physiological study of the plant (*A. belladonna* could be employed: as an antispasmodic in cases of coughing, asthma, epilepsy; as an inhibitor of gastric secretion in cases of ulcer; as a regulator of the vagus nerve in cases of nausea and vomiting) followed by a presentation of the appropriate dosage of *Belladone Dausse*. The final section (called "Pharmacology") was a long discussion of the various preparations combining all the forms and receipts mentioned in the Codex.

The form of research associated with this professional/pharmacological perspective was marginally chemical. There were two in-house laboratories in the 1920s. The first was a chemical laboratory involved in testing activities such as the evaluation of the quality of the raw material, while the second conducted pharmacological experimentation. Their role was not to produce new substances, but to assess the properties of plants already being used and to contribute to the surveillance of production. Assays of purity were not based on tests of chemical structure or composition but on what Dausse's scientists called "assays of identity," combining observation of the forms and colors of the plants, microscopic examination, qualitative chemical reactions, and measurement of dried weight for separate but, nonetheless, mixed principles. The measurement of physiological effects was included in the panoply but mentioned only when toxicity and potency were closely related, raising the delicate problem of establishing a dosage that would avoid heavy side effects, or of course acute poisoning. This was the case for extracts of Digitalis, for which in-house standardization became a routine procedure based on Dausse's adaptation of the classic assay on a frog's heart developed by academic physiologists.

Up to World War II, most innovations were associated with the quest for more productive plant preparations, mechanization remaining the favorite target. Dausse's management thus aligned their products on the most convenient and "modern" presentation of drugs, expanding their facilities to transform plants into pills or ampoules rather than vials. As pointed out by Michèle Ruffat in her history of Sanofi-Synthélabo (Ruffat, 1996), this culture would not resist the postwar transformation of French pharmacy: Dausse's activities in the 1950s and 1960s reveal a gradual alignment on the screening model of innovation, a slow transition toward a regime dominated by the characterization and manipulation of therapeutic *molecules* rather than preparations.

Madaus and its *biologische Heilmittel*: The industrial regulation of German alternative pharmacy

The legal status of plant preparations in Germany was in many respects similar to the French situation. Pharmacists benefited from a monopoly on sales of all extracts included in the pharmacopoeia, which, ideally, they prepared themselves. In addition, a standing commission composed of physicians and apothecaries was in charge of listing prescription remedies (Wimmer, 1994). Two specificities are, however, worth stressing. The first is that in Germany, fewer plants required a prescription to be sold than in France, where herbalists could moreover only sell the so-called *simples*, thus excluding all plants considered as toxic at a given dosage. In Germany only roughly one-third of the plant preparations listed in the *Arzneibuch* fell into the jurisdiction of prescribing physicians.

A second specificity is that in Germany, after 1900, administrative control of "galenic" remedies – those originating in changes in either the mode of preparation or in the combination of drugs already included in the pharmacopoeia – became a permanent source of conflict between the pharmaceutical profession and the industry. This point is illustrated by a 1906 debate on the possibility of the government's passing a "Galenika-Verordnung." Concerned about the multiplication of specialties, which were not "new" remedies but classic drugs made on an industrial scale, the national pharmacists' union then pleaded for a government decree that would specify that "Galenika" must always be prepared by a pharmacist. Opposed by Cepha, the alliance of the German chemical–pharmaceutical industry, this proposal was never turned into law, the Ministry of Commerce arguing that such a measure would threaten German exports.

This was also the case in France, but industrialization of plant extracts in Germany was not to be easily halted, and the number of preparations sold by industrial producers grew rapidly. In the 1920s, a medium-sized firm such as Riedel was producing 339 out of the 376 "Galenika" described in the German pharmacopoeia, Gehe was producing 256, and Merck 171.

Specific patterns remained to be analyzed, but the books these companies published to inform physicians reveal a pharmacological culture close to that discussed in the case of Dausse.[7] There were major differences, however: the emphasis placed on chemical purity, and the absence of strong reference to the physiological effects and the value of plant complexity, that of mixtures of substances, or to anything comparable to the development, control, and standardization of *intraits*. To find practices echoing this aspect of Dausse's culture and regulatory role, we need to look at firms that did not seek to occupy the "chemical" segment of the drug market but rather sought other forms of innovation. The most fascinating example is that of the Madaus pharmaceutical company, which had research and production sites in Saxony. Madaus operated at the boundary between academic and popular medicine, and participated in a movement for more natural and holistic therapeutic practices, which became increasingly visible and influential during the 1920s and 1930s (Jütte, 1996; Dinges, 1996; Rothschuh, 1983). Madaus was a powerful protagonist in the popular-medicine movement, otherwise composed of a vast array of associations including lay participants and professional healers, the latter nearly equaling academic doctors in numbers in many regional cities. The natural-medicine movement grew in strength after the national socialists seized power in 1933 as "the art of healing" gained institutional recognition. As mentioned above, the 1936 regulation gave *Heilpraktiker* a status. In parallel, institutes for alternative medicine were opened in a few universities. Madaus's way of making, producing, and regulating plant preparations differs from the French configuration not only as a consequence of its role in the alternative medical market, but also as an effect of the firm's commitment to the industrial regulation of drugs, its tools, values, and legitimate forms of evidence.

As recounted by C. Timmermann, Madaus was founded in 1919 by the three sons of a church minister's wife who practiced medicine herself as a nonlicensed healer. Benefiting from the then widely debated "crisis of medicine," the firm's production combined herbal and homoeopathic remedies. Success was quick to come (Timmermann, 2001). In 1925, Madaus had already opened branches in Amsterdam, Barcelona, and in several German cities in addition to its headquarters in Dresden. On the eve of World War II, its employees had reached 800 in number. Madaus strongly promoted "natural" healing methods. Its production included a wide range of "biological" therapeutic agents, from classical homeopathic preparations of metals to hormones, enzymes, or combinations of bacterial antigens. The bulk of the firm's ready-made specialties were, however, composed of various types of plant derivatives.

Madaus maintained a complex and contradictory relationship with the German movements for alternative healing practices and their strong interest in "natural" and "biological" therapeutic agents. On the one hand, the firm and its main figure, Gerhard Madaus, were strongly associated with the critique of "academic medicine," writing articles and pamphlets, sponsoring events, publishing journals, leaflets, and books promoting the discussion among *Heilpraktiker* and spreading their views on diseases or remedies to the public at large. On the other hand, Gerhard Madaus pushed the company on the path of industrial and mass production of homeopathic and plant remedies, arguing for the modernization of practices, for more experimentation and evaluation. The aim was to revisit, modify, and improve the corpus of known recipes and traditional preparations in order to optimize and generalize their uses, bringing them back into the arsenal and purview of local practitioners as well as academic physicians and pharmacists.

The broad palette of documents and publications on the firm's products and activities left in its archives in Dresden is associated with two discourses. The first is a medical discourse based on a global vision of the body, recommending attention to the systemic and multiple dimensions of diseases, pleading for an ecological understanding of the relationship between people and their environment, arguing for the need of more natural or biological means of intervention. The luxurious *Jahrbücher*, published by the firm for its clients, and, more strategically, for local doctors, featured eloquent discussions of this holistic and natural perspective.

A few examples suffice to illustrate the point. The 1932 *Jahrbuch* included a long essay on the evolution of life that sought to modernize and update the old idea of "signature," stressing the affinities and correspondences among the various realms of nature (Madaus, 1932). An ecological relationship like the oxygen cycle exemplified global ecological dependency, while therapeutic considerations were grounded on morphological and functional analogies between plants and humans, such as the structural relationship between chlorophyll and hemoglobin or that between blood pressure and plant-juice pressure.

Similarly, in a famous textbook on biological medicine, published in 1938, G. Madaus argued for some medical modesty given our lack of knowledge about bodily interactions:

> Virchow's discovery of the cellular nature of pathologies is still of decisive importance even if this localization is nowadays taken as a symptom... In the end, it is of rather secondary importance to know whether a given pathology is limited to a small number of organs and systems. A specific diagnosis showing that one secretory gland or another, the immune system, the circulatory apparatus, or the reticuloendothelial system are affected does not say much about the way in which the entire organism is actually concerned. The question of the interactions between systems, of their relations to the pathological course within one isolated organ lies at the very center of modern research on constitutions and is far from having been clarified. (Madaus, 1938)

In the late 1930s, the firm also edited a "pocket book" on biological medicine that would help local practitioners compare the advantages of various forms of medicine. The booklet presented "general" therapies such as a controlled diet, as well as combinations of preparations considered helpful in specific indications. In the case of diabetes, for instance, recommendations combined a healthy life style – meaning a lot of sun, fresh air, and physical activities – with prescriptions of Madaus ready-made plant combinations that minimized the excretion of sugar. The firm also produced and sold insulin, which was viewed as a necessary, but insufficient means of controlling the disease:

> Diabetes is not only a pathology of the pancreas. It is better understood as a general disturbance of metabolism that also affects the liver and various muscles besides the pancreas. Another essential etiological factor is the vegetative nervous system. As a consequence, a pure replacement therapy is not the ideal treatment, as has been known for a long time in medical research and medical practice. Due to the lack of a better alternative, however, substitution therapy is not to be interrupted.[8]

The "biological" medicine and "biological" remedies Madaus advocated thus did not imply – in contrast to the views of many *Heilpraktiker* – that isolated and homogeneous preparations were simply to be abandoned. As employed by G. Madaus, the idea of "biological" is a boundary idea, which helped to articulate and link entities and practices originating in heterogeneous worlds, that is, popular medicine and its uses of entire plants, nineteenth-century *materia medica*, laboratory-based physiology, and its molecular effectors like hormones and vitamins.

The second discourse documented in the firm's literary production is that of industrialization. Up to World War II, Madaus was a flourishing enterprise and the merits its owners placed in the virtues of mass production show little difference with the defense of industrial-specialty producers such as that advanced by Merck or Gehe, which did not support the alternative medicine movement but mined the pharmacopoeia. The stated goal was to produce more for less money in order to enlarge the market and to facilitate access to remedies. Quality also came into play since industrial extracts were thought to increase medical efficiency by delivering products granted with stable composition, longer conservation, better storage, and easier transportation. Homogeneity, standardization, and quality control were key concepts opposing essential ideas in the culture of alternative practitioners such as the bodily constitution of patients, the variability of diseases, or the adaptation of formulations and preparations to the needs of individuals.

In spite of this opposition, the two discourses met in the strong critique of purity G. Madaus and its company associated with the promotion of complex extracts. The firm did not only place its products in line with the traditional "art of formulating," it gradually elaborated a local understanding of "scientific preparations" stressing biological mixtures and combinations. Several arguments were at play, namely that plant extracts are composed of many ingredients, often dozens, that efficacy is not to be attributed to a single active principle, that clinical efficacy may be linked to the synergies among these constituents, that the potency of a preparation is not to be attributed to the isolated effect on a single organ, that pathologies have to be seen in a global life context including the influence of environment and habits, that the power of plant extracts is related to the life of the plants themselves, to their mutual relationship, and finally to the dynamics of ecological systems.

As a consequence of these claims, the industrialization of plant preparations at Madaus focused on three registers of practices, which were rarely combined in other companies: (a) the mechanization of processing; (b) the standardization of products and the expansion of quality testing; (c) the domestication and mass culture of medicinal plants. These corpuses of practices were rooted in three ways of knowing that the firm sought to develop in its experimental sites, which included a chemical laboratory, a biological institute, and a farm.

Madaus's innovation strategy was primarily, as in the case of Dausse, linked to the mechanization of processes (Kuhn, 1936). For instance, the firm engaged in modifications of the rules for making homeopathic dilutions that aimed at better conservation and were far from being accepted by homeopathic physicians (Madaus, 1935; Balz, 2010). The most important changes, those for which Madaus submitted several patent applications, had to do with the machine-based treatment of fresh plants.[9] The company continued to sell dried extracts of entire plant organs, but gradually transformed their presentation into sugar-coated pills. In the eyes of Madaus, the so-called *Teep* (for *Tee-Pulver*) preparations had the immense advantage of providing such pills without any isolation or purification steps. As proclaimed in the *Teep* advertisements (Figure 3.2), the Teep kept all the plant constituents, "they did not eliminate proteins or plant hormones and are therefore of a much higher quality than classical homeopathic decoctions and tinctures." *Teep* were obtained by mincing and grinding freshly collected plant parts with sugar (Figure 3.3), followed by drying under warm air. This raw "0" preparation was stored in dry storage rooms for later mixing (or "dilution") with additional sugar in order to get the dosage of plant material sought for the Teep pills, which were produced with exactly the same type of machinery used in all pharmaceutical companies: grinders, mixers, dryers, and the most important pill-compounding machines.

Madaus considered that Teep preparations were better, not only because they were less expensive and more stable than the classic "soft" plant preparations, but more importantly because the Teep process preserved the mixture of substances found in the fresh plants. Numerous experiments

Figure 3.2   Advertisements for Madaus Teep plant preparations. Reproduced from
Biologische Heilkunst, Dresden, 1932

were therefore conducted at the company's chemical and biological laboratories in order to document the quality of Teep preparations. A series of
analyses presented in 1935, for instance, compared various Teep with alternative presentations of the plant material, that is, simple drying to make
Tee, homeopathic tinctures, or extracts obtained following a treatment by
organic solvents. For all the families of substances investigated (hormones,
proteins, vitamins, enzymes, saponins, oils, cellulose, pigments, carbohydrates, pectin, and waxes) the results suggested that the Teep procedure
retained the largest amount of the original material (Madaus, 1935b) (Figure
3.4). Harnessed to holistic medicine, chemical science thus demonstrated
improvements over the procedures prescribed either by the homoeopathic
Arzneibuch or by the academic pharmacopoeia.

Turning popular medicine into scientific–industrial medicine, however,
required important connections with the practices of "school medicine."
This is best exemplified with the second and dominant aspect of industrialization: standardization and the research supported to control the quality
and effects of large quantities of plant products. The sensitive experience of
plant connoisseurs with their knowledge of forms, odors, texture, and tastes

Figure 3.3 Preparation of Madaus Teep pills. Reproduced from *Madaus Jahrbuch,* 1937

was not given up, but rather systematized and complemented with laboratory testing. Industrial standardization was thus connected with multiple activities at the same time: (a) writing and implementing formal production protocols; (b) systematized control of the quality of incoming raw materials; (c) measurement of the potency of the final plant preparations; and (d) occasional evaluation of their clinical side effects. The chemical laboratory under A. Kuhn's guidance was central to this effort. A few examples will suffice to give a sense of its specificity (in comparison with the testing laboratory of pharmaco-chemical firms), of its roots in the professional culture of preparation, and of the changes originating in industrialization and scaling up.

Kuhn and his colleagues spent considerable amounts of time investigating the composition of Teeps made of various plants such as *Mentha, Valeriana, Viscum, Digitalis, Lycopodium, Oleander, Aloe, Arnica,* and so on. Much of this investigation consisted in straightforward quality control. For instance, in December 1939, following complaints that a given lot had acquired a suspicious color, Kuhn and his colleagues determined the content of various *Arnica* Teeps. Although the quantity of oils that could be extracted with ether was highly variable, all the preparations analyzed presented a normal arnica smell and the same yellow deposit.[10] The grinding protocol followed in the production department was in this case not changed.

1. Alkaloide und Glykoside:

Auch bei vorsichtigem Trocknen leidet der Gehalt an obigen wirksamen Stoffen. Der größte Teil (z. B. durchschnittlich 80% Alkaloide und Glykoside) bleibt erhalten.

So ging die Wirkung von Aconitum durch das Trocknen von 3100 F.D. pro g auf 2500 zurück[1]), bei Digitalis von 600 F.D. auf 400 F.D.[2]). Besonders auffällig ist das Unwirksamwerden bei Helleborus, Paeonia, Aconitum, C o n i u m, Sabina[3]).

2. Amara:

Die Bitterstoffe leiden nur wenig beim Trocknen; nur bei einigen Pflanzen, wie Bryonia, gehen sie verloren.

3. Acria:

Hierunter verstehen wir pflanzliche Stoffe, die hautreizend und blasenziehend wirken. Sie gehen beim Trocknen verloren.

Z. B. bei Bryonia, Ranunculus bulb. et acer, Pulsatilla, Clematis vitalba et recta, Cochlearia, Cepa, Daphne Mezereum[4]), Alisma plantago[5]). Ebenso gehen manche Säuren verloren, z. B. Helvellasäure.

4. Enzyme:

Als charakteristisch für den Frischzustand einer Pflanze kann man die Peroxydase betrachten, die häufig vorkommt. Sie geht beim Trocknen in den meisten Fällen vollständig verloren, z. B. in Spiraea ulm., Rheum, Polygonum.

In seltenen Fällen ist sie sehr schwach noch vorhanden, z. B. bei Urtica. Proteasen verändern sich durch Trocknen der Drogen nicht.

5. Hormone:

Die F o l l i k e l h o r m o n e sind sehr beständig und bleiben erhalten.

Die Auxine (Hormone der Zellstreckung) bleiben erhalten[6]).
Die M e r i s t i n e (Hormone der Zellteilung) sind wahrscheinlich identisch mit den Wundhormonen.

Die W u n d h o r m o n e gehen durch Trocknen nicht verloren, wohl aber die Antiwundhormone und wahrscheinlich die Nekrohormone.

T h y r e o t r o p i n e werden durch Trocknen unwirksam[7]).
Glukokinine sind ziemlich trockenbeständig.
P f l a n z e n s e k r e t i n e sind in der Lufttrockendroge unter schwacher Einbuße 8 Jahre haltbar[8]).
Wir schätzen den Verlust auf 10%.

[1]) Madaus Jahrbuch 1934, S. 53.
[2]) Eigene Beobachtung 1933, bisher nicht veröffentlicht.
[3]) Schulz, i. Hufelands Journal, Bd. 70, V., S. 89.
[4]) Vgl. [3]).
[5]) Thoms, Handb. d. prakt. u. wissensch. Pharmazie, Bd. V, S. 457.
[6]) Kiem, Handb. d. Pflanzenanalyse, Bd. 4, 2, S. 1017.
[7]) Vgl. [6]); S. 1034.
[8]) Vgl. [6]); S. 1033.

Schematische Darstellung der Verluste beim Trocknen von Heilpflanzen.

Die roten Felder zeigen den Verlust an, auf Grund von Durchschnittsberechnungen und Schätzungen.

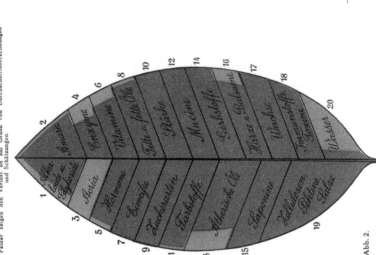

Abb. 2.

1 Alkaloide u. Glykoside
2 Amara
3 Acria
4
5 Hormone
6 Enzyme
7 Eiweiße
8 Vitamine
9 Zuckerarten
10 Fette u. fette Öle
11 Farbstoffe
12 Stärke
13 Ätherische Öle
14 Mucine
15 Saponine
16 Gerbstoffe
17 Harze u. Balsame
18 Wachse
19 Zellulosen, Pektine
20 Wasser
Fermentenzyme, Lavaine
Salze

Figure 3.4 Experimenting on the value of Teep preservation of plant components. Dr. Madaus Jahrbuch 1935

Controlling the plants collected in the wild or harvested at the Madaus farm was a demanding activity that could occasionally result in changes in the production practices. One example of this is the case of *Hypericum* extracts. While developing the Teep procedure for this plant, Madaus chemists actually noticed that one of the active components, an alkaloid called hypericin, occasionally disappeared during the final mixture with sugar. This was traced back to a rapid loss of solubility during extraction, for reasons that remained unknown but were attributed to a physical or enzymatic alteration of hypericin. The Teep-preparation protocol was therefore modified to accelerate the mixture, no longer leaving the brew of minced plants to stand for hours in the open air.[11]

Madaus scientists focused their studies of extract composition not on isolated substances but on broad categories of interacting substances, that is, alkaloids, oils, proteins, carbohydrates, and so on. Nonetheless, even if for Madaus the relationship of these combinations to the botanical understanding of specificity and classification was essential, the firm's use of orthodox/ academic pharmacology was far from marginal. This is eloquently testified by Kuhn's attempts to standardize the composition of *Belladonna* extracts on the basis of specific molecular analysis. Recognizing that the potency of the plant was due to the mixture of atropine and a few related alkaloids, Madaus chemists screened the pharmaceutical literature to design an innovative combination of extraction steps and physical measurements, which resulted in a quantitative assessment of the three most important alkaloids of the plant, that is, scopolamine, hyoscyamine, and atropine.[12] Published in pharmaceutical journals, the procedure was systematically used to investigate the composition of seeds and roots from different varieties of *Belladonna*. The main conclusion of this inquiry – in contrast to received pharmacological knowledge – was that the most potent varieties were not enriched with atropine but with another alkaloid, scopolamine. A chemical strategy would then have sought to purify both components for the production of separate drugs or of a standard combination. Madaus's approach of botanical complexity led to a different route. The Teep preparation of *Belladonna* was modified to contain a balanced association of the three alkaloids on the basis of a combination of different varieties and different parts of the plant, with proportions established on the basis of repeated chemical analysis.

Chemical testing was important, but the local quality control measurements relied even more strongly on the practices of biological standardization. The latter is often viewed as a pragmatic and incomplete substitute to molecular analysis; something indispensable when the composition and structure of substances is not known; something to be later replaced by more specific and accurate chemical tests. Within this perspective, a biological or physiological assay is the choice of a surrogate marker, a biological phenomenon that can be quantified and considered specific enough of the drug's potency to be equated with the amount of active principles supposedly contained in the preparation. Historically, physiological and biological controls were, however, not necessarily preliminary steps toward a molecular redefinition of pharmaceutical activity; they were valued for their own sake as the best way to operationalize and objectify the complexity of bodily relationships and drug effects. Madaus' biological standardization worked this way. In contrast to the producers of sera and vaccines, who considered the results of biological testing and the definition of potency in terms of animal (guinea pig, rat, or rabbit) units as direct indications of composition doctors could use to adjust dosage, the results of biological testing at Madaus had an internal rather than external value.

The owners of Madaus shared academic pharmacists' fascination for the laboratory, which in their eyes ranked above hospital and clinical settings as a knowledge-production site. Biological testing was accordingly a way to "mine" the corpus of popular and alternative medicine in order to make new mixtures and find new indications. Standardized animal models were therefore used in three different ways to deal with the complexity of plant extracts: (a) to evaluate and control the variability of collected plants and adapt mixtures to ensure some stability of effects based on the global composition; (b) to survey the implementation of production protocols although the issue was less the surveillance of operators and the detection of errors than the adaptation of practices to highly variable materials; (c) to provide experimental information on new specialties and/or indications.

Madaus's scientific workers did not just employ classic techniques like the frog assay to estimate the potency of *Digitalis* extracts that was described in all pharmacy textbooks. The firm's biological institute also developed numerous alternative procedures. In the case of *Digitalis* Teeps, evaluation mobilized remote biological connections. The frog assay was completed by observing the changing pace of the pulsation of vacuoles in the microscopic organism *Daphnia* (Kuhn & Penz, 1937). The exposés written for registration also document this permanent use of locally designed biological assays. A text written in 1941 describing a preparation called "Lycocyn," which consisted of a "standardized" extract of fresh *Lycopus europeus* to be employed to treat thyroid disorders, thus presented a brand new test during which the preparation counteracted the side effects of thyroxine, the thyroid hormone inducing lowered resistance to low oxygen pressure in guinea pigs. The test did not target a specific substance, but the specific effect, measured in ad hoc biological units, of a wholesome extract, the purification of which was of no interest (Kuhn & Penz, 1937).

Two examples illustrate the more general value of animal experimentation and modeling. The first one is the use of *Vascum album* (mistletoe) that Madaus sold as powder and Teep for its antispasmodic and hypotensive properties. Accepted uses included cases of epilepsy, hypertension, atherosclerosis, and headaches. Popular plant-medicine manuals also mentioned its use against cancer (Madaus, 1938). Madaus restricted its publicly advocated indications to hypertension and atherosclerosis. Cancer, however, became a topic for in-house experimentation. As reported in *Biologische Heilkunst,* the German reference journal for biological medicine, these experiments did not consist in healing animals affected with tumors but in studies in plant physiology. Madaus biologists showed that mistletoe extracts could inhibit the growth of plant cells and block the normal healing of wounds in trees. This property could explain not only long-term natural parasitism but also more general inhibition of cell growth, which might play a role against cancer even if the company never endorsed such indication (Madaus, 1933). The second example is that of garlic (*Alium sativum*), a plant widely sold by Madaus in the form of *Alliocaps* pills to treat rheuma, inflammation, and atherosclerosis in conformity with the German pharmacopoeia. The role of the laboratory was not to find new indications but a new model in that case mice were made atherosclerotic following the administration of high-dosage compositions of vitamin D, which could be cured with parallel ingestion of *Alliocaps* (Koch, 1936).

Pointing to laboratory work at Madaus is not to say that the "alternative" context left no trace in the firm's ethos of measurement, efficacy, and productivity. For instance, when arguing that hormonal diseases were not pathologies caused by the absence of a peculiar chemical but originated in global regulatory malfunctioning, G. Madaus advocated a much smaller dosage, in line with the homoeopathic principle that hormones will help cure the disorders they induce in high concentration (Kuhn & Jurek, 1933). More generally, hormones and vitamins held a central position in the experimentalized space of biological medicine. Viewed as highly potent effectors, they were also valued for their common roles in all living beings, plants and animals alike. This community in turn explained why sex hormones could affect the growth and differentiation of plants and, more importantly, why plants like rosmarin, which for a long time had been associated with the treatment of "women's diseases," could affect the human reproductive system. As a consequence, inquiries combining classic extracts and the experimental systems of steroid biochemistry were not only doable, but highly recommended for the industrial modernization of biological healing (Madaus & Kunze, 1933).

As boundary objects between alternative medicine, classic *materia medica,* biochemistry, and the industrial production of extracts, biological therapeutic agents had a critical importance with translation processes working in several directions, that is, rationalizing popular or classic practices, but also contributing to the molecularization of plant extracts. The best example of such a move is the work Madaus scientists pursued on vitamins. Between 1940 and 1942, within the context of the war and scientific and industrial mobilization, vitamin C became a central topic for A. Kuhn and the chemical laboratory. They launched a long series of investigations on alternative sources of vitamin C, beginning with the needles and bark of various coniferous trees. While looking for naturally rich local plant material that could replace oranges and lemons, Kuhn abandoned all discussions on synergies and mixtures, focusing Madaus testing only on ascorbic acid. The initial choice of needles from firs and spruces did not prove successful. The decision not to market the resulting syrup had nothing to do with holism or opposition to "molecular reductionism," but originated in failed attempts at neutralizing the very bitter taste of the firm's preparations. The limits of molecularization are, nonetheless, visible in the fact that Madaus never discussed the possibility of marketing pills of isolated and pure vitamin C.[13]

The third dimension of industrialization at Madaus was the long-term work devoted to the cultivation of key medicinal plants. The annual harvest was a recurring source of worries regarding the quantity and the quality of the material collected in the wild, which usually stopped at summer's end with bitter complains about shortages.[14] The dreamed response was cultivation as a means to minimize collections in fields and forests. A few species like *Digitalis* were actually domesticated, but they remained exceptional and they raised numerous questions regarding the consequences of cultivation on plant physiology and on the composition of extracts. A good example is provided with Madaus research on *Belladona*, which confirmed the idea that its agricultural production, coupled with the use of fertilizers, had increased the toxicity of dried preparations (Madaus, 1938).

A significant part of the research conducted within the chemical laboratory, the greenhouses, and the experimental farm was therefore targeting this issue and exploring the conditions that would make domestication successful. What is left of the protocols reveals two types of experimentation. One research line focused on the impact of agricultural practice, for instance, looking at the effect of various fertilizers on both the global composition of the plants and their therapeutic properties (or better said their laboratory surrogates). In the case of *Atropa belladona*, such trials led to the conclusion that the extracts made out of plants cultivated on sandy soils contained more essential oils (extracted with ether) than those cultivated on clay with the consequence that a combination of the two would average and stabilize the potency of the extracts (Koch, 1936).

The second line of research was more original as it focused on the ecology of medicinal plants. Relations among all living entities were central to the vision of plants as containing active elements that affected life processes in animal and human bodies. Madaus's botanical science was holistic and communitarian, taking plants not as isolated and self-sufficient organisms but as integral members of stable communities maintained through relations of synergies and antagonisms among species. Although G. Madaus never used the concept of ecosystem, he, nonetheless, insisted on associations among plants as characteristics of a given environment, rooted in badly known physiological exchanges like the circulation of nutrients between roots (Madaus, 1938). Theoretically, a good culture system should replicate the most important relations between the targeted medicinal plant and its accompanying species. Experiments were therefore conducted to document and select meaningful combinations. Only those performed in the biological laboratory and the greenhouse have left archival traces. In the case of *Viola odorata* (the extracts were used against skin diseases, eczema in particular) the association with wheat and barley gave opposite results as the latter stimulated growth and germination while the former inhibited them (Madaus, 1934). The cultivation scheme of *Digitalis* was modified following similar experimentation showing that cocultivation with *Galega officinalis* (a plant used against diabetes) increased the amount of alkaloids (Madaus, 1935). It seems, however, that most cultivation processes, either in the field or in greenhouses – as was the case for plants of tropical origin like *Cactus grandifolia*, which was needed for a very successful cardio-tonic called *Goldtropfen* – were organized without any reference to this ecological approach of plant physiology.

A final component of the regulatory activities of Madaus worth discussing in this chapter is related to the definition of the proper medical uses of the firm's products, and more broadly, the standardization of herbal and homoeopathic treatments. Madaus followed the path of many companies that sought to "discipline" the practice of physicians by disseminating their own scientific information. As mentioned in the introduction to this chapter, "scientific marketing," that is, the simultaneous organization of research and promotional activities, became an important tool in building pharmaceutical markets in the 1930s.[15] For Madaus, this development resulted in two different forms of "propaganda." The oldest and most classic of these used advertisements in professional and lay journals that focused on the name of the firm and built its image through slogans and trademarks. The second approach focused on the mobilization of laboratory and clinical knowledge and took the form of numerous articles, leaflets, exposés, and textbooks written for regular physicians and biological healers. A typical publication was the above-mentioned *Biologische Heilkunst,* a journal of academic format circulated among therapists, supported by the firm through many advertisements, edited by a board including G. Madaus, with himself and his collaborators signing many articles.

Among the new tools of scientific marketing, Madaus's *Lehrbuch der Biologischen Heilmittel* is worth considering as it shows a structure similar to that of Dausse's pharmacological textbook, including more than 400 plants and counting more than 3,000 pages. Each plant had its specific chapter listing its names, locations (with maps), morphology, composition, parts of interest, physiological effects, toxicological symptoms, therapeutic indications, and modes of preparation and conservation. Differences were, nonetheless, significant: (a) indications were rooted in long-term history with systematic references to old medical treaties dating back to the sixteenth century; (b) lay therapeutic experience was given a significant place; (c) toxicology did not mean animal experimentation for modeling the dose–response relationship but the reporting of clinical cases; (d) homoeopathic conditions were central to the definition of proper uses.

The chapter on *Atropa belladonna* thus includes everything that could be found in a pharmacopoeia but also all sorts of facts about the ecology of the plant and a lengthy development on its therapeutic history, with references to treatises written by Discorides, Gesner, Albert the Great, or Boerhave. Madaus, for instance, discussed a "Bulgarian cure" against epilepsy, shaking, or muscular seizures advertised in German newspapers by a certain Iwan Raeff as a secret remedy well-known in Italy, where it benefited from the Royal Court's promotion. G. Madaus evaluated the "cure" severely on the basis of its putative composition as defined by the academic pharmacists who had tried to analyze the mixture on the one hand, and on the other hand of what he thought about the interactions between *Atropa belladonna's* alkaloids and the rest of the material, that is, animal charcoal, roots of *Acorus calamus,* and nutmeg. The rest of the pharmacological section of this chapter actually did not differ from what could be read on *Atropa belladonna* in any pharmaceutical textbook. Even dosage was not an exception since the "homeopathic" dilutions (D3 to D5) Madaus recommended did not conflict with those proposed in the official German pharmacopoeia.

The relations to local practices and local healers was, however, not only top-down, it was bottom-up as well. In contrast to the emerging industrial regulation, the ethos of *Heilkunde* valued the knowledge of local practitioners highly. Their expertise was taken as an indispensable source of insights to be developed and incorporated into the corpus of the new medicine, and whenever possible turned into industrial practice. It was therefore expected that Madaus scientists should pay significant attention to the insights and suggestions patients and local physicians would send.

A good example of using this kind of information can be seen in the attempts to exploit the healing potential of wild fruits and vegetables, something that would gain a new momentum during the war in a context marked by the general quest for "ersatz." The chemical laboratory gave serious attention to letters proposing one or another line of investigation. A certain Dr. Wirz from the *Hauptamt für Volksgesundheit* in Munich thus recommended the hypoglycemic influence of wild berries, blackberries in particular, and their use in difficult pregnancies:

> Several sorts of berries play a significant role in the popular art of healing; for example the fruits of elder tree, the redcurrant, and blackberries themselves. Redcurrant has in particular been successfully employed in the treatment of troublesome pregnancies or in cases of lactation failure. I know that many gynecologists employ them. Now, Dr. Griefd, the director of the diabetes hospital in Berlin, who is also a chemist, has started to investigate the activity of the darkening substances of these berries and has obtained very interesting chemical and pharmaceutical results.
>
> To me, the most important of them are indications that the coloring matter has an effect on blood sugar. In case of pathological increase, the

dyes may act as substances reducing blood sugar. If confirmed, this effect would account for an important element of popular beliefs as many problems in the course of pregnancies are associated with increased blood sugar. Without waiting for the final results, I have acted to secure part of the rich harvest of berries in the Netherlands and have let prepare juices that could be distributed to women, midwifes, and national–socialist associations. The positive results already observed are astonishing. The most impressive is the case of a woman in her fifth pregnancy who had been advised to abort in her sixth month because of a very serious lethal risk linked to heavy protein excretion. After 14 days of treatment with redcurrant, the excretion ceased. The pregnancy led to the birth of a strong boy at full term.[16]

Dr. Wirz was in return assured that the pigments of the various wild berries had been carefully catalogued and that the influence of anthocyanins on blood glucose levels would be closely looked at in the firm's laboratories. The promise was actually kept, but the results were too mixed to make it a high-priority issue and a source of new products.[17]

In the Madaus perspective, interesting practitioners were healers as well as school doctors and pharmacists. The most significant undertaking to mobilize their local know-how was a survey the company organized in 1935 with a widely disseminated questionnaire on the use of medicinal plants. The 12,000 responses collected by the firm were analyzed to produce aggregated data, which were merely frequencies of use and were presented as state-of-the-art knowledge emanating from a collective built on practical work with herbal drugs (Madaus, 1936). This material was used first in the Madaus compendium, where it provided the basis for two different listings. Each chapter thus included a figure showing the most frequent indications treated with extracts of the plant under consideration, while a final section of the book listed the most frequently used plants for a given pathology, for instance, *Lobelia, Ephedra, Datura,* or *Eucalyptus* in the case of asthma, without any further comment or comparison with the textbooks of herbal medicine, leaving readers to draw their own conclusions.

What is to be taken from this brief description of the fate of plant extracts at Madaus is less that the latter's preparations differed from those of Dausse either in their role in popular medicine or in their standardized nature than the fact that in the 1930s and 1940s Madaus followed a system of drug production that combined the professional, the industrial, and the "lay" forms of regulation. Elements of the first are found in the incorporation of decisive elements of the pharmacological culture, starting with the interest in active (rather than pure) substances found in the usual receipts of *materia medica,* to be used in moderate quantities according to the experts' understanding of the loose boundary between efficacy and toxicity. That Madaus always referred to local physicians and recognized healers as the main agents and most knowledgeable actors of herbal medicine is another element of this professional order, which echoed the quest of French herbalists for a full legal status as health practitioners. As a consequence, the part played by "lay regulatory" actions, though real, remained limited. The greatest power lay in the role granted to the market, with direct sales of many herbal and homeopathic remedies, which allowed patients, their families, and all potential users to buy (or not) Madaus products, thus granting to final users and not just to the gatekeepers a measure of power in shaping the market. In contrast to Dausse's maneuvering with the aims and tools of professional regulation, the most important elements in Madaus's operations were in the end those originating in the industrial regulatory culture and in the implementation of mass production, quality control, and standardization. Biological tests of potency, homogeneous protocols, advanced division of labor and mechanization, as well as the control of collected and raw materials are the landmarks of an evolution that also led to hybrid forms of marketing. If the *Madaus Jahrbücher* were luxury goods, easy to read, and richly illustrated, to be placed in the doctor's waiting room, other strata of publications sponsored by the firm were proper scientific journals in biological medicine that published original research results, essays, and book reviews in a format that mimicked that of academic journals.

Concluding remarks

Contemporary France has the reputation of being a country that invented a unique alliance between private industrial firms and state administration. Labeled as "Colbertism," this conjunction took shape during the first industrial revolution. Through the twentieth century, it supposedly led French high-ranking civil servants trained in the country's major engineering schools – such as the École des Mines or the École Polytechnique, which staffed the state's technical bodies – to work in close connection with private capitalistic entrepreneurs in order to further industrial investments, scale up production, protect the national markets, and rescue, if needed, threatened strategic enterprises or banks (Minard,1998; Woronoff, 1994; Caron, 1997). Had this pattern characterized the history of the local pharmaceutical industry, we should have observed in early twentieth-century France, or even earlier, the emergence of strong forms of administrative regulation, possibly combined with burgeoning elements of industrial surveillance and control of drugs. Echoing previous work showing the slow transformation of the French pharmaceutical firms into large corporations, this chapter, with the case of Dausse and its plant preparations, actually documents a different pattern, linking a scattered economic landscape with diverse forms of industrialization and innovation (Gaudillière, 2005b). When it comes to the fate of therapeutic agents, the country's major temptation was less to increase the state's regulatory power than to preserve and rely on the expertise and power of the medical and pharmaceutical professions. On this basis, two patterns – one specific and the second one unspecific to the French context – are worth discussing.

The first regards the nature of interventions. Strong professional regulation resulted in the absence (until the 1941 establishment of the *visa* system) of any form of premarketing evaluation organized under the authority of the state. As many observers of French medicine have noticed, the central state health administration was anemic and did not have much power (Murard & Zylberman, 1996). Drug regulation was in effect delegated to the medical corporations. The trajectories of plant extracts and organ therapy discussed here confirm what has been documented when comparing the regulation of sera in France and Germany (Gradmann & Hess, 2008). Critical regulatory actions were on the one hand the writing of the Codex by elite pharmacists and on the other hand the approval of new specialties by the Académie de médecine. True, biological therapies using sera or hormones, due to their novelty in the pharmacopoeia, their variability, and their potency, were granted special status. In France, however, this special status was limited to a system of preliminary authorization – with or without inspection – of the *production facility*. Rather than being a means for drug surveillance, this control reflected the current understanding of professional autonomy. When acting as experts, physicians and pharmacists were left alone to decide what drugs were worth producing and prescribing, whereas when acting as producers, their legal responsibility was the same as that of any industrialist, that is, they were liable for fraud, misbranding, or the sale of poisonous substances.

A second, less expected dimension of this professional regulation focuses on the type of knowledge associated with the critical regulatory function granted to the pharmacopoeia. Historians of pharmacy have often pointed to the chronological conjunction of the chemical revolution and the stabilization of pharmacists' professional orders in the first decades of the nineteenth century. The assumed consequence is then that a chemical paradigm centered on the purification, the structural description, and, when possible, the synthesis of therapeutic substances dominated the making of drugs in parallel with the pharmacological model of the relationship between doses and effects mentioned in the introduction to this volume. In contrast to this assumed connection, the case of Dausse's *intraits* analyzed here suggests that until late in the twentieth century, chemical entities barely played a role in pharmacists' practice of preparation. Up to the 1920s, the receipts of the pharmacopoeia did not favor making pure, molecular entities, but rather making stable, reliable compositions of medical matters, a majority of which originated in living (mostly plant) bodies.

A pharmacist's culture of preparation that owed little to the model of purification and synthesis thus dominated the early industrialization of drugs in continental Europe. This form of innovation placed value on the art of combining or the art of presenting known – and often complex – substances.

As shown by the two examples discussed here, such industrialization mobilized chemical tools as a means for concentration, control, and standardization, as well as marginally as a source of isolated substances. Purity could be a plus, but it was not mandatory or even necessary. The alternative was to rely on associations and synergies. Complexes seemed especially valuable and important to preserve when plant extracts came under consideration. The industrialization of plant and organ extracts therefore relied on mechanics on the one hand and physiology on the other. The model of professional regulation advanced in our introductory chapter should therefore be amended to take into account this diversity of know-how beyond the mere mobilization of pharmacological modeling. As illustrated here, biological testing systems were not just important elements in the industrial practice of standardization and quality control. In parallel, academic pharmacists used them to perform physiological functions and make them manifest, meaning that they became tools to explore and signal the synergies and complexities that remained central to the culture of preparations.

If there is a caricature of German contemporary history to parallel the image of the French industrial state, it is the idea that a rapidly growing chemical industry colonized the entire pharmaceutical sector after the 1890s. One major interest of a comparative history of plant extracts is to show the importance of these practices, which made a subset among German firms comparable to their French counterparts living off the exploitation of specialties registered in the pharmacopoeia. The history of Madaus thus reveals a culture of preparation that shares many aspects with the practices at Dausse, including the organization of plant collection and breeding, the importance of mechanical innovations, or a deep interest in physiological tools, and research. Madaus was, however, not Dausse. The social and intellectual landscape within which the firm blossomed was not the French professional order, but a rare combination of industry and (institutionalized) alternative medicine. What was decisive here was less G. Madaus's holistic approach of the living, which nurtured a system of correspondences among plants, animals, and human beings, than the integration of the so-called biological therapies – homeopathy, as well as the use of plants and organ extracts – into the industrial regulatory order with its values of mass marketing, productivity, standardization, and homogeneity, all of which were taken as synonymous with quality and effectiveness. The consequences were not only the prominent role attributed to mechanics and processing or the expansion of advertisements, but the mobilization of pharmacology *and* chemistry for quality control. As a company looking for a more scientific form of popular medicine, Madaus paradoxically engaged in the development of as many standards and assays as more molecularly oriented firms like Schering, Merck, or Hoechst.

Madaus highlights the commonalities of the industrial regulation of drugs. It developed in-house research facilities focusing on physiology, using biological assays as privileged tools of intervention, inventing relations with physicians and local practitioners that blended science and marketing. This was not only an in-house challenge. The correlate of standardization and quality control within the firms was a pattern of state interventions that echoed and reinforced entrepreneurial interventions (a situation powerfully illustrated with the state regulation of sera in Germany) but did not constitute an autonomous administrative way of regulating. While substantiating the idea of an industrial way of regulating, the parallels between Madaus's regulatory activities and those of firms like Boehringer or Gehe, which mass-produced biologicals, do not make the differences between local innovation cultures less real. Even when investigating cellular metabolic pathways, Madaus's ambition was to make the biological complexity visible, relying on ecological investigations, *mélanges*, and the mobilization of local healers' therapeutic experiences. The tensions brought about by this transformation of biological therapeutic agents previously associated with forms of medical practices stressing the individual and constitutional nature of disease into mass-produced and prescription-ready pills are easy to perceive.

Madaus's handling of these tensions illustrates an important feature of drug regulation during the twentieth century. As stressed in the introduction to this volume, the various ways of regulating that can be distinguished are not entities granted with structural or functional ontologies; they rather summarize different grammars of action. As a consequence, even if there is a chronology of their emergence, they often operated in conjunction, overlapping in time and place. The boundaries between them are therefore not given, but an effect of the activities performed by and within each regulatory setting whatever its nature: Codex commission, patent office, drug bureau, or enterprise. In contrast to the case of Dausse where industrial regulation was peripheral, the industrialization of plant extracts at Madaus shows how this "boundary work" resulted in a quasi-internal/external separation between the industrial and the professional ways of regulating. The first was played at the level of in-house laboratories, control settings, and production units, revolving around the technological qualities of products and the use of standard processes. Its bearing on Madaus propaganda was general rather than specific, and took the form of broad claims regarding the stability and the preserved composition of Teeps. Professional regulation dominated the relationship with health practitioners, including the massive circulation of information that characterized Madaus's scientific marketing. Typically, the multiple recommendations or advice regarding indications, combinations, and conditions of use included in the firm's literature (academic journals included) did not only pay due respect to the autonomy of physicians (as was the case at Dausse) but argued for the decisive value of local experiences in clinical effects as well as in the nature of preparations all sorts of practitioners, that is, healers, physicians, and pharmacists, could accumulate. Within the peculiar context of interwar German biological medicine, drug regulation thus implied a kind of bottom-up type of aggregation, which after World War II would become incompatible with the generalization of screening and clinical trials (Gaudillière, 2010).

In the end, the pervasiveness of industrial regulation in Germany (at least in comparison with France) provides a possible explanation for what is otherwise difficult to understand, namely, the contrasted fate of herbal medicine in both countries. The 1936 reform granting *Heilpraktiker* a legal status is usually understood as an example of profession building that mimicked the history of school medicine, benefiting from a peculiar political window originating in the Nazi's initial love for alternative, supposedly popular health practices, where healers were able to negotiate partial but effective institutionalization. France did not offer the same opportunity. Hence, even if after World War I French herbalists expressed their acceptance of a status as second-rank drugmakers, the pharmacists' corporate bodies blocked the path toward the recognition of a full-fledged profession, pleading for the 1941 ban. The need to take into account cognitive and material practices when analyzing drug regulation, however, points to a different path of interpretation. Pharmacists' and herbalists' ways of understanding the nature of drugs as well as their mode of preparation were much less dissimilar than what is traditionally assumed, a feature that reinforced the administrative nature of their competition within the system of professions. The fact that the German reform persisted once its national–socialist birth context had vanished might therefore be seen as a consequence of both the declining influence of professional regulation and the successful industrialization of "alternative" therapies in the interwar period. Following this line of analysis, the success of French pharmacists might well be rooted in the herbalists' delayed industrialization, which left the former to think they were the only legitimate actors in trading and regulating medicinal plants. The opposite fate of French homeopathy after World War II points in the same direction: once this form of therapy was turned into an industrialized production of standard pills, it was officially recognized by the health-insurance system.

Notes

1. Decree law of September 11, 1941, article 59, paragraph 1.
2. "Stopverordnung," February 11,1943. Bundesarchiv Berlin, B142, 1432 Arzneimittelgesetzgebung.
3. See C. Bonah's chapter (Chapter 10).
4. See Table 0.1.
5. *Bulletin de l'Association amicale des anciens élèves de l'École Nationale d'Herboristerie,* "Le stage en herboristerie," N° 2, pp. 1–2, 1932.
6. *Les intraits Dausse et les sucs végétaux,* Laboratoires Dausse, Paris, 1910.
7. See, for example, *Gehe's Codex der Beziehungen von Arzneimiteln,* Schwarbeck, Gehe, 1926.

8. Dr. Med. Madaus, *Taschenbuch für die Biologische Praxis*, 5. Auflage 1938. Dr. Madaus & Co, Radebeul/Dresden, Sächsisches Hauptstaatsarchiv Dresden, 11610, Nummer 114.
9. Deutsche Reich Patent 129588 and 131837.
10. Aktenordner Nummer 127, Pharmazeutische Untersuchung Berichte, 1937–1942. "Bericht Dezember 1939."
11. Sächsiche Staatsarchiv Dresden, 11610 Firma Dr. Madaus. Aktenordner Nummer 141, Pharmazeutische Untersuchung Berichte, 1942–44. "Bericht Oktober 1942."
12. Sächsiche Staatsarchiv Dresden, 11610 Firma Dr. Madaus. Aktenordner Nummer 127, Pharmazeutische Untersuchung Berichte, 1938–1942. "Bericht September 1939."
13. Sächsiche Staatsarchiv Dresden, 11610 Firma Dr. Madaus, Aktenordner 124 and 138, Pharmazeutische Untersuchung Berichte, 1940–42.
14. Sächsiche Staatsarchiv Dresden, 11610 Firma Dr. Madaus, Aktenordner 127 and 128, Pharmazeutische Untersuchung Berichte, 1937–42.
15. See U. Thoms and J-P Gaudillière (eds), *Pharmaceutical Firms, Drug Uses and Scientific Marketing*, Special issue of *History and Technology*, forthcoming.
16. Sächsiche Staatsarchiv Dresden, 11610 Firma Dr. Madaus, Aktenordner Nummer 138, Pharmazeutische Untersuchung Berichte, 1942–44, Brief Kuhn an Dr. Koch, November 4, 1942.
17. The remaining reports only include one later mention of the wild-berries study, in the fall of 1943.

Bibliography

Balz, V. 2010. *Zwischen Wirkung und Erfahrung – eine Geschichte der Psychopharmaka. Neuroleptika in der Bundesrepublik Deutschland, 1950–1980*. Bielefeld: transcript.

Brinkmann, M. & Franz, M. (eds). 1982. *Nachschatten im weißen Land: Betrachtungen zu alten und neuen Heilsystemen*. Berlin.

Brissemoret. 1908. *Essais sur nos préparations galéniques*, Paris : Laboratoires Dausse.

Caron, F. 1997. *Les deux révolutions industrielles du XXème siècle*. Paris : Albin Michel.

Chauveau, S. 2005. "Le statut légal du médicament en France. XIXème-XXème siècles" in *Histoire et médicament*, C. Bonah and A. Rasmussen (eds.). Strasbourg : Glyphe, 87–113.

Dausse, A. 1836. *Mémoire sur la préparation de tous les extraits pharmaceutiques par la méthode de déplacement au moyen d'un appareil approuvé par la société de pharmacie contenant un procédé nouveau pour faire les extraits des plantes aromatiques suivi d'un tableau donnant exactement les quantités d'extrait fournies par chaque plante.* Paris.

Dinges, M. (ed.). 1996. *Medizinkritische Bewegungen im Deutschen Reich* (ca. 1870–ca. 1933). Stuttgart.

Gaudillière, J.P. 2005. "Drug trajectories: Historical studies of science, technology, and medicine" *Studies in History and Philosophy of the Biological and the Biomedical Sciences,* 36: 602–12.

Gaudillière, J.P. 2005. B. "Une marchandise pas comme les autres: Historiographie du médicament et de l'industrie pharmaceutique en France au XXème siècle" in *Histoire et médicament,* C. Bonah et A. Rasmussen (eds.). Strasbourg: Glyphe, 115–58.

Gaudillière, J.P. 2010. "The singular fate of screening in twentieth century pharmacy: Some thoughts on drug standardization and drug regulation." In C. Bonah,

C. Masutti, A. Rasmusse and J. Simon (eds) *Harmonizing Drugs: Standards in twentieth Century Pharmaceutical History*. Strasbourg: Glyphe.

Gradmann, C. & Hess, V. (eds). 2008. "Vaccines as medical, industrial and administrative objects" special issue of *Science in Context*, 21(2): 145–310.

Haug, A. 1985. *Die Arbeitsgemeinschaft für eine Neue Deutsche Heilkunde*. Husum.

Joanin, A. 1921–1939. *Les Remèdes Galéniques*. Paris: Laboratoires Dausse. 1er fascicule 1921, dernier fascicule 1939.

Jütte, R. 1996. *Geschichte der alternativen Medizin: Von der Volksmedizin zu den unkonventionelleren Therapien von Heute*. Munich.

Koch, E. 1936. "Biologische Untersuchungen," *Jahrbuch Dr. Madaus*. 42–58.

Kuhn, A. 1936. "Die Aufbereitung von Arzneipflanzen," *Jahrbuch Dr. Madaus*. 29–36.

Kuhn, A. & Jurek, E. 1933. "Zur Neugestaltung des homöopathischen Arzneibuches," *Jahrbuch Dr. Madaus*. 67–69.

Kuhn, A. & Penz, H. 1937. "Auswertungsversuche von *Digitalis* an Wasserflöhen," *Jahrbuch Dr. Madaus*. 74–78.

Madaus, G. 1932. "Die Lehre der Signatur," *Jahrbuch Dr. Madaus*. 8–28.

Madaus, G. 1933. "Versuche über Wundheilung bei Pflanzen," *Biologische Heilkunst*, 14: 332.

Madaus, G. 1934. "Unsere Heilpflanzen heilen auch die Erde," *Jahrbuch Dr. Madaus*. 14–23.

Madaus, G. 1935a. "Die Wirkung von *Digitalis purpura* und ihrer Abhängigkeit von der Düngung," *Jahrbuch Dr. Madaus*. 21.

Madaus, G. 1935b. "Die Scädigung von Heilpflanzen bei ihrer Verarbetung zum Heilmittel," *Jahrbuch Dr. Madaus*. 22–31.

Madaus, G. 1936. "Die Bedeutung der Heilpflanzen in der modernen Therapie," *Jahrbuch Dr. Madaus*. 15–25.

Madaus, G. 1938. *Lehrbuch der biologischen Heilmittel*. Leipzig: Georg Thieme Verlag. vol. I, 12.

Madaus, G. & Kunze, H. 1933. "Einige Ergebnisse der Untersuchungen in unserer biologischen Versuchstation im Jahr 1932" *Jahrbuch Dr. Madaus*. 11–24.

Minard, P. 1998. *La fortune du colbertisme. État et industrie dans la France des Lumières*. Paris : Fayard.

Murard, L. & Zylberman, P. 1996. *L'hygiène dans la République. La santé publique en France ou l'utopie contrariée, 1870–1918*. Paris: Fayard.

Rothschuh, K.E. 1983. *Naturheilbewegung, Reformbewegung, Alternativbewegung*. Stuttgart.

Ruffat, M. 1996. *175 ans d'industrie pharmaceutique. Histoire de Synthélabo*, Paris: La Découverte.

Timmermann, C. 2001. "Rationalizing folk medicine in interwar germany: Faith, business, and science at dr. Madaus & co," *Social History of Medicine*, 14: 459–82.

Wimmer, W. 1994. *"Wir haben fast immer etwas neues" Gesundheitswesen und Innovationen der Pharma-Industrie in Deutschland*. Berlin: Duncker & Humblot.

Woronoff, D. 1994. *Histoire de l'industrie en France du XVIème siècle à nos jours*. Le Seuil.

4
Making Risks Visible: The Science, Politics, and Regulation of Adverse Drug Reactions

Harry M. Marks

Cured yesterday of my disease
I died last night of my physician

Matthew Prior (1664–1721)[1]

A popular statin used to lower cholesterol is associated with an unusual breakdown of muscle tissue.[2] An analgesic prescribed for arthritic patients is linked to heart attacks. The scenarios involved in the recent cases of Baycol and Vioxx should be familiar. A novel drug passes intense regulatory scrutiny. The drug's makers heavily promote it. Following widespread use, a previously unnoticed side effect is observed. Investigative journalists then trumpet the drug's fall from grace, revealing a "back story" in which the warning signs of harm were ignored or suppressed. The drug's makers defend their product and their integrity while medical reformers and social scientists condemn corporate cupidity. Members of a bewildered public wonder about drug safety while injured patients and outraged politicians call for remedial action.

Psychically gratifying as such "histoires morales" are, they are of limited analytical value. The problem of identifying and regulating "adverse" drug effects is chronic, not acute, long-standing rather than recent (USGAO, 2006:2). The drug in such scenarios might be Baycol (2001) or Vioxx (2004), but it might as easily be the antibiotic chloramphenicol (1952), the oral contraceptive Enovid (1962), the diabetes drug Tolbutamide (1969), or various nonsteroidal anti-inflammatory drugs (NSAIDs) (1980s) (Maeder, 1994; Marks, 1999; Marks, 1997; Greene, 2005; Abraham, 1995). One might even argue that the problem of adverse drug reactions coincides with the creation of modern therapeutics: the serum sickness associated with the diphtheria antitoxin; the toxicities of Ehrlich's antisyphilis arsenicals (Marks, 1997). Yet if manifestations of drug toxicity appear at the dawn of modern immunotherapy and chemotherapy, the technical and regulatory systems for identifying and managing such toxicities are more recent in origin.

My argument in this chapter is twofold. First, that the handling of adverse drug reactions ultimately derives from a regulatory system that at all costs preserves medical autonomy. For the US case, the problem is ultimately rooted in the statutory language of the 1938 drug law, which authorized the Food and Drug Administration (FDA) to regulate the claims drug manufacturers could make about the safety and therapeutic uses of their products. With a few notable exceptions (Daemmrich, 2002; US GAO, 1990), the FDA regulates drug labeling, not drugs. The resulting regulatory system allows for maximum physician autonomy in drug prescribing.

Second, within such a regulatory system, "making risks visible" poses both technical and political difficulties (Chateauraynaud & Torny, 1999). Voluntary physician reporting of "unusual" drug reactions is inefficient in all health polities but the highly fragmented US health care system makes detection of adverse effects especially difficult (Finney, 1971; Tansey & Reynolds, 1997; Marks, 1999). Proving a causal link between drug and injury presents further difficulties: a slightly elevated risk for a common medical condition requires studying many more patients than demonstrating an increased risk for an unusual side effect (Brewer & Colditz, 1999). The burden of proof required will vary, according to current notions of evidence and the severity of the contemplated regulatory action. When the prospect looms of adding new warnings to a drug's labeling, drug companies enter a war of position, seeking – up to a point – to defend the market for their product. While the adversarial character of these negotiations vary with changes in the political and ideological climate, the result is invariably delay (Marks, 1995, 1997; Abraham, 1995; Maeder, 1994).[2]

In the following section, I describe the structural premises of the modern drug regulatory system in the United States. I follow with a discussion of two paradigmatic cases: chloramphenicol and diethylstilbestrol (DES), drawing on the work of Thomas Maeder and Susan Bell. I then analyze the contested history of administrative and technical efforts in the 1960s, 1970s, and 1980s to make drug hazards visible.

A modern system of drug regulation, 1938–2007[3]

On June 25, 1938, the US Congress enacted legislation authorizing the US FDA to review the safety and composition of new drugs. The new law required drug firms to demonstrate that drugs were "safe for use under the conditions prescribed, recommended or suggested in the proposed labeling thereof [my emphasis]."[4] This statutory language incorporated the twin premises of contemporary drug regulation: (1) that drug safety must be defined in risk–benefit terms (a drug that was unsafe for treating colds might be safe for treating pneumonia or influenza); (2) that drug regulation was largely informational in character, intended to instruct but never to command physicians' actions. For the most part, the FDA regulates drug labeling, not drugs or medical practice.[5]

This informational approach to regulation was a product of law, medical culture, and politics. By law, most of the FDA's regulatory powers lay in their authority to regulate labeling. This informational approach was characteristic of much New Deal regulation, which sought to improve markets by improving information.[6] Both the risk–benefit standard for judging drugs and the attempt to direct physician behavior via information were characteristic of the preceding 30-plus years of professional self-regulation by the American Medical Association's (AMA's)Council on Pharmacy and Chemistry (Marks, 1997, 22–41). J. J. Durrett and Theodore Klumpp, the medical officers in charge of the FDA's drug regulation efforts, were well-acquainted with this professional milieu. Both men were graduates of Harvard Medical School: Durrett was an FDA veteran; Klumpp joined the agency in 1936 from teaching at Yale Medical School, where he maintained many contacts.[7]

Klumpp and Durrett worked out the details of the FDA's regulatory approach via their handling of sulfapyridine, the first novel therapeutic compound to come up for review under the new law. Sulfapyridine was the newest of the sulfonamides, anti-infective drugs that had been introduced into medical practice only a few years before (Bovet, 1988). The sulfonamides were remarkable drugs, capable both of quelling life-threatening infections and of inducing life-threatening anemias, depressed white cell counts, and other serious blood disorders (Long & Bliss, 1939; Dowling, 1977). Sulfapyridine, the latest sulfonamide, was proposed for use in treating pneumonia, an infection against which sulfanilamide, the original sulfa drug, was useless.

Approving sulfapyridine was a matter of assessing its risks in relation to its clinical benefit. As Durrett put it:

> If the drug that killed one person in 10,000 was of only minor use therapeutically it might still be judged to be unsafe, whereas the drug which killed one in a thousand persons if it had marked and undisputed therapeutic value, such as the drug under question [sulfapyridine], it would still be a safe and valuable drug.[8]

To evaluate the drug, Klumpp and Durrett reviewed data on over 2,000 clinical cases, along with preclinical toxicological studies submitted by the drug's manufacturer, Merck.[9] They interviewed each of the clinical investigators Merck had supplied with the drug, along with an additional 45 research physicians. Those contacted constituted a virtual who's who of prominent clinical investigators and infectious disease specialists (Marks, 1997, 84–91).

The researchers consulted agreed that the drug was likely to prove of considerable benefit in treating pneumonias. Questions remained, however, about when and how it was safest to use. Clearly, "the drug was not killing many people" but at the same time, there was "no uniformity of opinion with respect to the harm which the drug might be capable of doing from one investigator to another."[10] The problem was compounded by the existence of serum therapy, a proven therapy for treating pneumonia. While difficult to use in community settings, serum therapy was standard treatment in major hospitals (Marks, 1997; Podolsky, 2006). Some researchers thought that the FDA should delay approval while more data was gathered about the relative risks and benefits of the two treatments. M. A. Blankenhorn (University of Cincinnati), David Rutstein (New York State Health Department), and O. H. Robertson (University of Chicago), fearful that physicians might use the drug casually, each wondered whether the FDA might approve sulfapyridine only for restricted distribution to "qualified clinicians" until more was known about its optimal (and safe) use.[11]

On March 9, 1939, the FDA approved sulfapyridine for treating pneumonia, while cautioning manufacturers to warn physicians about the need to closely monitor patients for signs of serious toxicity.[12] The FDA's Klumpp acknowledged academics' concerns that sulfapyridine "will undoubtedly be abused by the unwise and ill-informed whenever it is put on the market." But physician misuse had nothing to do with the "intrinsic safety or danger of the drug." Such misuse was a "problem of medical practice." Disciplining "unwise" or ignorant practitioners lay outside the FDA's authority.[13] As another agency official would write, the prospect of physicians using drugs "contrary to the recommendations in the labeling is unfortunately a matter which we are not permitted to consider in connection with the new drug applications."[14]

The philosophy articulated by Durrett and Klumpp rests at the heart of US drug regulation into the twenty-first century. Despite major changes in the FDA's statutory authority, and in the burden of evidence required to market a new drug, informational labeling remains central to US drug regulation. At various times, academic physicians have suggested that the FDA restrict a drug's distribution to selected specialist groups (Marks, 1997, 94–95). For the most part, such suggestions have fallen on deaf ears.[15]

Conflicts over drug regulatory policy consequently emerge as conflicts over particular uses of the FDA's authority to regulate drug labeling. When the FDA has sought to regulate drug claims more aggressively, industry has challenged the FDA in the courts or in Congress. Along with its medical allies, industry has invoked physician autonomy, medical need, and freedom from governmental ukases as inviolable principles (Marks, 1995). When, in 1948, the FDA attempted to remove glandular extracts from the market, on the grounds that an inactive and hence ineffective drug could not, by definition, be safe, Abbott Laboratories' Medical Director, George Hazel, claimed that the ruling,

> invokes a principle of profound importance in a free society as to whether medical therapeutics should be limited by administrative decrees, or be decided by the physician in charge of the individual case and be based upon his evaluation of the case in the light of his training and experience in clinical medicine. (Hazel, 1949: 37)[16]

Industry issued similar legal and political challenges when the FDA attempted to implement efficacy reviews of drugs it had approved prior to the 1962 Drug Amendments.[17] More recent FDA officials who have taken a strong regulatory stance on drug claims have had their wrists slapped by Congress and/or the courts. David Kessler, FDA Commissioner from 1990 to 1997, attempted to curtail corporate promotion of so-called off-label uses – clinical indications for an existing drug that had not been reviewed by the FDA (Kessler, 1990, 1991). Congress responded with legislation allowing companies to distribute "scientific" data so long as they ultimately intended to seek regulatory approval for the off-label uses. These generous provisions of the 1997 Food and Drug Modernization Act were not sufficient for some free-market advocates, who successfully challenged the FDA's new regulations by invoking the First Amendment rights of corporations.[18]

Although negotiations over the precise wording of new drug labels are routine, some labeling conflicts become deeply politicized. Not surprisingly, such intensified conflicts develop around reports of the adverse effects of previously approved drugs. FDA attempts to add warnings to the labeling for the antibiotic chloramphenicol, the oral hypoglycemic tolbutamide, the oral hypoglycemics, the antidepressant SSRIs (selective serotonin release uptake inhibitors), various NSAIDs, or the analgesic Darvon have generated lengthy combats over both specific labeling language and regulatory authority (Abraham, 1995; Leslie, 2005; Maeder, 1994; Marks, 1997: 216–29; Marks, 1999; Soumerai, 1987; Smith, 1979). Are the effects noted "truly" a product of the drug or of unusual circumstances (individual sensitivity and/or inappropriate usage)? How much attention should be paid to particular effects in revised labeling? How directive should revised labeling be in instructing physicians how to use a particular product?[19] What should the agency do about manufacturers who undercut their own warning labeling in other promotional activities?[20] While the FDA has the legal power to order a drug removed from the market, they rarely invoke it. Protracted regulatory conflicts therefore focus on the justification for specific changes in informational labeling.

The moralized histories that accompany the recent episodes of adverse drug reactions (e.g., Vioxx, Baychol, SSRIs) are themselves a product of this regulatory framework. Such accounts focus on the motives of historical actors, judging actions in moralized terms: Have FDA guardians relaxed their vigilance? Have drug companies suborned the integrity of the scientific process, via misrepresentation or suppression of evidence? Have physicians abdicated their professional responsibilities? These narratives reflect a long-enduring conflict within the medical profession itself, between reformers who regard commercial influences as inherently corrupting – "nearly all abuses arise because someone profits thereby" – and practitioners committed to the conjoined principles of therapeutic innovation and professional autonomy.[21] For the latter, the issues at stake in controversies over drug warnings are equally plain: "excessive" regulation endangers the flow of pharmaceutical innovation and threatens the inherent right of physicians to use drugs as they see fit. Both sides invest regulatory language with enormous powers to direct action, yet even reformers seem uncomfortable with more directive regulation that might limit physician autonomy.[22] Given the consistent evidence that many physicians ignore drug warnings and frequently use drugs without regard to their approved indications, such deference to professional autonomy seems indefensible (Melville & Mapes, 1980; Radley *et al.*, 2006; Lasser, 2001; Jones, 2001).[23] But the core issue here is not individual, corporate, or even regulatory morality, but the structures of drug regulation. Short of removing a drug from the market, the FDA's ability to alter physician practice is limited to written directives transmitted through labeling changes, medical alerts, and other forms of information management.

Making risks visible

As we all know, thanks to Mary Douglas, in a world full of dangers, only some hazards are certified as "risks." The recognition and interpretation of danger involves cultural work (Douglas & Wildavsky, 1982; Schlich & Tröhler, 2006). That drugs, ill-used, could cause harm is an ancient proposition. That the therapeutic instruments of modern scientific medicine could cause serious harm was readily acknowledged. Not long after the introduction of diphtheria antitoxin, clinicians and researchers recognized that some patients reacted badly to the injection of foreign protein. The effects of "serum sickness" and "anaphylaxis" were, nonetheless, manageable – first, by caution in the initial administration of antitoxins and vaccines, with the physician advised to look for signs of allergic sensitivity or hyper reactivity before administering a full dose and, second, by engineering changes in the production of biologics (Dowling, 1977, 38–39; Moulin, 1991, 141–45; von Pirquet & Schick, [1905]; Marks, 1997, 60–62).[24] A similar logic was translated to the domain of chemotherapy. Toxic reactions to Paul Ehrlich's Salvarsan and its analogues were well-known. Physicians were duly cautioned and pharmacologists worked to devise less toxic arsenicals (Marks, 1997: 54–55). Reports in the 1930s that the early sulfa drugs could engender life-threatening anemias evoked a similar response: physicians using these drugs must be on the lookout for signs of anemias or hepatic damage, which were, it was thought, the product of "idiosyncratic" reactions to the drugs (Marks, 1997: 78–82; Long & Bliss, 1939: 266–92).[25]

The FDA's policies after 1938 on mandatory premarket toxicological testing did not substantially alter the fundamental premises of regulation. Toxic effects must be identified but they were not necessarily a barrier to drug approval. As Susan Bell has argued, few products were as closely scrutinized as the synthetic hormone DES. Laboratory researchers had identified some of the drug's side effects: uterine bleeding and possible carcinogenic action. High rates of nausea and vomiting in early clinical tests generated further concern. As with sulfapyridine, the FDA held protracted discussions with researchers, going beyond the data submitted by the manufacturers. The clinical specialists they surveyed, nonetheless, considered that the drug's substantial benefits outweighed its risks. The FDA approved the drug for a narrow set of indications, in the hopes that this would discourage inappropriate use (Bell, 1980, 1994, 1995). The drug was, nonetheless, widely used for a variety of gynecologic and obstetric complaints, including prevention of miscarriage (Apfel & Fisher, 1984). In 1969, Arthur Herbst, an astute clinician working at a referral hospital noted that he had seen six cases of vaginal adenocarcinoma in adolescent women within the past two years. A mother of one of these patients recalled having used DES during her pregnancy. Herbst and his colleagues followed up with a case-control study that indicated that the mothers of 7 of 8 cases had taken DES during pregnancy, while no mothers of the 32 control cases had used the drug. Vaginal adenocarcinomas were sufficiently rare in young women that the FDA felt justified in warning against using DES during pregnancy (Herbst, 1970, 1971; Langmuir, 1971; DHEW, 1971). Despite the warning, some physicians continued to use the drug to improve pregnancy outcomes (Gillam & Bernstein, 1987).

In many ways, the DES story is typical. Evidence of some toxicity is subsumed within a risk–benefit calculus that leads to widespread drug use. Further evidence of substantial and serious harm (cancer in daughters of mothers who took DES during pregnancy) leads to a labeling change specifically warning physicians against the drug's prenatal use. Some physicians, nonetheless, continue using the drug in such circumstances. Yet DES was unusual in the relative ease with which its effects were detected and verified. As statisticians would later argue, a relatively rare clinical event due to drug use is easier to detect than a drug-related increase in the occurrence of a common condition such as coronary heart disease in middle-aged men (Brewer & Colditz, 1999). It is also much easier to prove: while some DES advocates remained skeptical (Reid, 1972), Herbst's careful study of eight patients was sufficient to convince the FDA.

As Thomas Maeder has shown, identifying and remedying the harm done by the antibiotic chloramphenicol (aka chlormycetin) was far more difficult. Developed and initially tested for its uses against scrub typhus, the drug was approved by the FDA on January 12, 1949 (Maeder, 1994: 73–102). Parke-Davis, the drug's sole manufacturer, promoted the drug heavily as a broad-spectrum antibiotic (Maeder, 1994). Over the next two years, two physicians described fatal cases of aplastic anemia in patients taking chloramphenicol (Maeder, 1994, 107–09). Then, in May, 1952, James Watkins, the nine-year-old son of Dr. Albe Watkins died of anemia developed after receiving several courses of antibiotic treatment. His son's illness and death launched Dr. Watkins on a 16-year campaign to alert medical groups and the FDA to the dangers of Parke-Davis' drug (Maeder, 1994, 20–45, 348–49).

The FDA did not ignore Albe Watkins' concerns. By late Spring, 1952, they were getting reports of blood dyscrasias from hematologists and other physicians around the country. In June the FDA organized a more formal survey, bringing the total number of cases of blood disorders in patients taking the drug to 410 (Maeder, 1994,109–24, 132–43). As with DES, the problem was not in detecting anemias in patients who by all rights should not have them. Rather, the problem lay, first, in interpreting the connection with chloramphenicol and, second, in deciding what to do about the problem.

The FDA's survey was by no means as informative or decisive as Arthur Herbst's case-control study of DES. Of the 410 cases identified, more than half (233) either did not involve chloramphenicol or provided no clear record of its use. In another 116 cases, chloramphenicol was not the only drug being taken. Only in 61 cases was chloramphenicol the only drug involved. Despite these and other methodological limitations, a panel of hematologists and infectious disease specialists convened at the National Research Council was fairly certain that the drug was responsible for producing some anemias (Maeder, 1994, 144–45).[26] The problem was in what to do: remove the drug from the market? restrict its use somehow? or modify the labeling to alert physicians to its dangers. Dr. Chester Keefer, who had managed the allocation of penicillin during World War II, summed up the sense of the meeting. Keefer

> felt that the evidence was reasonably convincing that chloramphenicol caused blood dyscrasias. He emphasized that it was not known how often these occurred and that a continuing study was needed. In considering the four alternatives proposed by the [Food and Drug] Administration he favored placing a warning on the label. He felt that after such a step had been taken it was the responsibility of each practising physician to familiarize himself with the possible toxic effects of the drug. He felt that it would be impracticable to specify in the labeling that the drug should be used only for typhoid and resistant infections. He believed that each physician would use it as he saw fit. He felt that there was not enough evidence to warrant taking the drug off the market and that a

combination of the first and second alternatives under consideration [e.g. to restrict use]...was impracticable.[27]

Several infectious disease specialists on the National Research Council (NRC) panel were prominently concerned with the misuse and overuse of antibiotics. Yet none thought the dangers of chloramphenicol justified restricting physicians' ability to use the drug "as they see fit."[28] Following their panel's advice, the FDA instructed Parke-Davis to issue new labeling warning physicians about the possibility of blood dyscrasias and the need to monitor patients who were taking the drug repeatedly or over long courses of treatment. The warnings were drowned out by the Parke-Davis president's instructions to his sales force:

> Chlormycetin has been officially cleared by the FDA and the National Research Council with no restrictions on the number or the range of diseases for which Chlormycetin may be administered. (Maeder, 1994, 160, 208–18)

Such corporate "spin" enabled the drug to recapture (and increase) its market share by the late 1950s, at which time new adverse effects emerged – in premature neonates given the drug prophylactically. These reports launched another round of efforts to police the drug's use via labeling changes. The NRC's experts rallied in support of the beleaguered drug, if anything more vociferously than before. They emphatically rejected proposals that the drug be restricted to specialist or in-hospital use, a strategy that

> would, in fact, be an attempt to regulate the professional activities of physicians. That enormous quantities of chloramphenicol are currently being prescribed is evidence in itself that the drug is being employed unwisely if not unnecessarily in many cases. How can physicians be taught or persuaded to employ this and other drugs properly. [...] The problem is an important one for which some reasonable solution should be sought. However, I doubt that the FDA can or should assume the function of policing physicians.[29]

FDA officials apparently agreed. In a press release announcing the labeling changes, they noted:

> Beyond this, there is need for the continuing education of the physician...This, of course, is a responsibility of the leaders of medicine and not of the Food and Drug Administration.[30]

Chloramphenicol remained in active use through the 1970s, surviving multiple labeling changes (Maeder, 1994, 84–189, 197–206, 218–24, 292–313,

358–62).[31] The problem of getting physicians to "employ" chloramphenicol or any drug "properly" would continue to vex NRC leaders.[32]

Chloramphenicol's continued survival was ensured by a confluence of factors: infectious disease specialists, eager to retain a drug they saw as valuable; an academic community reluctant to constrain practitioners' clinical autonomy; regulators whose principal regulatory tool was drug labeling; and a company willing to take advantage of these circumstances to heavily promote their product. The difficulty of getting reliable, comprehensive estimates of the anemias caused by chloramphenicol was, at best, a minor theme.[33] It was, however, soon to assume greater importance.

The identification of an adverse drug effect is a complex process. First, an observant physician must notice (and report) an unusual and otherwise inexplicable event, for example, Arthur Herbst's run of vaginal adenocarcinomas in adolescent women. Next, the strength of the association between drug use and the event must be measured. If the event is sufficiently unusual, then a very small study (like Herbst's case-control study) may be enough to securely establish that drug x is producing so many excess cases of condition z. The defects in red blood cell production caused by chloramphenicol are not common, but there are many possible causes of blood dyscrasias. The hematologists consulted in individual chloramphenicol-related cases may have been convinced about those cases, but without an estimate of increases in the rate of chloramphenicol-caused blood disorders, it was difficult to say how much of a public health problem the drug represented. And in the absence of either a disease registry or an adverse-effects surveillance system, such estimates were hard to come by (as the FDA found out).

Methodological excursus. There are multiple strategies for analyzing a putative adverse drug effect. A case-control study like Herbst's will provide an estimate of relative risk, the increase in the proportion of adverse events associated with a drug exposure. Case-control studies provide prima facie evidence of causation, but without additional information will not tell you much about the extent of the problem. The more common the suspected complication, the larger such the study must be (Jick, 1977). Disease registries provide a way of tracking the absolute number of adverse events, but cannot ensure that these are all drug-related. Herbst and his colleagues set up such a registry for DES-related adenocarcinoma; a similar registry for all suspected cases of drug-related blood dyscrasias was established by the AMA's Council on Drugs in 1955 (Erslev & Wintobe, 1962).[34] Finally, one may have an ongoing surveillance system, capable of tracking both drug use and adverse events. In the aftermath of problems with chloramphenicol, thalidomide, and the oral contraceptives, researchers at various institutions began discussing and initiating pilot surveillance programs. Early surveillance programs were established at the Peter Bent Brigham Hospital (1963), the Johns Hopkins Hospital (1964), Philadelphia (1964–66), in Boston (1966), and San Francisco (19??). Other institutions and networks followed

Table 4.1 Chronology of drug reaction reporting systems in the United States, 1952–72

1952–55	AMA Council on Drugs establishes Registry on Blood Dyscrasias
1963	Mandatory adverse drug reaction reporting established at Peter Bent Brigham Hospital, Boston
1964	Leighton Cluff establishes surveillance system at Johns Hopkins Hospital for analysis of adverse drug events
1964–66	Greater Philadelphia Adverse Reaction Reporting Registry involving five Philadelphia teaching hospitals.
1966	Hershel Jick establishes drug surveillance program at Lemuel Shattuck Hospital, Boston. Expanded to 22 hospitals later in the decade.
196?	Kaiser-Permanente establishes drug surveillance/adverse effects monitoring for outpatients
1972	Yale-New Haven Medical Center Outpatient Program
197?	University of Florida (Leighton Cluff)

their example in the late 1960s and early 1970s (Table 4.1). The history and operation of these programs remains fragmentary, much like the programs themselves.[35] Some, like the Kaiser-Permanente project, were part of a larger institutional program to computerize medical records.

Others were the initiative of individual researchers interested in the epidemiology of adverse drug reactions: Leighton Cluff at Hopkins and Florida; Hershel Jick in Boston; Sam Shapiro at Boston University. Yet others (e.g., Philadelphia) may have been a response to the anxieties of academic physicians about the growing publicity afforded adverse drug effects (Durant, 1965). Most programs were organizationally unstable, dependent on local entrepreneurs and a mixture of local and outside funding. When entrepreneurs moved on or funding dried up, the programs disappeared.[36]

Registries, surveillance programs and case reporting are usually discussed in terms of their methodological properties – how well each performs in identifying and then verifying adverse drug events in various epidemiological circumstances (Anello, 1977; Brewer & Colditz, 1999; Strom, 1989, 1994, 2000).[37] Such discussions take place in the open, at professional meetings, and in the scientific literature. Yet lurking behind such public discussions were a set of deep, privately held, concerns about how adverse drug effects, if found, should be managed. These behind-the-scenes discussions were conducted in a highly charged political atmosphere, engendered by the FDA's implementation of the 1962 Drug Amendments. In addition to operationally defining the standards for judging a new drug "effective," this legislation charged the FDA with determining the therapeutic value of the hundreds of drugs evaluated for safety between 1938 and 1962, and with setting the rules for ethical human experimentation (Marks, 1997; 2000b). The FDA's incursions into the previously unregulated terrain of medical research and therapeutic practice were often on the minds of those trying to address the problem of adverse drug reactions.

Medical ambivalence and disease registries: The hidden politics of adverse drug reactions, 1960–1975

The reluctance of medical specialists to advocate aggressive intervention in the chloramphenicol case reflects a larger ambivalence within academic medicine toward the federal government, which deepened during the 1960s.[38] Discussions at both the National Research Council and the AMA's Council on Drugs regarding adverse drug reporting reflect this ambivalence. With passage of the 1962 Drug Amendments, FDA officials had issued regulations requiring investigators to obtain written informed consent from patient subjects in experimental studies. Drug industry officials proved adept in mobilizing the discontent these regulations provoked in the academic community.[39] The academic members of the AMA's Council on Drugs saw a clear link between the informed consent issue and that of managing adverse drug reactions:

> All agreed that the FDA is exceeding the letter of the [1962] law and two examples (informed consent now means "written consent" and the proposed restriction of drugs for use according to the specialty of the physician were) were offered as proof of this.[40]

These political tensions hampered efforts to create a uniform registry for reporting adverse drug reactions. At issue was the question of who would control the data and how it would be used. At various times, a registry under joint FDA–AMA auspices, an independent AMA-managed registry, and an adverse drug surveillance system under the auspices of the Joint Commission on the Accreditation of Hospitals (JCAH) were discussed. The fortunes of any one proposal rose and fell with the general relations between organized medicine and the FDA. Academic physicians were crucial intermediaries in this process. Increasingly, as the decade wore on, tensions between the staff of the AMA Council on Drugs and academic physicians like John Adriani colored discussions of registry proposals. Adriani, Chair of the Council of Drugs, like other academic researchers, was suspicious of the increasing ties between the AMA and the drug industry. In working on the creation of an adverse drug event monitoring system, Adriani found himself repeatedly undermined by AMA staff, especially Dr. Jean Weston who worked "for" the Council.[41] Over time, Adriani seems to have put increasing hopes in an adverse events monitoring system to be promoted by the JCAH.[42]

Weston spoke for the rank and file physicians in the AMA's governing House of Delegates, who objected to FDA attempts to remove from the market various drugs approved prior to 1962. The FDA was required by law to remove such drugs if no substantial evidence of therapeutic efficacy could be found (Marks, 1997: 218).[43] Medical societies around the country reprinted copies of an article by Pharmaceutical Manufacturers Association president C. Joseph Stetler, attacking the FDA for interfering in the practice of medicine (Stetler, 1972).[44] For Weston, regulatory attempts to address adverse drug reactions reflected a similar sort of unnecessary inference in professional territory.[45] In an article published after she left the AMA, Weston reviewed the experiences of the various drug registries and surveillance programs established earlier in the decade. According to Weston, only a handful of studies at academic institutions had revealed high rates of adverse drug events, and there was reason to doubt that these were representative of medical practice at large. Perhaps, she suggested, the problem of adverse drug reactions was not, after all, so great (Weston, 1968).[46]

The AMA did announce a registry for spontaneous drug reporting in 1966, and somewhat less ceremoniously closed it down in 1970. The principal vehicle for spontaneous drug reporting thereafter was provided by the FDA (Daemmerich, 2004).[47] The FDA surveillance system, greatly expanded and modernized, remains central to the detection of adverse drug effects (Rossi, 1983; Leappe, 2002; Wysowski & Swartz, 2005).[48] The FDA's efforts are supplemented by a number of registries run by specialty societies, and by an ever-increasing number of hospital and health care network-based systems to detect and evaluate suspected adverse effects (Strom, 1994, 2003; Stern & Bigby, 1984).[49]

A detailed analysis of adverse effects surveillance over the past twenty years would require another paper. However, certain broad outlines seem clear. Despite substantial increases in both administrative and computing capacity for adverse drug surveillance, there remains a debate between advocates of such monitoring systems and proponents of more traditional methods for identifying the risks and benefits of drug therapy. Such debates extend to within the FDA itself, between members of the Center for Drug Evaluation and Research, responsible for premarket testing of drugs, and staff in the Office of Drug Safety, responsible for postapproval surveillance (Ferris, 2005; US GAO, 2006; Leappe, 2002). Similarly, despite some innovations in statistical methods, the basic dilemma remains unchanged: many effects appear only when a drug is used on large numbers of patients, and proving the connection between a drug and an effect is not a straightforward matter. No wonder, then, that there is delay in recognizing and acting on adverse drug effects (Lasser *et al.*, 2002; Ladewski *et al.*, 2003).[50]

Important as improvements in administrative and analytic capacity for monitoring adverse drug effects may be, they do not in themselves get to the heart of the problem. In writing about the history of adverse drug effects, I have emphasized the centrality accorded to clinical autonomy. The way physicians use drugs is regulated by language – drug labeling.[51] Since the 1940s (at least), whenever US physicians see the FDA as intruding upon the privilege of clinical autonomy, they have gone on the attack (Marks, 1995, 1997). (Industry has not been slow in exploiting this disposition, to their own advantage). I do not say this to attack the principle of clinical autonomy, but to open a discussion about its justification. As often as it is invoked, rarely is clinical autonomy deliberated.

The central argument for clinical autonomy in recent decades is that such autonomy is crucial in an era of rapid technological innovation.[52] Only if physicians are allowed to experiment with new, unproven uses of drugs will we arrive at the knowledge of what those drugs can do and how best to use them. How else would we have learned that calcium-channel blockers, approved for treating angina, can be used to treat hypertension? (Gelijns, 1998) Would one really want to shackle those physicians who figured out the precise mix of drugs and drug scheduled to contain drug resistance in immuno-compromised patients? Or, to take a nondrug example, to constrain the clever clogs who made his rabies-infected patient hypothermic so as to inhibit the slow-growing rabies virus from killing her? Against the benefits offered by a culture of innovation are the as-yet-unmeasured costs of clinical autonomy. A great deal depends on whether we think that these costs are down to the handful of physicians who use new drugs promiscuously and seemingly with very little constraint, or rather to far greater numbers of practitioners who use drugs inappropriately and injudiciously, with little appreciation for the sorts of evidence needed to demonstrate the therapeutic value of a new drug. Either argument could be supported from existing socio-medical studies of prescribing, but absent more systematic study of this issue, it would be hard to conclusively prove either (Jonville-Bera *et al.*, 2005; Melville & Mapes, 1980; Inman & Pearce, 1993; Armstrong, 2002). Yet our continued tolerance for clinical autonomy, for a less directive approach to regulating therapeutic practice and for the seemingly inevitable delays that accompany the adjudication of newly discovered adverse drug reactions, depends on what we think the result of this calculus would be. Has a labeling-oriented approach to drug regulation provided us with a steady stream of innovative and beneficial therapeutic practice, or has it put us at risk for unnecessary exposure to the harmful effects of ill-used drugs? An extended discussion of that question would seem more productive than another round of exchanges about the motives and conduct of drug companies and regulatory authorities.

Notes

* The chapter is as Harry left us, including the introducing poem.

1. Cited in Moser, 1969: 267.

2. It is generally argued (Daemmerich, 2004; Abraham, 2002; Abraham & Davis, 2006) that the US regulatory system is tougher on adverse effects and signs of drug toxicity than other regulatory systems, especially the United Kingdom and Germany. That does not obviate the point I am emphasizing: that the US approach of regulating through labeling is extraordinarily protective in allowing for clinical discretion in the use of drugs.
3. This section draws heavily on Marks, 1997, 1994, 1995. The reader is referred to these publications for additional detail and documentation.
4. Federal Food, Drug and Cosmetic Act, chap. 5, sec. 505 (d). This clause sets the standards for new drugs; sec 502 (j) authorized the secretary to treat as misbranded *any* drug that was dangerous to health when used in the dosage and conditions recommended in the labeling.
5. That the FDA from the onset of the 1938 law sought to regulate claims of therapeutic benefit is clear, despite subsequent political and legal challenges from industry that forced the FDA to get specific statutory authority to examine claims of drug efficacy. See Marks, 1997, 1995; compare Temin, 1980; Daemmerich, 2002.
6. Legislation creating the Securities and Exchange Commission is the most obvious example of an attempt to fix capital markets by regulating information about those markets. There are elements of a similar approach even in more interventionist economic legislation, such as the National Recovery Administration, so dependent on industry-specific "codes" for regulating industrial competition.
7. Durrett graduated HMS, 1914, Klumpp, HMS, 1928. See Jacques Cattell, ed., *American Men of Science*, 10th edn (Tempe, AZ: Jacques Cattell Press, 1960–62), vol. I, p. 1021; Council on Pharmacy and Chemistry, *Bulletin* 65 (December 4, 1935), 774; "Changes in the Food and Drug Administration," *JAMA* 111 (September 17, 1938), 1116; Who's Who in America, 1982–1983, 42nd ed. (Chicago: Marquis Publishing, 1982), vol. I, p. 1838.
8. J. J. Durrett, Memorandum of Interview with Perrin H. Long and E. Kennerly Marshall [Johns Hopkins], December 5, 1938, 88–69A_2099, Box 1, NDA 90, vol. 1, Washington_National Records Center, Suitland, MD [Hereafter W-NRC].
9. Theodore G. Klumpp, Memorandum for Mr. Campbell, February 23, 1939, 88–69A_2099, Box 1, NDA 90, vol. 3, W_NRC.
10. J. J. Durrett, Memorandum of Interview with Dr. Joseph Rosin [Merck] and D. W. Richards [Columbia University], December 1, 1938; in Durrett's view, "there was no way to dispute the value of this drug." See J. J. Durrett, Memorandum of Interview with Perrin H. Long and E. Kennerly Marshall [Johns Hopkins], December 5, 1938. Both references 88–69A_2099, Box 1, NDA 90, vol. 1, W_NRC.
11. See the detailed discussion in Marks (1997): 84–92, and M. A. Blankenhorn to W. G. Campbell, February 2, 1939; O. H. Robertson to W. G. Campbell, February 7, 1939; David D. Rutstein to W. G. Campbell, February 7, 1939. See also Hugh Morgan to W. G. Campbell, February 6, 1939; Memorandum of Interview with Harris S. Johnson, February 3, 1939; W. H. Carroll to W. G. Campbell, February 9, 1939; L. H. Schmidt to W. G. Campbell, February 9, 1939. All 88–69A_2099, Box 1, NDA 90, vol. 3, W_NRC.
12. W. G. Campbell to Dr. Joseph Rosin [Merck], March 9, 1939, 88–69A_2099, Box 1, NDA 90, vol. 3, W_NRC. Five additional manufacturers had submitted applications for sulfapyridine since the original Merck filing.
13. Theodore G. Klumpp, Memorandum for Mr. Campbell, February 23, 1939, 88–69A_2099, Box 1, NDA 90, vol. 3, W_NRC.

14. Hunter F. Kennedy to Perrin H. Long, August 27, 1942, 88–59A_2736, Box 107, F 511.07 (August), W-NRC.

15. Recently, in a handful of cases, the FDA has imposed distribution requirements for selected drugs whose unrestricted use might pose particular hazards (Karwoski, 2006; Seligman, 2002). The FDA has required elaborate conditions on physicians who prescribe the drugs thalidomide and accutane and the patients who take them. Both are capable of causing birth defects. For accutane, see Shulman, 1989; Cuzzell, 2005. The registration system was put into place only after repeated failures of efforts to warn physicians through labeling and medical alerts. For thalidomide, see Timmermans & Leiter, 2000; Daemmerich & Krücken, 2000.

16. George W. Calver, Physician to the US Congress, made virtually the same plea on behalf of the Congressional wives he treated: George W. Calver to the Pure Food and Drug Administration, July 1, 1948, 88–59A-2736, Box 582, folder 5126.12, W-NRC.

17. The 1962 Amendments, which explicitly authorized the FDA to examine the efficacy of new drugs, also required the agency to review the evidence for the efficacy of drugs it had approved between 1938 and 1962 under the "safety" statute. The FDA, overwhelmed by the task, initially delegated the reviews to expert physician panels convened by the National Academy of Sciences. Industry challenged these initial determinations on the grounds that the FDA had unconstitutionally delegated its regulatory authority to a nongovernmental body. The other contested issue was whether the FDA was allowed to insist on evidence from "well-controlled" clinical trials as a condition of reapproval. See "Drug Efficacy," 1971, 195–222; Gardner, 1973a, 1973b.

18. Remarkably, neither these aspects of the 1997 law or the subsequent legal challenges have received serious political analysis. For a summary, see Blackwell & Bent, 2003.

19. In the case of oral contraceptives, Lara Marks emphasizes the FDA's decision to focus on labeling warning patients, not physicians (Marks, 2000). See also Watkins, 2002.

20. One or more of these issues arose in each of the examples cited above, and are dealt with in greater detail elsewhere. In the oral hypoglycemic case, the challenges to labeling changes ultimately reached the Supreme Court (Marks, 1997: 216–28). For chloramphenicol, see Maeder, 1994; for SSRIs, see Leslie, 2005; for Darvon, see Soumerai, 1987; for the NSAIDs see Abraham, 1995.

21. Robert Hatcher to Torald Sollman, November 25, 1936. Torald Sollman papers, Archives, Cleveland Health Science Library, Cleveland Ohio. On the long history of anti-commercial in therapeutic reform, see Marks (1997, 2000a). For a recent elaborate example of anticommercialism, see Angell (2004), although the points should be familiar to regular readers of the American press and to Congress watchers.

22. As Nicholas Rasmussen has argued, even "therapeutic reformers" were involved in an extensive network of research and promotion for drug companies as early as the 1930s (Rasmussen, 2005, 2006). The proposition that the United States lags behind other countries in drug innovation was key to ideological debates over regulation from the 1970s on; however, the argument can be found much earlier.

23. For an excellent overview and study of prescribing with reference to drug labeling warnings, see Wagner *et al.*, 2006.

24. See Wilson, 1967, for a fairly thorough survey of earlier reports of reactions to vaccines and other biologics. A more detailed survey of the interplay between the clinical recognition/management of reactions and improvements in vaccine production would be very helpful, as would suggestions to discussions of this topic by historians that I've missed.
25. As von Pirquet & Schick [1905] emphasize, these are acquired sensitivities that usually manifest themselves clinically after an initial use. Long & Bliss, though less interested in the biological phenomenon, would agree. The theme of "idiosyncrasy" in the face of any new adverse reaction would continue to recur.
26. Interestingly, a proposal by two National Research Council statisticians to do a case-control study to measure the extent of the damage caused by chloramphenicol was not pursued. See Gilbert W. Beebe and Seymour Jablon, Use of Medical Records Systems of the Army and Veterans Administration in the Study of the Risk of the Production of Aplastic Anemia by the Therapeutic Use of chloramphenicol. July 30, 1952 [MED: Ad Hoc Conference on Chloramphenicol, 1952, National Academy of Sciences Archives, Washington DC. Hereafter NAS archives].
27. Ad Hoc Conference on Chloramphenicol, August 6, 1952. Med: Ad Hoc Conference on Chloramphenicol, 1952, NAS archives.
28. Only Harvard hematologist William Dameshak seems to have stood strongly for restrictions on the drug's use. Harry Dowling and Maxwell Finland, both critics of the way antibiotics were promoted and used, thought the drug's many benefits outweighed its risks. See Ad Hoc Conference on Chloramphenicol, 1952, NAS archives; Maxwell Finland to Robert Stormont [AMA Council on Pharmacy and Chemistry], August 27, 1953, Box 12, Maxwell Finland papers, Francis Countway Library of Medicine, Harvard Medical School, Boston, MA. Scott Podolsky has emphasized that Finland had other more pressing concerns: the promotion and use of "combination antibiotics" and the related rise of antibiotic resistance (Personal communication, July 27, 2006). For background, see Whorton, 1980, and Marks, 2000a.
29. Remarks of C. Lockard Conley (Johns Hopkins) in Replies to Questionnaire on Chlormycetin, December 1, 1960, MED: Requests for Advice Chloramphenicol, 1960–1962, NAS archives. Max Finland again thought that far too much attention was being paid to chloramphenicol – a result, he suggested, of a "very personal and emotional interest of one physician" – presumably a reference to Able Watkins. See handwritten reply [n.d.] of Maxwell Finland to an inquiry from R. Keith Canaan [NRC], December 1, 1960. Box 4, Finland papers, HMS. I am indebted to Scott Podolsky for calling Finland's remarks to my attention. For similar concerns regarding FDA impingement on the doctor–patient relation, see Food and Drug Administration, Subcommittee on Chloramphenicol of the Medical Advisory Board, Minutes. February 26, 1968, Minutes. February 20, 1969, Box 34, John Adriani papers, National Library of Medicine.
30. FDA Press Release, January 26, 1961. MED: Requests for Advice, Chloramphenicol 1960–1962. NAS archives.
31. Of the specialists consulted by the NRC, only infectious disease specialist Harry Dowling and hematologist Maxwell Wintrobe spoke up for restricting the drug to hospital use. Despite his concerns over the drug's dangers and misuse, William Dameshak joined the majority in rejecting any moves by the FDA in the direction of restricting the drug's distribution to hospitals – hospital physicians were equally capable of using the drug inappropriately and/or without

proper precautions. Replies to questionnaire on Chlormycetin, December 1, 1960, MED: Requests for Advice Chloramphenicol, 1960–1962, NAS archives. For Dameshek's concern over the harm done by the drug, see Maeder, 1994: 343. For subsequent discussions of changes in chloramphenicol labeling, see also FDA, Ad Hoc Committee on Chloramphenicol, Minutes, February 26, 1968, February 20, 1969. Box 34, John Adriani papers, NLM.

32. See R. Keith Cannan [NRC] to William S. Middleton, March 2, 1964, March 24, 1964, Drug Efficacy Study: Middleton: Drug Research Board: Correspondence, NAS archives.

33. On the lack of estimates, see the remarks of Gilbert Beebe, Chester Keefer, and Harry Dowling in Ad Hoc Conference on Chloramphenicol, August 6, 1952, NAS archives. For proposals to set up registries for chloramphenicol related anemias, see Beebe, *ibid.*; and William Dameshek to John Dingle, September 5, 1952 [MED: Ad Hoc Conference on Chloramphenicol, NAS archives].

34. For methodological criticism of the AMA registry, see Drug Research Board (1970), 11.

35. This section draws on Anello, 1977; Ruskin & Anello, 1980; Stewart *et al.*, 1977; "Meeting the problem," 1966. For the United Kingdom, see Finney, 1965, 1971; Inman, 1980; Tansey & Reynolds; 1997.

36. The longest-lived of these programs, the Boston Collaborative Drug Surveillance Program (BCDSP), began in 1966 and continues today. The core of the original program was a cooperative effort among Boston-area hospitals to collect data on in-patient adverse drug reactions. The collaborative was especially productive in testing hypotheses about putative drug-linked events via case-control studies. This large database enabled investigators to routinely explore possible drug effects that smaller institutions could not evaluate (Lawson, 1980). Like several other surveillance programs, the Boston Collaborative Drug Surveillance Program was partially funded by the FDA, which was looking for a way to supplement its principal source of data on adverse reactions: spontaneous reporting from physicians (Anello, 1977). It does not appear, however, that surveillance programs were especially useful in replacing spontaneous reporting as a way of finding new suspected drug effects (Venning, 1983; Rossi *et al.*, 1983; Wysowski & Swartz, 2005).

37. See Finney, 1965, 1971. The British biostatistician, David Finney, was the first to take these issues up in a serious way.

38. See Marks, 1992, for a discussion of the longer-range ambivalence of academic researchers towards the federal government.

39. For histories of the informed consent regulations, see Curran, 1970, and Marks, 2000b. Examples of the industry organizing campaign against the informed consent regulations include Walter A. Munns, Smith Kline and French, May 24, 1961 [memorandum on efficacy requirements]; Louis S. Goodman to Maxwell Finland, June 7, 1961; both in A. McGehee Harvey Papers, Chesney archives. Lee van Antwer [Assistant Medical Director, G.D. Searle] to Harry Gold, September 20, 1962; Lawrence B. Hobson, Squibb Institute for Medical Research] to Harry Gold, August 31, 1962; Harold L. Upjohn [???] to Harry Gold, September 19, 1962; all in "FDA Guidelines 1962," Harry Gold Papers, New York Hospital; Henry K. Beecher to George P. Larrick, December 22, 1964, FDA Docket, Section 130.37, Consent for Use of Investigational New Drugs on Humans. FDA records. I have seen additional examples of this mobilization in the papers of numerous clinical investigators and pharmacologists of the era.

40. AMA Council on Drugs, Minutes. Executive Committee, August 31, 1967. See also Council on Drugs, October 6–8, 1966. Report of Bethesda Conference on Relationship of the Clinical Investigator to the Patient, Pharmaceutical Industry and Federal Agencies. Both in Box 27, John Adriani papers, MS 453, National Library of Medicine. A third friction point was the implementation of the Drug Efficacy Study. The FDA had commissioned the National Academy of Sciences/National Research Council to review the efficacy of all the drugs the FDA had screened for safety between 1938 and 1962. The recommendations of the NAS panel were strongly contested both by industry and by the medical rank and file (Bazell, 1971).
41. On uses and control of adverse reaction data, see Drug Research Board, Minutes. Sixth Meeting. October 11, 1965, MED: Drug Research Board Minutes, NAS archives; "Meeting the problem," 1966; various correspondence, Box 30, John Adriani papers, NLM. On tensions between academic physicians and AMA, see R. Keith Canaan to William S Middleton, September 30, 1963, Drug Efficacy Study: Drug Research Board, William Middleton Correspondence, NAS archives; AMA Council on Drugs, Minutes, September 17–19, 1970; John Adriani to John Curry, March 25, 1969, Box 32; Maxwell Wintrobe to Harry Shirkey, February 9, 1967; Adriani to Wintrobe, May 6, 1966; Wintrobe to Adriani, June 7, 1966, Box 34 [all in Adriani papers, NLM]. By comparison with the analysis offered here, Daemmrich (2004: 120–25) offers a relatively bloodless account of these developments.
42. John Adriani to Herbert L. Ley, Jr [FDA] February 24, 1969, provides a good history of the various registry proposals. Box 26, Adriani papers, NLM.
43. The FDA delegated this massive assessment task to the National Academy of Sciences/National Research Council, which enlisted specialty panels of academic physicians to evaluate the efficacy data on each drug. The reports issued by the Drug Efficacy Study (DES) provided the basis for the FDA's actions in removing these drugs. Now that the DES papers at the National Academy of Sciences are open, there is little excuse for historians continuing to avoid a full analysis of this program, central to understanding the shifting relations between academic medicine, the drug industry, and the medical profession in the 1960s and 1970s.
44. The article was also printed in the *Ohio State Medical Journal* 68 (1972) and *The Journal of the Medical Association of Alabama* 42 (1972), among others. Not all versions identified Stetler as the source/author of the piece, which attacks the FDA for suggesting that it might require statements of relative efficacy in drug labeling.
45. See Jean K. Weston to James L. Goddard [FDA], March 31, 1966, Box 27, John Adriani papers, NLM.
46. Dr. Weston, previously on the staff of the AMA Council on Drugs had moved by 1968 to the National Pharmaceutical Council, a drug industry trade group.
47. The starting date is a little ambiguous; some publications from 1966 imply that the registry was already up and running (de Nosaquo, 1966).
48. According to Louis Lasagna (1983), the FDA began experimenting with formal postmarketing surveillance in 1970, with levodopa. So-called Phase IV trials were officially recognized in the 1997 FDA Modernization Act, though the practice by then was well-established.
49. The major novelty here are the various European surveillance systems developed in the 1980s and 1990s, which potentially provide another source of information to US regulators. See Strom, 1994, 2003, for a survey.

50. See Evans 2000; Kaufman & Shapiro, 2000, for discussions of recent methodological advances.
51. I here disagree fundamentally with Peter Temin, who sees the history of drug regulation as moving in the direction of an increasing "command and control" logic (Temin, 1980).
52. In earlier decades, the central argument lay more around the notion of clinical judgment – the physician's capacity to know what is best in treating *your* individual case. The innovation argument, however, was never absent. I would suggest, speculatively, that they have reversed their relative ideological importance in recent decades. See, for example, the debates in the 1970s about the off-label use of the beta-blocker practolol [Practolol correspondence in *JAMA* (1974), 227, 201–02; *Annals Internal Med* (1973) 79, 752–54; Morelli, 1973; Mundy *et al.*, 1974].

Bibliography

J. Abraham (1995) *Science, Politics and the Pharmaceutical Industry* (London: UCL Press).

J. Abraham (2002) "Transnational industrial power, the medical profession and the regulatory state: Adverse drug reactions and the crisis over the safety of Halcion in the Netherlands and the UK," *Social Science & Medicine* 55, 1671–90.

J. Abraham and C. Davis (2006) "Testing times: The emergence of the practolol disaster and its challenge to drug regulation in the modern period," *Social History of Medicine* 19, 127–49.

C. Anello (1977) *Identification of Adverse Drug Reactions to Marketed Drugs in the United States and the United Kingdom* (Washington: Biometrics and Epidemiology, Bureau of Drugs, FDA).

M. Angell (2004) *The Truth about the Drug Companies: How They Deceive Us and What to Do about It* (New York: Random House).

R. J. Apfel and S. M. Fisher (1984) *To Do No Harm: DES and the Dilemmas of Modern Medicine* (New Haven: Yale University Press).

D. Armstrong (2002) "Clinical autonomy, individual and collective: The problem of changing doctors' behavior," *Social Science & Medicine* 55, 1771–77.

R. J. Bazell (1971) "Drug efficacy study: FDA yields on fixed combinations," *Science* 172, 1013–15.

S. E. Bell (1980) *The Synthetic Compound Diethylstilbestrol (des) 1938–1941: The Social Construction of a Medical Treatment* (Brandeis University Ph.D. thesis).

S. E. Bell (1994) "From local to global: Resolving uncertainty about the safety of DES in menopause," *Research in the Sociology of Health Care* 11, 41–56.

S. E. Bell (1995) "Gendered medical science: Producing a drug for women," *Feminist Studies* 21, 469–500.

A. E. Blackwell and J. M. Beck (2003) "Drug manufacturers' First Amendment right to advertise and promote their products for off-label use: Avoiding a pyrrhic victory," *Food Drug Law Journal* 58, 439–62.

Daniel Bovet (1988) *Une Chimie Qui Guerit: Histoire De La Decouverte Des Sulfamides* (Paris: Payot).

T. Brewer and D. Colditz (1999) "Postmarketing surveillance and adverse drug reactions," *JAMA* 281, 824–29.

F. Chateauraynaud and D. Torny (1999) *Les Sombres Précurseurs. Une Sociologie Pragmatique De L'alerte Et Du Risque* (Paris: École des Hautes Études en sciences sociales).

W. J. Curran (1970) "Governmental regulation of the use of human subjects in medical research: The approach of two federal agencies" in Paul A. Freund (ed.) *Experimentation with Human Subjects* (New York: George Braziller), pp. 413–14.

J. Cuzzell (2005) "FDA approves mandatory risk management program for isotretinion," *Dermatology Nursing* 17, 383.

A. Daemmrich (2002) "A tale of two experts: Thalidomide and political engagement in the United States and West Germany," *Social History of Medicine* 15, 137–58.

A. Daemmrich (2004) *Pharmacopolitics. Drug regulation in the United States and Germany* (Chapel Hill: University of North Carolina Press).

A. Daemmerich and G. Krücken (2000) "Risk versus risk: Decision-making dilemmas of drug regulation in the United States and Germany," *Science as Culture* 9, 505–34.

de Nosaquo (1966) "The American Medical Association Registry on adverse reactions," *Annals of Internal Medicine* 64, 1325–27.

DHEW [Department of Health Education & Welfare] (1971) "Certain estrogens for oral or parenteral use," *Federal Register* 36, 21527–38.

M. Douglas and A. Wildavsky (1982) *Risk and Culture. An Essay on the Selection of Technical and Environmental Dangers* (Berkeley: University of California Press).

H. F. Dowling (1977) *Fighting Infection. Conquests of the Twentieth-Century* (Cambridge: Harvard University Press).

"Drug Efficacy and the 1962 Drug Amendments" (1971), *Georgetown Law Journal* 60, 185–224.

Drug Research Board, National Research Council (1970) *Report of the International Conference on Adverse Reactions Reporting Systems* (Washington: National Academy of Sciences).

T. M. Durant (1965) "Drug problems and the Philadelphia Plan," *JAMA* 192, 131–34.

A. J. Erslev and M. M. Wintrobe (1962) "Detection and prevention of drug-induced blood dyscrasias," *JAMA* 181, 114–19.

S. J. W. Evans (2000) "Pharmacovigiliance: A science or fielding emergencies," *Statistics in Medicine* 19, 3199–209.

M. Ferris (2005) "Does the FDA adequately protect the public?" *CQ Researcher* 15, 1–31.

D. J. Finney (1965) "The design and logic of a monitor of drug use," *Journal of Chronic Diseases* 18, 77–98.

D. J. Finney (1971) "Statistical aspects of monitoring for dangers in drug therapy" *Methods of Information in Medicine*, 10, 2–8.

J. Gardner (1973a) "Supreme Court will decide outcome in FDA, industry drug-effectiveness battle," *National Journal* 519–26.

J. Gardner (1973b) "Supreme Court Rule on ineffective drugs gives FDA sweeping regulatory powers," *National Journal* 963.

A. C. Gelijns, N. Rosenberg and A. J. Moskowitz (1998) "Capturing the unexpected benefits of medical research," *New England Journal of Medicine* 339, 693–98.

R. Gillam and B. Bernstein (1987) "Doing harm: The DES tragedy and modern American medicine," *Public Historian* 9, 57–82.

J. A. Greene (2005) *The Therapeutic Transition: Pharmaceuticals and the Marketing of Chronic Disease* (Harvard University: Ph.D. thesis).

A. L. Herbst and R. E. Scully (1970) "Adenocarcinoma of the vagina in adolescence: A report of 7 cases Including 6 clear-cell carcinomas (so-called mesonephormas)" *Cancer* 25, 745–57.

A. L. Herbst, H. Ulfelder and D. C. Pskanzer (1971) "Adenocarcinoma of the vagina: Association of maternal stilbestrol therapy with tumor appearance in young women" *New England Journal of Medicine* 284, 878–81.

W. H. W. Inman (1980) "The United Kingdom" in W. H. W. Inman (ed.) *Monitoring for Drug Safety*, (Boston: MTP Press Ltd), 9–48.

W. H. Inman and G. Pearce (1993) "Prescriber profile and post-marketing surveillance," *Lancet* 342, 658–61.

H. Jick (1977) "The discovery of drug-induced illness," *New England Journal of Medicine* 296, 481–85.

J. K. Jones, D. Fife ; S. Curkendall ; E. Goehring ; J. J. Guo ; M. S. Shannon (2001) "Coprescribing and codispensing of cisapride and contraindicated drugs," *JAMA* 286, 1607–09.

A. P. Jonville-Bera ; F. Bera ; Autret-Leca (2005) "Are incorrectly used drugs more frequently involved in adverse drug reactions? A prospective study," *European Journal of Clinical Pharmacology* 61, 231–06.

C. B. Karwoski (2006) *Practical Experience with risk management plans in the US*, Paper presented at the 42nd annual meeting of the Drug Information Association.

D. W. Kaufman and S. Shapiro (2000) "Epidemiological assessment of drug-induced disease," *Lancet* 356: 1339–43.

D. A. Kessler and W. L. Pines (1990), "The Federal regulation of prescription drug advertising and promotion," *JAMA* 264, 2409–15.

D. A. Kessler (1991) "Drug promotion and scientific exchange," *New England Journal of Medicine* 325, 201–03.

L. A. Ladewski *et al.* (2003) "Dissemination of information on potentially fatal adverse drug reactions for cancer drugs from 2000 to 2002: First results from the research on adverse drug events and reports project," *Journal of Clinical Oncology* 21, 3859–66.

A. Langmuir (1971) "New environmental factor in congenital disease," *New England Journal of Medicine* 284, 912–13.

L. Lasagna (1983) "Discovering adverse drug reactions," *JAMA* 249, 2224–25.

K. E. Lasser *et al.* (2002) "Timing of new black box warnings and withdrawals for prescription medications," *JAMA* 287, 2215–20.

K. E. Lasser et al. (2006) "Adherence to black box warnings for prescription medications in outpatients," *Archives of Internal Medicine* 166, 338–44.

L. L. Leape (2002) "Reporting of adverse events," *New England Journal of Medicine* 347, 1633–38.

L. K. Leslie, T. B. Newman, P. J. Chesney and J. M. Perrin (2005) "The Food and Drug Administration's deliberations on antidepressant use in pediatric patients," *Pediatrics* 116, 195–204.

P. H. Long and E. A. Bliss (1939) *The Clinical Use of Sulfanilamide and Sulfapyridine and Allied Compounds* (New York: Macmillan Co).

T. Maeder (1994) *Adverse reactions* (New York: William Morrow).

H. M. Marks (1992) *Leviathan and the Clinic: Academic Physicians and Medical Research Policy, 1945–1955*, Paper presented at History of Science Society Meetings, Dec. 27–30.

H. M. Marks (1994) *Playing It Safe: Federal Drug Regulation after 1938*, Presented at the Organization of American Historians, 1994 Annual Meeting, Atlanta, GA, April 14–17.

H. M. Marks (1995) "Revisiting 'The Origins of Compulsory Drug Prescriptions'," *American Journal of Public Health* 85, 109–15.

H. M. Marks (1997) *The Progress of Experiment. Science and Therapeutic Reform in the United States, 1900–1990* (New York: Cambridge University Press).

H. M. Marks (2000a) "Trust and mistrust in the marketplace: Statistics and clinical research, 1945–1960," *History of Science* 38, 343–55.

H. M. Marks (2000b) *Where Do Ethics Come From? The Role of Disciplines and Institutions*, Conference on Ethical Issues and Clinical Trials, University of Alabama at Birmingham, February 25–26.

L. Marks (1999) "'Not just a statistic': The history of USA and UK policy over thrombotic disease and the oral contraceptive pill, 1960s–1970s," *Social Science & Medicine* 49, 1139–55.

"Meeting the Problem. Panel discussion of Experiences and Problems Involved in Report Adverse Drug Reactions" (1966), *JAMA* 196, 421–28.

A. Melville and R. Mapes (1980) "Anatomy of a disaster: The case of practolol" in Roy Mapes (ed.) *Prescribing Practice and Drug Usage* (London: Croom Helm), 121–44.

H. F. Morelli (1973) "Propanolol," *Annals of Internal Medicine* 78, 913–17.

R. H. Moser (1969) *Diseases of Medical Progress: a Study of Iatrogenic Disease* 3rd edn, (Springfield, IL: Charles C. Thomas).

A. M. Moulin (1991) *Le dernier langage de la médecine. Histoire de l'immunologie de Pasteur au Sida* (Paris: Presses Universitaires de la France).

G. R. Mundy *et al.* (1974) "Current medical practice and the Food and Drug Administration: Some evidence for the existing gap," *JAMA* 229, 1744–48.

S. H. Podolsky (2006) *Pneumonia before Antibiotics. Therapeutic Evolution and Evaluation in Twentieth-Century America* (Baltimore: The Johns Hopkins University Press).

D. C. Radley ; S. N. Finkelstein and R. S. Stafford (2006) "Off-label prescribing among office-based physicians," *Archives of Internal Medicine* 166, 1021–26.

N. Rasmussen (2006) "Making the first anti-depressant: Amphetamine in American medicine, 1929–1950," *Journal of the History of Medicine and Allied Science* 61, 288–323.

N. Rasmussen (2005) "The drug industry and clinical research in interwar America: Three types of physician collaborator," *Bulletin of the History of Medicine* 79: 50–80.

D. E. Reid (1972) "A controversy in fetal ecology," *American Journal of Obstetrics and Gynecology* 114: 419–21.

A. C. Rossi, D. E. Knapp, C. Anello, Robert T. O'Neill, Cheryl F. Graham, Peter S. Mendelis, and George R. Stanley (1983) "Discovery of adverse drug reactions: A comparison of selected Phase IV studies with spontaneous reporting methods," *JAMA* 249: 2226–28.

A. Ruskin and C. Anello (1980) "The United States" in W. H. W. Inman (ed.) *Monitoring for Drug Safety* (Boston: MTP Press Ltd), 115–28.

T. Schlich and U. Tröhler (2006) *The Risks of Medical Innovation: Risk Perception and Assessment in Historical Context* (London: Routledge).

P. J. Seligman (2006) "Effects on medical practice of regulatory actions." http://www.fda.gov/cder/Offices/OPaSS/WHORegDec1. accessed September 10, 2006.

S. R. Shulman (1989) "The broader message of accutane," *American Journal of Public Health* 79, 1565–68.

R. J. Smith (1979) "Federal government faces painful decision on darvon," *Science* 203, 857–58.

S. B. Soumerai, J. Avorn, S. Gortmaker, S. Hawley (1987) "Effect of government and commercial warnings on reducing prescription misuse: The case of propoxyphene," *American Journal of Public Health* 77, 1518–23.

R. S. Stern and M. Bigby (1984) "An expanded profile of cutaneous reactions to nonsteroidal anti-inflammatory drugs," *JAMA* 252, 1433–37.

C. J. Stetler (1972) "Relative efficacy: When the government decides what drug should be prescribed, is the patient better served?" *Journal of the Mississippi State Medical Association* 13, 427–32.

R. B. Stewart, L. E. Cluff and J. R. Philp (1977) *Drug Monitoring: a Requirement for Responsible Drug Use* (Baltimore: Williams and Wilkins).

B. Strom ed. (1989) *Pharmacoepidemiology* (New York: John Wiley & Sons).

B. Strom ed. (1994) *Pharmacoepidemiology*. 2nd edn (New York: John Wiley & Sons).

B. Strom ed. (2003) *Pharmacoepidemiology*. 3rd edn [electronic book] (New York: John Wiley & Sons).

E. M. Tansey and L. A. Reynolds (1997) "The Committee on Safety of Drugs," *Wellcome Witnesses to Twentieth Century Medicine* 1, 103–32.

P. Temin (1980) *Taking Your Medicine: Drug Regulation in the United States* (Cambridge: Harvard University Press).

S. Timmermans and V. Leiter (2000) "The redemption of thalidomide: Standardizing the risk of birth defects," *Social Studies Science* 30, 41–71.

U.S. Government Accounting Office (1990), "FDA drug review. Postapproval risks 1976–1985." GAP/PEMD-90–15.

U.S. Government Accounting Office (2006), "Drug safety. Improvement needed in FDA's postmarket decision-making and oversight process," GAO-06–402.

G. R. Venning (1983) "Identification of adverse reactions to new drugs," *British Medical Journal* 286, 199–202, 289–92, 458–60, 365–68.

C. F. von Pirquet and B. Schick [1905] *Serum Sickness* (Baltimore: Williams and Wilkins) [1951 translation of Die Serumkrankheit].

A. K. Wagner *et al.* (2006) *Pharmacoepidemiology and Drug Safety* 15, 369–86.

E. Watkins (2002) "'Doctor, are you trying to kill me?' Ambivalence about the patient package insert for estrogen," *Bulletin of the History of Medicine* 76, 84–104.

J. K. Weston (1968) "The present status of adverse drug reaction reporting," *JAMA* 203, 89–91.

J. Whorton (1980) "'Antibiotic abandon': The resurgence of therapeutic rationalism," in John Parascandola (ed.) *The History of Antibiotics: a Symposium* (Madison: American Institute for the History of Pharmacy), 125–36.

G. N. Wilson (1967) *The Hazards of Immunization* (London: Athlone Press).

D. K. Wysowski and L. Swartz (2005) "Adverse drug event surveillance and drug withdrawals in the United States, 1969–2002," *Archives of Internal Medicine* 165, 1363–69.

5
Regulating Drugs, Regulating Diseases: Consumerism and the US Tolbutamide Controversy

Jeremy A. Greene

Most existing scholarship on pharmaceutical regulation has focused on legal, political, economic, and organizational dimensions of the regulatory process.* Relatively little attention has been paid to the relationship between the regulation of drug products and the epidemiology and definition of the disease categories to which they are necessarily linked. As drugs have increasingly come to define diseases, and diseases to define drugs, the regulatory nexus connecting the two has served to make the informational economy of pharmaceutical risks and benefits increasingly crucial to the decisions and health practices that doctors and patients make on a daily basis. Over the late twentieth century, pharmaceutical policy because a key contested terrain in the regulation of medical practice and medical markets. In this chapter I will present a case study that succinctly illustrates some of the successes, failures, and lingering tensions surrounding these efforts.

Early in the afternoon of May 20, 1970, a report was leaked over the Dow Jones newswires that Orinase (tolbutamide), "a drug used to lower blood sugar in diabetic patients," might be harmful (Knowles, 1977: 152). Orinase had been Upjohn's showcase success and sales leader for nearly a decade, and news of its possible toxicity spelled disastrous things for the company and its investors. Before the New York Stock Exchange closed that afternoon, Upjohn's stock had fallen in heavy trading. The next morning, the *Washington Post* reported the preliminary findings from the federally funded University Group Diabetes Program (UGDP) – the largest, longest, and most definitive study of diabetes therapy yet performed – with the implication that at least 8,000 patients a year may have already died as a result of Orinase consumption. As the story was picked up by the Associated Press, the Food and Drug Administration (FDA) hastily issued a press release that provided only the briefest abstract of the study and pronounced that the agency intended to revise the labeling of tolbutamide and other oral hypoglycemic drugs (Food and Drug Association, 1970). In the meantime, all hell broke loose.

At that time, hundreds of thousands of Americans were taking Orinase every day for mild (asymptomatic) diabetes, largely on the premise that the pill reduced their long-term risk for diabetic complications and heart disease. Over the next few days, FDA commissioner Charles Edwards would receive hundreds of phone calls and letters from patients concerned to find that the drugs they were taking to reduce their health risks might actually be increasing them. The question was particularly troubling for the asymptomatic diabetics taking Orinase, like "C.P.," a Virginian man who wrote:

> I have a mild diabetes – the kind that shows up in blood tests only it does not show in normal urine tests. For the past two years have been taking 2 Pills daily Upjohn *Orinase* and using saccharin ... Last week both were pointed out as dangerous to use as reported in the Wash Post. My Doctor thinks he should have more authoritative information before advising discontinuing or curtailment of these items. Could your office please advise on continued use of these items in view of frightening reports of Wash Post newspaper. I am 65 years old.[1]

C.P. and the rest of the *Washington Post's* readership were among the many Americans who learned of Orinase's putative toxicity before their physicians did (Mintz, 1970). This "premature announcement" of Orinase's toxicity, before the FDA had issued any warning to physicians and before the UGDP study was published in the clinical literature, would unleash a public debate over risk and asymptomatic disease that would last over a decade, create rancorous divides between advocates, researchers, and regulatory agencies, and leave hundreds of thousands of diabetics, their families, and their physicians in a muddle of uncertain practice, contested information, and strained trust (Kolata, 1979).

If patients like C.P. were disturbed by the news of Orinase's toxicity, the news hit their physicians twice as hard. Those who learned of the controversy through the newspaper were relatively lucky compared to the thousands of physicians who first learned of the debacle through their agitated patients. Commissioner Edwards subsequently received the following letter from a "Poor Practitioner," who complained hotly of the difficulty he was thrown into due to the study's untimely publicity and the regulatory and epistemological uncertainty that followed:

> Dear Dr. Edwards:
>
> Now that my nurse, receptionist, and bookkeeper are no longer tying up the three telephone lines to discuss with patients who are extremely worried and apprehensive about the Orinase situation, I am able to obtain a free line to dictate this letter to your attention.

I sincerely believe that the "public leak" by the FDA to the newspapers, and Walter Cronkite in particular, is not only a very stupid and indiscreet action on the part of your agency, but I firmly also believe this transgresses any and all medical ethic. This is a scare tactic to the general public who are using an ethical and adequate drug program and in so doing this, you are disrupting control balance of relationship of physician to patient and all other such relationships. You are further dictating by fiat medical practice and to make matters worse, you are using an unpublished study which has no logic, inadequate statistics and improper evaluation.

I deplore such action. I trust that it will not occur in the future over this or any drug. In the past, your direction has been to have the drug company release a news letter to physicians regarding dangerous or untoward side effects of drugs when they have been proven. In this case, you have done neither. I would hope very much that the FDA would retract publicly its stand and correct this situation and future ones as they may occur.

Sincerely yours,
DAVID L. ROBERTS, M.D.
Poor Practitioner [2]

The poor doctor Roberts and the hapless patient C.P. are but two of the thousands of minor actors in the public drama that would become popularly known as the Tolbutamide Controversy. As it unfolded over the full course of the 1970s, this fight over Orinase and the UGDP trial would prove to be one of the ugliest conflicts in the history of therapeutic investigation, and would come to involve a set of Congressional hearings, an FBI investigation, and a court ruling challenged all the way up to the US Supreme Court (Kolata, 1979). Eminent clinicians, typically reserved in public comments, took to calling each other "snake-oil salesmen," "unbridled sensationalists," and "drug-house whores" (Pressley, 1991: 447). Although the court proceedings had ended by 1984 and the dispute gradually disappeared from the pages of medical journals and popular newspapers, the debate never would reach a point of resolution.

The focus of this chapter is not to resolve whether Orinase reduced or increased the cardiovascular mortality of its consumers – now that Orinase has been replaced by newer generations of antidiabetic agents – this question has become largely irrelevant. Nor am I interested in tracing the terms of the debate as a conflict over clinical trials knowledge, as others have already done (Finkelstein *et al.*, 1981; Meinert and Tonascia, 1986; Pressley, 1991; Marks, 1997). Rather, I would like to suggest that the trials of Orinase in the 1970s offers an important perspective in the history of drug regulation that is frequently overlooked: the contested postmarket regulation of a drug in practice. Read alongside medical, industry, and popular literatures of the time, the hundreds of Orinase-related letters collected in the FDA archives during the controversy permit a unique case study in how the regulation of therapeutic indications for pharmaceutical products, and the changing definition of patients as consumers, became entwined in a crisis over medical authority in late twentieth-century America.

The relationship between the regulation of pharmaceuticals and the regulation of medical practice is central to this narrative. A drug is not regulated merely as a good in its own right, but always as a commodity tied to a set of knowledge regarding its production and its consumption. As we shall see, these forms of knowledge are far easier to regulate before a drug hits the market. Once a drug has developed broad use in practice, the multiplicity of economic, social, and cultural forms called into being by the daily practice of pharmaceutical consumption form an unruly terrain that is much more difficult for any single regulating agency – whether state-based, profession-based, or industry-based – to control. Granted, this case is narrated in a late twentieth-century American context, which implies a strong state-based regulatory agency coupled with a rhetoric of individual liberty and a politics of consumer advocacy that is not necessarily applicable in other national contexts (Daemmrich, 2004; Carpenter, 2010). Nonetheless, the contested status of postmarket regulation is a phenomenon much broader than the postwar American context, and my hope is that several themes can be generalized from this study.

Four important themes run through this narrative. First, that postmarket pharmaceutical regulation is a far more contested – and less formally governed – terrain than premarket regulation. Second, that unlike premarket regulatory debates, which tend to focus on the qualities of the drug itself as concrete and the patient populations as hypothetical, the theoretical purity of postmarket debates is always muddled by the existence of actual consumers who have been taking the medications at hand, and who now have incorporated the pharmaceutical consumption into their identities as patients. Compared with premarketing regulatory decisions, postmarketing regulatory decisions are more concerned with effectiveness than efficacy, and they are more concerned with the external validity of trial results rather than the internal validity of trial design. Third, that at stake in these debates are not only the availability or withholding of a drug product, but also of the information and the avenues to knowledge surrounding a drug product. Finally, and perhaps most evidently, the narrative of diabetes and Orinase in the 1960s and 1970s highlights the role of mutually defining relationships between drugs and diseases in the multiple registers of therapeutic regulation.

Origins of the Tolbutamide Controversy

The difficulty of regulating drugs for new diseases – or for diseases with changing boundaries – is central to this story. In the decade before the

UGDP trial, the emergence of Orinase and other oral hypoglycemic drugs had helped catalyze the transformation of diabetes from a symptom-bound disease into numerical diagnosis treated on a preventive basis. For the most part, as late as 1950, it was difficult to be diagnosed with diabetes without exhibiting symptoms (polydipsia, polyuria, autophagia) or signs (glycosuria) of the disease. As I have written elsewhere, by the late 1960s the numerical diagnosis of "mild" or asymptomatic diabetes – detected on the basis of blood sugar levels alone – had become common in practice due in part to the influence of oral hypoglycemics (Greene, 2007).

Not all diabetologists supported this new system of asymptomatic treatment, however: prominent skeptics portrayed the treatment of laboratory values in the absence of symptoms as an unhelpful and potentially unhealthy by-product of precision measurement that had nothing to do with good clinical practice (Tolstoi, 1953). Criticisms of the use of oral antidiabetic agents came from a surprising set of positions within the medical profession. On the one hand, many supporters of strict blood sugar control, including Boston's Eliot Joslin, were concerned that the ease of use of oral medications would lead diabetics to abandon the temperate discipline of right living that had long been the essence of good diabetic care (Joslin, 1959; Feudtner, 1995). They were joined, from the other end of the spectrum, by a "new school" of therapeutic moderates who saw rigid control of blood sugars in the absence of symptoms as a sort of physiological Puritanism, an unhelpful and potentially unhealthy by-product of precision measurement that had nothing to do with good clinical practice (Tolstoi, 1950). The two camps could not have been more different in their approach to diabetes control, but both agreed that the widespread use of tolbutamide was questionable. Even as late as the 1960s, the unsettled issue of the treatment of diabetes on the basis of blood sugar level was considered one of the great controversies in internal medicine, though all arguments had taken place largely within the polite context of the clinical literature (Bondy, 1966).

In this context, in the winter of 1958–59, the National Institute of Arthritis and Metabolic Diseases (NIAMD) designed a large-scale clinical trial to test the benefit of treating asymptomatic diabetes and to assess the long-term benefit of the recently released Orinase. The study would address three-layered questions: (1) did tolbutamide have a favorable impact on vascular disease? (2) did lowering blood sugar levels help decrease the risks of vascular disease? (3) what methods were useful in clinical trials for diabetes? By 1961, the study protocol was approved with seven different research sites; subjects with diabetes newly diagnosed by screening laboratory tests were recruited and randomly assorted into four treatment arms.

A control arm (PLAC) would receive diet therapy and a placebo pill; this would be compared against a tolbutamide arm (TOLB) that would receive diet plus 1,500 mg of Orinase daily; a "standard insulin" arm (ISTD) that would receive diet plus sufficient insulin for symptomatic control of blood sugar, and a "variable insulin" arm (IVAR) that would receive diet plus a more vigilant insulin regimen to insure strict glycemic control. A fifth arm (PHEN) was added in 1962 to evaluate phenformin (trade name *DBI*), a newer type of oral diabetes agent. By 1963, five more sites had been added, and by 1965 the full patient complement of 1,027 patients – roughly 200 to each arm – had been enrolled (UGDP, 1970). In essence, the four active arms of the trial compared two therapeutic strategies: a set of interventions producing strict glycemic control regardless of symptomatology (TOLB, IVAR, and PHEN) versus more *symptomatic management* (ISTD), both compared against placebo (PLAC).

The study was expressly *not* designed to assess tolbutamide for toxicity or added mortality. Six years into the study, coordinators noticed that more deaths were appearing in the tolbutamide group than in any other group, including placebo, chiefly from cardiovascular causes. Nothing in the previous literature had prepared the UGDP researchers for such a result, and the study coordinator was at first unsure how to proceed. His first response was to search for baseline differences that might explain the difference in mortality. As the trend continued to widen, he became increasingly anxious as to the ethics of continuing the tolbutamide arm, and alerted the other UGDP investigators of the early results in 1967. By that point, the study coordinator was convinced that even if tolbutamide was *not* harmful, there was no longer any possibility of demonstrating benefit; therefore the study arm could not be continued in an ethical fashion. Other investigators were not convinced but by June 1969 the majority of investigators voted 21 to 5 to stop the tolbutamide arm, and to immediately notify the FDA and Upjohn of the findings.

As evidenced by the vote, a minority of UGDP investigators continued to question the strength of this evidence. Differences in medical management among the study's sites, some thought, might account for the differences in mortality. This argument was compounded by the observation that the excess mortality in the tolbutamide arm was almost exclusively concentrated in 4 of 12 clinics, suggesting that the finding might be due to confounding factors in the study populations or variations in clinical practice or protocol implementation. Nonetheless, the majority of investigators were convinced of the strength of the findings, and the tolbutamide arm of the study was formally discontinued on October 7, 1969.

Much of the contemporary and subsequent discussions of the UGDP study have focused on the contested internal validity of the study design (i.e., whether the conclusions the investigators drew were valid), the contested external validity of the trial design (i.e., whether the conclusions of the UGDP study, valid or not, had any relevance beyond the universe of the study itself), and whether the decision to discontinue the tolbutamide arm was hasty or necessary at all. A great deal has already been written about these debates, and it is not my intent to either "get to the bottom of it" or provide an epistemological account to explain the ruptures. But I would suggest that the more profound historical lessons to draw from the UGDP lie not in the internal handling of trial evidence, but in the external, or public, management of disputed health information.

After the tolbutamide arm stopped in June 1969, closed-door sessions occurred between UGDP, FDA, and Upjohn to determine how to best publicize the study results. It was agreed that the next annual meeting of the American Diabetes Association (ADA), scheduled for the following June 1970, would be the best possible moment for public statements; allowing one year to fully evaluate the study data, plans were made to reconvene in Bethesda on May 21, 1970, to finalize professional and public communications and label revisions. Unfortunately for those concerned, the press leaked news of the results one day before their scheduled meeting.

Crisis, Publicity, and the Regulation of Drug Information

The Tolbutamide Controversy achieved publicity at a pivotal moment in the relationship between therapeutic research and its multiple publics. The years through which this controversy smoldered were a period of crisis for the paternalistic model that had characterized relations between the American medical profession and its patient public for at least a century (Rothman, 1991; Mintzes and Hodges, 1996). For advocates of greater transparency, the *Washington Post*'s early release of trial data was no "leak" but rather was an important step toward openness in communication at a time when increasing amounts of media coverage were being devoted to the topic of consumer health (Katz, 1984). For many physicians, discomfort with the media's handling of the University Diabetes Group Project was directly linked to fears that an era of relative professional autonomy and uncontested authority had ended. A central subtext of the Tolbutamide Controversy regarded the appropriate level of circulation of pharmaceutical-related information, defined in quite different terms by industrial, professional, and consumer stakeholders, all of whom portrayed a different vision of the proper role of the state in restricting or expanding access to drug-related knowledge. The proper role of "publicity" was here tightly bound to the precise definition of which "public" was understood to be the rightful audience of UGDP-related information.

Industrial regulation of information:
The financial/pharmaceutical public

As we now know, on the afternoon of May 20, as the participants in the FDA Expert Advisory Committee were readying their presentations and making their way to the Washington area, news of the UGDP results were broken over the Dow Jones ticker. That the first public news of tolbutamide's risks occurred not as a general press release or article in the science section of the newspaper, but as a report over the financial newswire, is highly significant. The first public for the Tolbutamide Controversy, then, was the broader financial community surrounding the pharmaceutical industry. Needless to say, this particular sort of publicity emphasized a concern for the welfare of the drug itself – as a product – that could be easily differentiated from a broader concern for the welfare of the diabetic patient. The Upjohn Company, a particularly interested public of the study, was able to insert itself swiftly into a counter narrative, mobilizing critiques from noted statisticians Alvan Feinstein and Stanley Schor that disputed the study's results.

In the early twenty-first century it has become common practice to see clinical trials results receive their first publicity in the business sections of newspapers, but in 1970 this was not yet a common occurrence. Pharmaceutical companies of the mid twentieth century had tended to produce a wide range of therapeutic agents that overlapped with other companies' offerings; the failure of any one product would not necessarily disturb the financial well-being of a company. But as a smaller number of exclusive, branded, multimillion dollar drugs came to dominate the interests of the industry, and as the industry grew to comprise a larger portion of the national economy, pharmaceutical news became big business news, and the impact of one clinical trial could affect the portfolios of thousands of investors. As a direct consequence of this transformation, by 1970 detailed information on ongoing clinical trials was eagerly sought by financial analysts, traders, and individual investors. Knowledge of the progress of a clinical trial itself became a valuable currency that circulated through a private-sector information economy.

As significant as the Dow Jones ticker was for the financial community, however, it was the next morning's *Washington Post* article – which representatives from the UGDP, FDA, and Upjohn read with their breakfasts before they made their way to their now pre-empted strategy meeting – that was instrumental in breaking the story to the broader clinician and consumer publics. If the excess mortality in the tolbutamide arm of the UGDP study had alarmed investigators regarding the fate of the 200 patients they had placed in tolbutamide treatment, this paled in comparison to the dilemma now facing America's physicians and the 800,000 patients estimated to be taking the drug (Mintz, 1970).

Professional regulation of information: The physician public

In addition to their bewilderment at learning that a pharmaceutical agent and therapeutic rationale that they had been recommending with confidence for over a decade were deemed worthless and potentially injurious on the basis of a single study, physicians were equally disturbed by the publicizing of the study through a media that sidestepped the traditional role of the physician as broker of health information. Physicians were accustomed to receiving their news from the FDA in a more direct manner – letters and bulletins – and understood themselves to be the vital link between the FDA and the consumer; a "first public" that would receive product warnings and transmit them to the ultimate consumers (Dowling, 1956). The immediate contestation of the study's results by Upjohn-associated statisticians, coupled with the fact that the study itself had not yet been published and therefore could not be evaluated by practicing physicians, further compromised the position of physician as knowledge-broker.

Nonetheless, as Dr. Roberts' letter suggested at the beginning of this chapter, physicians needed answers immediately, and some wrote letters in the earnest hope that the FDA might supply guidance in time of crisis. Many, like Philadelphia physician Norman Knee, politely appealed to the FDA for more information and offered their clinical reasoning for the continued usage of tolbutamide in asymptomatic diabetics based on the primacy of clinical experience.[3] The tone of many letters from physicians to the FDA at this time was still cordial, information-seeking, suggesting that in spite of the "unfortunate publicity" of the study, the FDA and the frontline physician might patch up this singular breach and resume their typical relationship in distributing health information.

M. J. Ryan, the director of legislative services for the FDA, quickly replied to these physicians in a letter that lamented the leak of the UGDP summary over the Dow Jones News Service as an unfortunate "case in which the financial community and newspapers got reports of this medical research before physicians, who might find themselves beset by troubled patients." After the *Washington Post* article broke the story to the nation, the FDA had felt obliged to issue a press release as soon as possible, and had not had the time to notify physicians first. Ryan apologized that the FDA had not developed a practical way of informing physicians in such instances before the information reaches the patient via the lay press, and assured physicians that the FDA would develop a better system in the future.[4] In a separate letter to physicians, Surgeon General Jesse Steinfeld apologized for the breach and reiterated that the government's hand had been forced by the early publicity.[5] By June 8 the FDA was taking steps to restore the primacy of FDA–physician relations, and had prepared the first in a series of letters to the nation's doctors, suggesting that although the individual physician must ultimately decide the utility of oral hypoglycemics, "they could no longer be given simply on the ground that they might help and could do no harm" (Schmeck, 1970a: 157). This basic logic, in which the warning was tied tightly to the product and still allowed the physician freedom to evaluate its relative merit, seemed a step in repairing the breach between physicians and the FDA by reinstating the physician as the central mediator of individual health information.

Consumer demand for information: The diabetic public

If physicians were put in a practical quandary as a result of the study's publicity, consumers of Orinase felt even more constrained, as they were trapped in a situation of limited information and vital consequences. Patients and family members inundated their physicians' offices with calls and filled the mailboxes of their senators, the Department of Housing, Economics, and Welfare (HEW), the FDA, and even President Richard M. Nixon. In the first few weeks of the affair, most letters retained a tone that was hopeful, if somewhat frayed with desperation, appealing to the FDA as a trusted authority that might help resolve a pressing and confusing issue. Many letter-writers stated that their own consumption of Orinase, like that of the bulk of the UGDP study subjects, took place in the treatment of mild, asymptomatic diabetes. As C.P. – the gentleman whose letter began this chapter – noted, they tended to have "a mild diabetes, the kind that shows up in blood tests only it does not show in normal urine tests."[6] A context of perceived urgency is manifest in the physical appearance of many of these letters, handwritten on torn sheets of papers or hastily typed with numerous errors. One Brooklyn patient "using *Oraniss* Tablets as prescribe by my Doctor" asked for the FDA to "let me know what are the Harmful effects, and what are the available Treatment for these patients who are considered (slight) Diabetic. I understand the death rate among these patient are very High."[7]

This pressing need for information was coupled with a sense of unease regarding the possibility of critiquing their own physicians' diagnostic and therapeutic decision making. Patients felt particularly uncomfortable about discussing the propriety of the prescription regimen with their physicians, and appealed to the FDA as a defender of the public interest. Even when they did speak to their physicians, patients' references to newspaper and radio programs were often dismissed offhand. "When we questioned our doctor about this news report," one patient complained, "he (1) knew nothing about it and (2) regarded our question as a personal offense. At best, he says he'll wait to see what the Food & Drug Administration says. Meanwhile, we don't know whether to allow continued use of the drug."[8] Another patient, whose son had been on Orinase for a year, was told by her doctor that the drug was perfectly safe to use since the Joslin Clinic continued to recommend it: "Will you kindly let me know whether the drug orinase is still safe to take," she asked the FDA "now this article has come out according to which there is proof against it ... why is it then still on the market? And why have not the doctors been advised not to prescribe it?"[9] Faced with a visible rift between public knowledge and professional practice, patients and their families appealed to the federal government to explain and mend this breach.

As much as these early letters suggest that patients felt they had recourse to federal protection of their consumer rights, their appeals to the FDA were frequently tinged with institutional mistrust. "J. F." wrote to the FDA in late May of 1970, "Could you please get me some more down to earth *100% correct* info on this problem. Unfortunately I am one of those people with a Diabetic Problem, and I have been taking *Orinase* for 3½ years. After reading the enclosed article I am very much upset and afraid."[10] As he appealed to his senator to make his case to the proper parties, including the Food and Drug Administration, however, J. F. warned of a more general unraveling of medical authority that included the FDA in its critique:

> If this article is true I would say that this drug killed more people than the Viet Nam War. I hate to be a conclusion Jumper But – It appears that the FDA, Upjohn Co., the Amer. Med. Assoc., and The Physicians Who Prescribed This Drug are to Blame...The American People deserve a better system of Protection From Harmful drugs than is at the Present Time in operation.[11]

Although J.F. included the FDA in the parties he blamed for lack of oversight, he, nonetheless, appealed to it as an entity capable of acting in the public interest, even if it failed to perform to the best of its duties in its past handling of tolbutamide. For other letter-writers, a taint of scandal indicted the FDA as possibly duplicitous in its responsibilities toward the welfare of the American people. Comparisons between a war on disease and the war in Vietnam analogized a distrust between citizens and the state in both cases. As another diabetic patient wrote a few days later,

> The FDA has now recommended Orinase be given only to a select group in case it is harmful, well, if it is, the harm is already done and they are 14 years to [sic] late. It is unbelievable a drug can be used that amount of years before possibly being declared unsafe, doesn't the Govt. care at all?...It does shake your faith in a country when you are told that such a thing could happen, it is not only a very expensive drug but could be like slow poisen [sic] to the Diabetic if not of any help...We can get to the moon and constantly be at war but our medical standards still rank lower than many other countries. Makes you wonder, doesn't it?[12]

It fell to Marvin Seife, then Director of the FDA's Office of Marketed Drugs, to respond to consumer queries. In a standardized letter, he replied to patients that the FDA had convened an expert advisory committee to review the results of the UGDP study, and that "despite a number of limitations in this study," both the FDA and the advisory committee agreed with the UGDP conclusions. "In the near future," Seife concluded "we will inform your doctor, along with all other practicing physicians, of the findings and medical implications of this study. In the meantime, we suggest that diabetic patients now taking tolbutamide or chemically similar agents continue on their current regimen until advised otherwise by their physician."[13] Seife self-consciously hedged his direct communications with consumers by referring them to their own physicians for the individually tailored health information they often demanded. While his agency was responsible for responding to consumer safety concerns, he was required to tread carefully lest he be accused of arrogating the role of the private physician in determining a course of therapy for an individual patient. All parties, it seems, were constrained by the cloud of uncertainty surrounding the relation of the UGDP trial data to public knowledge and clinical practice.

Publicity and the maintenance of controversy

Once the study became public, however, it was no longer an entity to be contained and managed by the FDA alone. Instead, after evidence of harm had been publicly presented, the defenders of oral antidiabetic therapy recognized that the production and maintenance of controversy through news media was perhaps the only strategy that could help to prevent any stable consensus from forming around the study results.[14] Physician and pharmaceutical defenders of tolbutamide use recognized that the production and maintenance of controversy through news media was a valid strategy to prevent consensus being formed around the study results, and that the continued publicization of controversy helped to maintain a sustainable space in which widespread usage of the oral antidiabetics could continue to be regarded as a legitimate therapeutics. This process was initiated by Upjohn public relations personnel and several leading diabetologists within hours of the Dow Jones report, and quickly found its way into public reporting of the UGDP results (Schmeck, 1970c). Upjohn mobilized critiques from noted statisticians Alvan Feinstein and Stanley Schor that disputed the study's results; in mid-November, a group of diabetologists wrote a letter to the *New York Times* describing the study as "worthless," and the Upjohn Company found multiple venues – in newspaper articles and journal advertisements – to make its case that the UGDP Study simply flew in the face of all other trials and all clinical experience.

In the fall of 1970, the Joslin Clinic – which would eventually become the nerve center for organized resistance against the extension of the UGDP Study results into clinical practice – held its first major publicity stunt. The annual meetings for the American Medical Association (AMA) were to be held in Boston at the end of the month, providing an ideal time to capitalize on the discomfort many physicians felt toward the AMA and FDA surrounding this issue. Physicians of the Joslin Diabetes Clinic, most notably Robert F. Bradley, Holbrooke Seltzer, and Peter Forsham, recruited 34 leading diabetologists from around the country to sign a statement dissenting from the AMA–FDA decision. Along with Cornell's Henry Dolger, the three leaders then held a press conference to publicize their discontent with federal interference into diabetes clinical practice that was dubbed its "Boston Tea Party" (Medical World News, 1970). The group named itself the Committee for the Care of the Diabetic (CCD), seizing rhetorical high ground by characterizing its adversaries as a set of distant technocrats who meddled in clinical realms they did not properly understand.

Publicity was a weapon vital to the armamentarium of all stakeholders in this controversy. The "Tea Party" paid off for its ringleaders: shortly after its initial press conference, the CCD was invited into negotiations with the FDA and UGDP investigators over labeling, a move that would ultimately delay the FDA for five years in making any headway on implementing its proposed changes and result in Senate hearings in the fall of 1974 and spring of 1975, and generating further rounds of public argument surrounding the FDA's action or inaction regarding the postmarket regulation of the drug. The Tolbutamide Controversy was not merely a controversy over diabetes care: had it been limited to therapeutic decision making, it would never have reached such proportions. Essential to understanding the publicity of the affair is understanding the public nature of the therapeutic agent at the center of it. A number of factors conspire to make a pharmaceutical like Orinase inherently more public than other therapeutics such as surgery, diet, or wound care. First, the pharmaceutical is a product, and therefore belongs automatically to the public world of goods, services, and trade. It is no accident that the first news of this study came over the Dow Jones newswire. Second, because of its identity as a commodity, the late twentieth-century pharmaceutical encodes within it a consumer-oriented approach to medicine. The pharmaceutical tablet is, in many ways, the perfect image of health care as commodity: a compact unit of therapy, portable, exchangeable across borders, universalized in shape to a nearly spherical form, its therapeutic value has nothing to do with the local circumstances of administration (unlike a surgical procedure or injection) and having everything to do with a highly abstract network of data, research, and therapeutic information of which it is both product and emissary. Finally, as a highly regulated consumer product, the pharmaceutical represented an early site for federal intervention in medical practice, and actions of the FDA have tended to define patients according to a consumer model. In extending the public duty of consumer protection into the private realm of diagnosis and treatment, however, the FDA would face a difficult new task.

Regulating disease

The Tolbutamide Controversy neatly illustrates the gap between premarket and postmarket therapeutic regulation of pharmaceutical products. In the wake of the 1962 Kefauver–Harris Act the FDA's role as gatekeeper in the approval of new agents had been strengthened with a formalized sequence of phased clinical trials and a broad set of regulatory powers over applicant companies. In contrast, once a drug is already "through the gate," the formal powers the state can bring to bear is markedly limited, and post-market regulation is characterized by a more complex matrix of informal regulation involving the medical profession, the pharmaceutical industry, and ultimately the pharmaceutical consumers themselves. Postmarket, the FDA's official actions over pharmaceutical manufacturers are largely limited to outright ban – rarely employed – or pressuring individual companies for voluntary withdrawals and/or labeling changes or promotional infractions.

In spite of patient and physician concerns to the contrary, the FDA never seriously considered a categorical ban on Orinase. The FDA was particularly constrained in its ability to regulate Orinase, which was one of a special subset of drugs that, by 1970, had been doubly approved. The first new drug application (NDA) for Orinase was approved in 1957, before the passage of the 1962 Kefauver–Harris bill that mandated proof of efficacy in addition to the proof of safety previously required for FDA approval. Along with 3,000 other drugs, Orinase was subjected to a retrospective efficacy review (termed DESI) by the National Academy of Sciences and the National Research Council between 1966 and 1969.[15] Upjohn's combination antibiotic Panalba had fared less well in the evaluation, but Orinase passed through the review process relatively easily.[16]

In addition to mandating efficacy, the Kefauver–Harris bill expanded the FDA's authority from the safety of the drug per se to a more formal evaluation of the appropriateness of the drug for a particular therapeutic usage, termed a "therapeutic indication." In a move with political signifi-cance for the later Tolbutamide Controversy, the FDA's attempts to set out a formal policy regarding clinical trials, therapeutic indications, and labeling change would be stalled for eight years largely by a series of lawsuits from the Upjohn Company regarding Panalba. The new process would only be formalized in May 1970, immediately before the UGDP controversy erupted. In the case of Orinase, the DESI review affirmed that Orinase was "an oral anti-diabetes agent which effectively restores blood sugar to normal ranges in selected diabetes patients."[17] In a proximate, short-term sense, Orinase had been proven both safe and effective. However, Upjohn's promotional claims for the drug had by 1968 extended to a second, long-term indication on an asymptomatic basis, and it was the safety and efficacy of this asymp-tomatic indication that the UGDP results challenged.

Moreover, the FDA had never received a clear mandate on how to regulate physicians' use of drugs, and the scope of its authority over therapeutic indi-cations had been murky ground since the passage of the Kefauver–Harris amendments (Lasagna, 1970). Prior to 1962, the FDA had restricted most of its regulatory responsibilities to the makers of pharmaceutical products. As the more formal therapeutic indication gained relevance in clinical prac-tice, the FDA began to issue "Dear Doctor" letters giving warnings on new adverse effects that came to light after drugs were launched; nonetheless, the FDA's authority was still tightly limited to the product itself, particularly over product labeling. Where the agency had an obligation to inform the physicians and the general public regarding product claims, federal law gave the FDA no direct jurisdiction over the physician's actions as prescriber (Dowling, 1971).

In its initial response to the new data on tolbutamide, in order to avoid dictating medical practice, the FDA turned instead toward the regulation of its consumers, delineating which populations of patients could be considered to have treatable disease. A day after the story broke in the *Washington Post*, the FDA announced that the use of Orinase was "no more effective than diet alone, and as far as death from heart disease and related conditions is concerned, may be less effective than diet or diet and insulin." The FDA statement was careful to point out that this warning *only regarded mild, adult-onset diabetics*, and that the drugs *might still be found to be useful in diabetics with more symptomatic disease*. The FDA recommended that Orinase and all other sulfonylureas should be used "only in patients with *symptomatic* adult onset diabetes mellitus who cannot be adequately controlled by diet alone and who are not insulin dependent," and a letter to physicians was sent out to that effect in June 1970 (FDA, 1970).

In an attempt to avoid dictating medical practice, the FDA had turned instead toward the regulation of its consumers, delineating which populations of patients could be considered to have treatable disease. As the *New York Times* noted, the FDA's move to restrict Orinase consumption to the symptomatic indicated a "tightening up of the criteria for using the oral diabetes drugs that might limit them to a relatively small group of patients" (Schmeck, 1970a: 157). Toward this effort the FDA cautiously recruited allies across physician associations such as the ADA and the AMA. Whereas the ADA had noted in June that "the evidence presented does not appear to warrant abandoning the presently accepted methods of the treatment of diabetes," by late October, the FDA could announce that the AMA and the ADA agreed with its intention to add a package insert warning for all oral antidiabetic drugs (Schmeck, 1970b: 42; 1970d). On October 30, 1970, the FDA further clarified its position on the symptom in a widely circulated bulletin, specifying that "the oral hypoglycemic agents are not recommended in the treatment of chemical or latent diabetes, in suspected diabetes, or in pre-diabetes" (Edwards, 1970). A joint press release of the AMA and FDA endorsed the bulletin, and the AMA's Council on Drugs agreed that the only legitimate use for the drugs was in the "symptomatic, maturity-onset diabetic" who could not be managed with insulin.[18] It seemed, at first, as though the medical profession would adopt the symptom as a site of risk differentiation and consumer protection (Schmeck, 1970b, 1970d).

But these early alliances would soon fade. After the AMA "Boston Tea Party" an increasingly vocal segment of physicians charged that attempts by the federal government to dictate who was and who was not a valid patient constituted an unprecedented "compromise of the physician's freedom to prescribe."[19] The CCD argued that the FDA's jurisdiction was limited to the *product* and not the definition of disease. Talks between the FDA and the CCD dragged on for several years and then broke down entirely.[20] Further events intended to resolve the controversy instead provided more fuel for the fires. The 1974 report of the Biometric Society – a group of statisticians called in to arbitrate the affair – was once again "leaked" to the press prior to publication in the medical literature. Newspaper reports across the country quoted the claims of the National Institute of Health's (NIH's) Thomas Chalmers that the use of these drugs was responsible for 10,000 to 15,000 unnecessary deaths each year in the United States. Chalmers' statements, publicized before any practicing physicians had access to the Biometric Society report, managed to both reagitate consumer anxiety and reactivate the hostility of a great number of practicing physicians (Moss, 1975: 206; O'Sullivan and D'Agostino, 1975).

In response, the AMA now demanded that all package inserts bear statements disclaiming that they were merely limitations on the parameters for pharmaceutical advertising, and "should not be considered a legal controlling influence over drug use by a given physician in the management of an individual patient." [21] Although a few physicians did write letters in support of the UGDP results, the vast majority of physicians who wrote to the FDA also criticized the proposed labeling as an infringement of the physician's role. [22] Many simply denied the ability of the FDA's decision to influence their practice. As one internist noted in a letter to the agency, "Whether so labeled or not, many conscientious and careful physicians will continue to use these drugs in selected patients until such time as the issue is finally decided."[23] And use them they did. By 1975, it had become clear that – in spite of a slight dip in sales in the immediate aftermath of the UGDP – oral antidiabetic prescriptions had continued to climb at a remarkable rate (Chalmers, 1975). "You must realize," another internist wrote, explaining the limits of the FDA's impact on therapeutic practice, "that I will not really be influenced by any warning that you eventually put on a label."[24] Amidst this bravado and bluster however, not all physicians were as sanguine about their insulation from the FDA's influence. This was especially true among those who had experienced firsthand the rising prevalence of malpractice litigation.

In the decade of the 1970s, malpractice litigation had become a tangible reality to practicing physicians on a scale heretofore unknown, and the courtroom had emerged as a de facto domain of pharmaceutical regulation that allowed consumers to individually or en masse use the indications on a drug label to press for greater standardization of medical practice (Hogan, 2003). As Neil Chayet – the CCD's most prominent attorney – had commented in the *New England Journal of Medicine* three years prior to the release of UGDP results, the pharmaceutical package insert was increasingly employed as a standard of practice to which physicians might be held legally accountable (Chayet, 1967). In the late 1960s, the first malpractice conviction regarding a package insert was issued against a dentist for using an anesthetic in a dosage unapproved in its labeling, and the prospect of package insert malpractice suits remained an open issue in the rest of the nation. Although physicians were empowered to use therapeutics in any way they saw fit, off-label usage of drugs opened physicians up to possible liability for adverse outcomes.

The asymptomatic nature of mild diabetes made the liability regarding Orinase prescription more complex than for other agents. Since the prescription of Orinase was partially based on a logic of decreased cardiovascular risk, and since the WARNINGS section on the proposed package insert was based on a logic of increased cardiovascular risk, the physician who prescribed Orinase could become a particularly broad target for litigation. As one internist noted in his written submission to the 1975 hearing:

> If this regulation passes, I have no doubt that a spate of new malpractice suits will arise. Whenever a diabetic on oral drugs has a CVA [cerebrovascular accident, or stroke] or an MI [myocardial infarction, or heart attack], his attorney will claim that the condition would not have occurred, had the patient not taken the medication. I suspect that the majority of such cases will be won by the defendant, but the stress of the case and the court costs and defense would remain. In view of the malpractice situation today, I am opposed to giving plaintiff's lawyers an additional tool. [25]

Other physicians a pointed out that these regulations would place an additional regulatory burden of malpractice risk on physicians. "If I am to follow the rules implied in the proposed labeling I will be required to discontinue the use of oral hypoglycemic agents entirely," a South Carolina physician sardonically observed, "or face the probable legal implications in a wrongful death action brought on by some enlightened family ... During this day and time of *suing the doctor* I feel that your proposed labeling ... put[s] the conscientious family level physician in an untenable situation. A damned if you do or damned if you don't position."[26]

The FDA did receive several letters from malpractice lawyers seeking to make cases. The following letter from an Alabama attorney is typical:

> I represent the estate of a deceased who died after being administered the drug Orinase. I am interested in gathering all information possible, concerning the effects of this drug, both good and bad, on humans. I would appreciate you forwarding me any information that your department has concerning Orinase.[27]

Although it is unclear how many such cases came to court, the concern over malpractice had an evident material basis. The FDA's response to such requests was to claim that the package insert was the only information on Orinase that it was legally allowed to share with the public due to confidentiality agreements.[28] At the same time, the looseness of causal association between asymptomatic diabetes and cardiovascular mortality, combined with the looseness of causal association between Orinase and cardiovascular mortality, meant that any strong wording on a package could be perceived as a legal trap. As one practitioner replied, "Any patient who dies of cardiac disease while on these agents will leave his physician open to suit based upon this labeling. The lawyers could have a field day."[29]

To characterize the violent reaction with which physicians received the FDA recommendations solely as a concern over malpractice, however, is to miss the more significant gap perceived between the knowledge required to label a drug safe or unsafe and the knowledge required to judge a therapeutic practice as effective or not effective. Many physicians pointed out that the FDA recommendations themselves illustrated the regulators' ignorance of the logic of diabetes therapy in practice. Physicians particularly objected to the FDA's suggestion that Orinase should be thought of as a third-line agent, for consideration only after both diet and insulin had failed. As one physician pointed out, Orinase was only really useful as a first-line agent, because it was much more difficult to convince a patient with no symptoms that the needle-based practice of insulin therapy was worthwhile.[30] As one Miami physician wrote, "It is a common observation that diet alone is sooner or later unsuccessful in treating diabetic patients with mild maturity onset diabetes... [t]hey are more easily controlled with the oral drugs, often refusing to consider self-administration of insulin."[31] Some insisted that many patients achieved better control of blood glucose with Orinase than with insulin or diet. They argued that the FDA's evaluation of risk was singularly one-sided, focusing on sins of commission while neglecting the sins of omission if patients denied oral therapy went untreated:

> Have they projected the statistical impact of such proposed labeling on oral agents as such directed labeling on oral agents as such directions will apply to larger numbers of people who are denied the use of oral agents? I feel that in my own family practice there will be deaths *due* to your proposals."[32]

Claiming the consumer

Several historians and sociologists have described the scientific debate over the UGDP trial as a case study of incommensurable dispute, in which each participant claimed to represent objectivity while deriding its adversaries as deluded by ideologies of self-interest. Simultaneous and parallel to this scientific debate, however, lay an equally important pragmatic debate regarding the continued day-to-day use of Orinase in a time of fractured medical opinion. Like the scientific dispute, the clinical contestation of tolbutamide in the decade of the 1970s was also marked by accusations of interest. In the latter debate, however, each side claimed not to represent objectivity, but rather to represent the best interests of the patient as consumer. Many physicians articulated a common paternalistic logic by which the medical profession was responsible for the consumption habits of patients; this was countered by an emerging radical consumerist lobby that argued that the state, and not the physician, had the responsibility for protecting consumers. But not all consumers felt themselves represented by radical consumerism, and several patients and physicians explicitly invoked the newer language of medical egalitarianism to argue for or against the FDA's actions in sometimes surprising ways. In this tangle of imputed self-interest and contested authenticity, the role of the consumer in the regulation of drug risks would be refracted through multiple lenses of stakeholder interest.

Each side attempted to portray its opposition as deluded by self-interest. Perhaps the most visible accusations of interest were levied against the pharmaceutical industry. Proponents of the UGDP liked to explain the tenacity of their opposition in terms of the deep pockets of the Upjohn Company. Thaddeus Prout, a UGDP investigator, suggested that the UGDP was unfairly persecuted by skilled industry public relations specialists, complaining that "one of the things that this controversy has brought out in the last 5 years, I guess – it seems longer – is the incredible way in which a group of physicians teamed up with industry to attack the only scientific evidence there is on the use of these agents, at a time when we sorely need it."[33] Many clinicians, in turn, accused the UGDP investigators of operating within a self-congratulatory incentive structure of bureaucratic and academic promotion with little regard for "clinical reality." As one clinician complained:

> It is always interesting to read the glowing reports of headline physicians who are responsible in part for many such infringements on the working physician's relationship with his patient. They write and report from a biased Ivy-Tower situation – certainly not from the day to day relationships with the vast majority of people seen by the grass-roots physicians of this country![34]

Direct allegations of financial interest were also made against the UGDP, particularly after a former UGDP researcher suggested that another had received bribes from a rival pharmaceutical firm, USV, which sought to divert market share away from Orinase toward its own oral diabetes drug DBI (phenformin).[35]

Other critics noted that federal bureaucracies such as the NIH and the FDA became interested parties in any trial representing a significant expenditure of the taxpayer's money. "Why should we quote this study when other studies published before and since seem to indicate just the opposite?" one physician noted, adding, that "I agree that we spent a lot of money on the study, but this does not make it a good one."[36] Another practicing internist complained:

> It seems to me a classic case of conflict of interest in which the Federal Government is approving the study which the Federal Government has funded, in spite of the voluminous exterior criticism. This is an irresponsible act upon your part and can serve only to alienate the medical profession even further than some of your past half assed actions have already done.[37]

All parties in this debate attempted to point out the self-interest of their opponents in order to present themselves as disinterested servants of patient welfare.

In their attempts to position themselves as the rightful representatives of the patient-consumer, stakeholders found themselves muddled by the peculiar nature of pharmaceutical consumption. As a result of a long history of professional and governmental regulations, every prescription drug can be seen to have at least two consumers: the health provider who chooses which pharmaceutical to prescribe and the patient who chooses whether or not to ultimately buy and consume the drug. For the greater part of the twentieth century, physicians had understood their own position as "mediate consumers" of pharmaceuticals to be a sort of natural type, a rightful interposition between pharmaceutical manufacturer and patient that was rooted in the paternalistic ethos of medical practice. To many physicians, the Orinase labeling proposal was part of a much broader trend of consumerism that was believed to be inimical to the patient's best interests. How could a federal agency understand the complexity of an individual patient's situation well enough to take on the responsibility of mediate consumer? Furthermore, in appealing directly to the patient as consumer, the FDA was making an egregious trespass into an area well-defined by the Hippocratic code. As one practitioner asked,

> Why should the patient be forced to make these decisions rather than relying upon the best judgment and advice of his physician, when that same patient can go to grocery store – or even a gasoline station – and pick up over counter drugs which are far more hazardous to his health, such OTC drugs which the FDA does not seem able to control? Rather your agency seems bent on a course which implies that the medical community is less conscientious in its use of medications than is the non-medically educated to prescribe for themselves.[38]

In a particularly acerbic response, another internist objected that the FDA's attempts to appeal directly to the consumer through labeling and remove the physician from the role of mediate consumer were simply absurd:

> You have hit a new high in bureaucratic label-pollution! ... Good Heavens! I would hope that diabetic pills would be sought by diabetics after consulting their physician who would indeed do all the thinking as to what's best for his particular patient's needs and weight the dangers of these medications against their benefit for the patient. What would you gain by putting such labels since the decision is the physician's and not the patient's? ... The FDA needs a label "Warning: liable to waste tax dollars.[39]

Not all physicians, however, saw consumerism as a corrupting influence on medical practice. In addition to a more general critique of paternalism within doctor–patient relations, the early 1970s also witnessed the flourishing of a radical consumerism movement as an agent of progressive political change, particularly after Ralph Nader's *Unsafe at Any Speed* demonstrated that consumers could form a political base to effect change with broad public health implications (Nader, 1965). Sidney M. Wolfe, one of the many physicians inspired by the political possibilities of a "republic of consumers," joined Nader to form the Health Research Group at Public Citizen, and came to represent the heart of the radical consumer movement in health care. Wolfe viewed the CCD with conspicuous disdain, regarding the court-required delay of warning labels as a move that had caused patients to remain on "dangerous, ineffective, and expensive drugs" for five extra years: the moral equivalent of mass manslaughter. "During this interval of irresponsible delay by the FDA," Wolfe noted, "approximately 250 million dollars worth of these drugs have been consumed in this country alone, and according to experts, 20,000–30,000 unnecessary deaths due to these drugs have probably occurred."[40] Unlike the physicians of the CCD, who saw the FDA's efforts as excessively intrusive, radical consumer groups such as Public Citizen argued that the agency's backpedaling on the issue of asymptomatic treatment had already allowed too much. "The problem is you are granting the indication," Wolfe's colleague Anita Johnson objected during the August 20 hearing, "Here you are granting an asymptomatic indication. Then you are holding your breath a little bit after you grant it. Our position is that it should not be granted at all."[41]

Between October 20 and November 12, 1975, the FDA received more than 200 letters from concerned patients, many of whom agreed with Wolfe and Johnston that the FDA should severely restrict the use of these "toxic agents." Family members of patients who had died of heart disease while on oral hypoglycemics wrote in to show support for a ban, as did other patients who had sustained reversible side effects who offered their individual testimonials as public evidence in support of regulation.[42] At one extreme, a Virginia woman detailed the decline and death of her husband from a pancreatic cancer that she insisted could only be the result of eight years of oral antidiabetic therapy. "The enclosed autopsy report," she added, "speaks for itself."[43]

Though these individuals – and the thousands who financially and materially supported Public Citizen – felt their identity as consumers was well-represented by Wolfe's position, many other consumers disagreed and characterized Wolfe's protectionism as merely another form of paternalism that ultimately misrepresented the voice of the consumer. The majority of letters from consumers found in the hearing dockets appear to be concerned that the FDA might *overly restrict* consumer freedoms, not that the agency needed to increase its regulatory activities as Wolfe suggested. "A.M.," a Wisconsin man diagnosed with "chemical diabetes" and treated on an asymptomatic basis first with Orinase and then with DBI (phenformin), wrote to Senator Gaylord Nelson early on in the UGDP controversy insisting that the consumer deserved continued access to risk-reducing treatments:

> Since consulting Dr. Parks, I have checked in to his office about every two months or so for blood sugar tests, and have been told each time that my blood sugar is at satisfactory levels. I have assumed that I would be on this medication indefinitely, and I personally have never felt any unusual symptoms…The alternative to taking this oral drug would probably be taking insulin by needle. I have no faith in my ability to administer insulin to myself without causing serious harm to my blood circulation system.[44]

Dozens of letters like A.M.s insisted that proper defense of consumer rights should focus on the freedom to take on risks involved in consuming a given product, rather than regulatory activities that restricted consumer choice.

Ironically, some physicians allied with the latter group of consumers and used the language of egalitarian patient–physician relations to criticize what they saw as Wolfe's overly paternalistic view of the patient as consumer. One physician remarked at the FDA hearings that overly protective regulation would rob the consumer of vital rights:

> [The patient] has the right to decide if he should make that kind of change. If he has made that kind of decision that he wishes to continue on his food habits, and that is incompatible with his diabetic manage-ment, it then, according to the package insert as it is being positioned here, remains for the physician to choose insulin.[45]

Restricting the use of these drugs, especially in the care of older patients, was portrayed as a tyrannical abuse of protective power that stifled the will of the patient. Another physician, specifically responding to Public Citizen's claims to speak for the consumer, wrote an angry note to the FDA that ended, "I hope that your final statement will not persuade successful patients to abandon the use of these drugs because of what they read in the newspaper. Who is responsible for them – you, me, or Dr. Wolfe?"[46]

Different factions could claim to represent the interests of consumers because the patient as consumer had no unitary voice. Was consumerism in medicine an authentic grassroots movement or was it an intrusion of the marketplace into the sacred space of doctor and patient? To engage in the Tolbutamide Controversy was also to come to terms with the multiple roles that risk had come to play in the regulation of therapeutic agents and diagnostic categories. Who was best qualified to balance the risks and benefits of Orinase for an individual person: the physician, the government, or the consumer himself or herself? Many of the consumers motivated to write to the FDA on this issue insisted that they should have the ability to perform their own risk analyses. T.H., a gentleman from Dallas, Texas, called the proposed labeling an "unnecessary and cruel ruling as it would submit us to the danger of insulin shock which could be much more dangerous than the possibility of heart involvement." T.H. argued that he was perfectly competent to perform his own risk–benefit calculations, adding he was "willing to take my chance on heart involvement in exchange for the comfort and convenience of being free from the danger of insulin shock or reaction."[47]

No party in the debate could properly claim to represent the interests of the consumer, but representing the consumer had become a political necessity in the changing health care climate of the 1970s. That the political value of consumer representation was evident to many patients themselves can be seen in the closing statement that A.M. added to a letter he sent to the FDA during the 1975 hearings:

> I also wish to say that I have no interest in any drug manufacturing firm and that my comment is entirely based on my fear that any FDA rule which would lead to Dr. Parks cutting off my DBI-TD prescription would be a direct threat to my life expectancy.[48]

The presence of this disclaimer suggests that the author's self-identification as a diabetic patient and hypoglycemic consumer was not sufficient, in his eyes, to elevate him from a situation of interestedness. The voice of the suffering patient had already been appropriated and mobilized by interested parties; and any authenticity of the patients voice had become complicated. Of the 200 letters the FDA received in the fall of 1975 in support of the drugs, many contained messages too similar in argument, tone, and metaphor for them to be merely independent occurrences. The following text was received verbatim by Edwin M. Ortiz (director of HEW's Division of

Metabolism and Endocrine Drug Products) in three different letters from "concerned friends" in three different geographic regions:

Dear Sir:

A number of my friends have diabetes, and are controlling the disease with oral medication (DBI, Dymelor) and diet. Insulin is neither required nor suitable for them. I am concerned, on their behalf, to learn that you are seriously considering depriving them of their rights by removing the oral medication from the market. The risk of long term side effects seems minute compared to the psychological and physical shock of ingesting insulin.

Please consider all the factors and do not remove these life-prolonging drugs from sale.

Very truly yours[49]

Given that both of the drugs named were Lilly products, it is overwhelmingly likely that these letters originated in the marketing or publicity offices of Eli Lilly & Co. The tactic of the "concerned friends" letter was perfect: untraceable, an air of authenticity and urgency, yet who could venture to demand proof of whether this writer really had any friends with diabetes? Though these ghost letters present an extreme example, it is clear that the struggle to claim the interests of the consumer had deeply complicated any simple, authentic, or unitary grassroots role of "the consumer" in the post-market regulation of American pharmaceuticals (Lynch, 2004).

Conclusion

The 1975 hearings ended in stalemate, as did a second public hearing in 1978; after a lengthy reanalysis of the data, these parties eventually yielded in 1984 to a labeling change for Orinase and all other drugs in its class. To read this final verdict as a vindication of the UGDP and its proponents, however, would be a misinterpretation. Supporters of oral hypoglycemic drugs had not lost the debate by 1984; they merely lost interest in continuing it. Orinase was by then off patent, and the newer oral diabetic agents, like Upjohn's *Micronase* (glyburide), had cleverly obtained an independent therapeutic class designation. These new "second-generation sulfonylureas," were categorically differentiated from drugs of Orinase's class and less affected by labeling changes or package warnings. Along with newer classes of oral antidiabetics, they remain a vital cornerstone in diabetes therapy today (Nathan, 2002).

Throughout most of the 1970s, while the Tolbutamide Controversy reappeared in newspaper headlines on a periodic basis, sales of oral hypoglycemics not only maintained themselves but continued to increase. To those who saw the UGDP Study as proof of the drug's toxicity, this was deeply puzzling. Mild diabetes and oral antidiabetic drugs were relatively recent phenomena in clinical medicine. Why did this potentially harmful new clinical practice persist?

To attempt to answer this question was to come to terms with the typically invisible forces and tensions that lurk within postmarketing regulation of drugs in practice. Postmarketing prescription drug regulation is complicated not only by a greater number of parties than premarket regulation, it is also complicated by the embodied phenomenon of clinical inertia. Inculcated over the past decade into believing and enacting a systematic program of early diabetic pharmaceutical prevention, prescribing physicians recognized their own agency and culpability if that system were to be overturned. As they complained about malpractice implications of changed labeling, physicians were not only thinking about lawsuits based on their present actions, but were also grappling with the theoretically far broader culpability for their past decade of participation within a therapeutic system now being examined as potentially harmful. After a decade of pharmaceutical therapy, it is difficult to tell a patient that they never really had a treatable disease without calling into question the entire edifice of medical knowledge and prior trust in the doctor–patient relationship.

Because pharmaceuticals represent a vital intersection of the private sector's ability to market goods, the federal government's ability to defend consumer safety, the medical profession's ability to determine therapeutic practice, and the will of the citizen as consumer, drugs like Orinase bring conflicts over medical authority and health information into sharp relief. Since the 1970s, the scope and frequency of UGDP-type occurrences has only increased, as evidenced recently by the widespread controversy over industrial and regulatory malfeasance in the detection of novel postmarket risks in drugs like the analgesic Vioxx (rofecoxib), the hormone-replacement agent Prempro (conjugated estrogen), multiple selective serotonin release uptake inhibitor (SSRI) depressants, and the more recent oral antidiabetic agent Avandia (rosiglitazone).

From the Tolbutamide Controversy onward, state-based and industry-based pharmaceutical regulation has increasingly involved the public, and the assembled structures of publicity, not only in the American context but across global markets. The definition of disease has broadened from a conversation formerly limited to doctors and patients into a very large-scale conversation indeed. The Tolbutamide Controversy began with a single question – were sulfonylureas safe and efficacious to use in patients with mild diabetes? – and fractured into a myriad of stubborn debates surrounding definitions of safety and efficacy of drugs, of the definition of populations of patients as diabetics, and the discourse and practices that tie them to clinical practice. The controversy and publicity surrounding the UGDP trial rendered visible the underlying struggle between professional, industrial, consumer-advocate, and state-based modes of postmarket regulation of pharmaceuticals and pharmaceutical-related information in the late twentieth century. In the early twenty-first century, as we find ourselves increasingly reevaluating the risk of contemporary blockbuster drugs in light of contested trial data, similar ruptures between these modes of regulation are continuously emerging, bursting open, and reforming. The persistence of these patterns of crisis and conflict within postmarket regulation only underscores the need for close historical scrutiny to the social dynamics of therapeutic regulation and therapeutic practice.

Notes

* Portions of this chapter appeared earlier as "Risk and the Symptom: The Trials of Orinase," chapter 4 of Jeremy A. Greene, *Prescribing by Numbers: Drugs and the Definition of Disease* (Baltimore: Johns Hopkins University Press, 2007). We gratefully acknowledge the Johns Hopkins University Press for granting us permission to reproduce this material in this volume.

1. C.P. to Food and Drug Administration, May 25, 1970, AF12–868, v.40, AF Jackets Collection, Food and Drug Administration, Rockville, Maryland. Note that all patients' names in this chapter are referred to by initials; this is a by-product of the Freedom of Information Acto (FOIA) request process required to gain access to the FDA AF Jacket collections. Physician and manufacturer correspondence is not considered to be protected by privacy law in the same manner and so physicians' names are considered public knowledge. In this chapter, I consistently use the monogram as a designation of patient/consumer letters while retaining the full names of physicians, researchers, policymakers, and lobbyists. For purposes of continuity, I have chosen to reproduce typographical errors faithfully without repetitive use of *sic*.
2. David L. Roberts to Charles C. Edwards, November 6, 1970; FDA AF Jackets: AF12–868, v. 42.
3. Norman S. Knee to Charles E. Edwards, June 1, 1970, AF12–868, v.40, FDA.
4. M. J. Ryan to Francis T. Collins, June 30, 1970, AF12–868, v.40, FDA.
5. Jesse L. Steinfeld to Morgan U. Stockwell, July 6, 1970 AF12–868 v.40, FDA.
6. C.C.P. to Food and Drug Administration, May 25, 1970, AF12–868, v.40, FDA.
7. J. S. M. to Food and Drug Administration, May 24, 1970, AF12–868, v.40, FDA.
8. M. M. to Henry E. Simmons, June 1, 1970, AF12–868, v. 40, FDA.
9. V. D. to Food and Drug Administration, June 6, 1970, AF12–868, v. 40, FDA.
10. J. F. to Senator Williams, May 28, 1970, AF12–868, v. 40, FDA.
11. Ibid.
12. R. R. J. to Senator William Proxmier, June 1, 1970, AF12–868, v. 40, FDA.
13. Marvin Seife to J. S. M., June 5, 1970; Marvin Seife to C. C. B., June 8, 1970, both found in AF12–868, v. 40, FDA.
14. Robert Proctor and Londa Schiebinger have recently labeled this practice – illustrated with great depth in the case of the tobacco industry's remarkable success at depicting the evidence linking cigarette smoking with lung cancer as "controversial" for several decades, and perhaps most visibly evident in the present attempts to discredit evidence surrounding global climate change – as

"agnogenesis": the orchestrated manufacturing of controversy around scientific evidence that is unfavorable to an industry or interest group. Robert N. Proctor and Londa Schiebinger (eds.) *Agnotology: The Making and Unmaking of Ignorance* (Palo Alto: Stanford University Press, 2008). The role of industry in the strategic production and maintenance of scientific controversy is best illustrated in the tobacco industry's decades-long fight to delay consensus on the health hazards of the cigarette; this is amply documented in the internal documents released in the aftermath of recent litigation and in Allan M Brandt, *The Cigarette Century* (New York: Basic Books, 2006). For a more general account of the use of industrial public relations to shape public debate in the postwar era; see Elizabeth Fones-Wolf, *Selling Free Enterprise: The Business Assault on Labor and Liberalism* (Urbana: University of Illinois Press, 1994).

15. The revision of tolbutamide's label following its DESI evaluation were in the process of finalization in late 1969 when the UGDP results were first announced to the FDA. John Jennings, Upjohn Company, to James O. Lawrence, n.d., received December 22, 1969, AF12–868, v. 39, FDA.

16. The recall of *Panalba* sparked a controversy in its own right that would prove the defining case of the DESI project. See Upjohn Company, "Drug Recall," company circular, March 19, 1970; N.H. to Richard M. Nixon, April 1, 1970; Perry C. Martineau to Richard M. Nixon, March 21, 1970; Charles E. Krause to Richard M. Nixon, March 24, 1970; H.E.W. to Charles E. Krause; April 29, 1970. All documents found in AF12–868, v.40, FDA.

17. Jennings to Lawrence, op cit.

18. HEW News press release, November 2, 1970, AF 12–868, v.42, FDA; "Oral hypoglycemic agents," *FDA Current Drug Information*, October 1970, 1, AF12–868, v.42, FDA.; AMA Council on Drugs, "Statement Regarding the University Group Diabetes Program (UGDP) Study, November 2, 1970," *Diabetes* 19(1970)2: vi–vii.

19. These clinicians' allegations that the FDA was overstepping its regulatory bounds were paired with allegations that the FDA should "think and act as physicians rather than administrators" in their regulation of other physicians. Robert F. Bradley to Henry Simmons, December 11, 1970, FDA Docket 75N-0062 v.2.

20. The CCD maintained that "controverting data" supported the usage of oral hypoglycemics, and insisted labeling must reflect a "fair balance" of scientific opinion. Ironically, the two sides had nearly reached a tentative agreement when the premature publication of the Biometric Society report in early 1975 reignited a public controversy that led to the public set of hearings. "Notice of Public Hearing, July 7, 1975," 10, 12, FDA Docket 75N-0062, v.1.

21. The AMA further objected to extension of the study's interpretation to the entire class of oral hypoglycemics, and to the requirement that physicians obtain informed consent prior to each prescription, "Comments of the American Medical Association Concerning Oral Hypoglycemic Drugs, Proposed Labeling Requirements, Published July 7, 1975," September 4, 1975, FDA Docket 75N-0062, v.1, quotation p. 2.

22. In the hundreds of letters the FDA received from physicians in 1975 regarding the tolbutamide controversy, only nine wrote in support of the FDA's decision. Five of the nine were identical form letters written by colleagues of one UGDP researcher at the Mayo Clinic.

23. Hans G. Engel to FDA Hearing Clerk, August 2, 1975, FDA Docket 75N-0062, v.1.

24. David H. Walworth to FDA, August 23, 1975, FDA Docket 75N-0062, v.1.

25. Hans G. Engel to FDA Hearing Clerk, August 2, 1975, FDA Docket 75N-0062, v.1.
26. James L. Bland to HEW, September 2, 1975, FDA Docket 75N-0062, v.1.
27. Clifford Emond, Jr. to Division of Public Information, Office of the Commissioner, FDA, June 30, 1970, AF12–868, v.40, FDA.
28. "Other than the approved labeling information, information contained in the NDA is regarded as confidential and may not be released. It is the approved labeling that sets forth the indications for use as well as appropriate cautionary and warning information made available to the prescribing physician." Carl Loustanis to Clifford Emond, Jr., July 26, 1970, AF12–868, v.40, FDA.
29. Peter M. Hartmann to HEW, September 2, 1975, FDA Docket 75N-0062, v.1. Hartmann was a Washington, DC, physician.
30. Morse Kochtitzky to FDA, September 3, 1975, FDA Docket 75N-0062, v.1.
31. L. M. Ungaro to Alexander Schmidt, December 16, 1975, FDA Docket 75N-0062, v.1.
32. James L. Bland to HEW, September 2, 1975, FDA Docket 75N-0062, v.1.
33. Testimony of Thaddeus Prout, *Transcript of Proceedings, FDA Panel on Hypoglycemic Drug Labeling*, August 20, 1975, 83–84, 75N-0062, v.11.
34. James L. Bland to HEW, September 2, 1975, FDA Docket 75N-0062, v.1.
35. As a practicing physician from New Jersey noted, "A doctor who was once a contributor to the UGDP study resigned from the study because one of the principals in the study was being paid off by the US Vitamin Corporation who was anxious to discredit other oral hypoglycemics and in particular the Upjohn Company. I feel that on the basis of this alone the UGDP study is not viable and should be discarded." Carl K. Friedland to FDA, October 2, 1975, FDA Docket 75N-0062, v. 8. The researcher in question later admitted to receiving funds from USV during the study, but insisted it had not impacted his work. Ironically, the UGDP study would later indict USV's drug *DBI* as harmful and the drug was subsequently pulled from the market.
36. Eugene T. Davidson to FDA, September 8, 1975, FDA Docket 75N-0062, v. 8.
37. B. Todd Forsyth to HEW, September 4, 1975, FDA Docket 75N-0062, v. 1.
38. J. R. Kircher to FDA Hearing Clerk, July 28, 1975, FDA Docket 75N-0062, v.1.
39. Sudah Deuskar to Food and Drug Administration, July 20, 1975, FDA Docket 75N-0062, v.1.
40. Sidney M. Wolfe and Anita Johnson to Alexander M Schmidt, June 10, 1975, FDA Docket 75N-0062, v.8., p.2.
41. Testimony of Anita Johnson, *Transcript of Proceedings, FDA Panel on Hypoglycemic Drug Labeling*, August 20, 1975, 25, FDA Docket 75N-0062 v.11.
42. J.W. to FDA, August 7, 1975; G. U. to FDA, August 18, 1975; L. Latour to FDA, July 28, 1975, all documents found in FDA Docket 75N-0062, v.1.
43. P.G.W. to FDA, July 24, 1975, FDA Docket 75N-0062, v.1.
44. A.M. to Alexander M Schmidt, August 25, 1975, FDA Docket 75N-0062, v.1.
45. Testimony of John Abeles, *Transcript of Proceedings, FDA Panel on Hypoglycemic Drug Labeling*, August 20, 1975, 121–22, FDA Docket 75N-0062 v.11.
46. David H. Walworth to FDA, August 23, 1975, FDA Docket 75N-0062, v.1.
47. T.W.H. to Food and Drug Administration, August 27, 1975, FDA Docket 75N-0062, v.1.
48. A.M. to Alexander M Schmidt, August 25, 1975, FDA Docket 75N-0062, v.1.
49. H.E.J. to Edwin M. Ortiz, September 26, 1975; B.T. to Edwin M. Ortiz, September 30, 1975; S.M.M. to Edwin M Ortiz, November 14, 1975, all documents found in FDA Docket 75N-0062 v.8.

Bibliography

American Medical Association Council on Drugs 1970, "Statement regarding the University Group Diabetes Program (UGDP) study, November 2, 1970," *Diabetes* 19(2): vi–vii.

Philip K. Bondy 1966, "Therapeutic considerations in Diabetes Mellitus," in *Controversy in Internal Medicine,* ed Ingelfinger *et al.* (Philadelphia: W.B. Saunders Co.), 499–503.

Allan M Brandt 2006, *The Cigarette Century* (New York: Basic Books).

Daniel P. Carpenter 2010, *Reputation and Power: Organization Image and Pharmaceutical Regulation at the FDA* (Princeton: Princeton University Press).

Thomas Chalmers 1975, "Settling the UGDP controversy," *Journal of the American Medical Association* 213 (6): 624.

N. L. Chayet 1967, "Power of the package insert," *New England Journal of Medicine* 277: 1253–54.

Arthur Daemmrich 2004, *Pharmacopolitics: Drug Regulation in the United States and Germany* (Chapel Hill: University of North Carolina Press).

Harry Dowling 1956, "The practicing physician and the food and drug administration," in *The Impact of the Food and Drug Administration on Our Society,* ed. Henry Welch and Felix Marti-Ibanez (New York: MD Publications), 28–29.

Harry F. Dowling 1971, *Medicines for Man: The Development, Regulation, and Use of Prescription Drugs* (New York: Alfred A. Knopf).

Charles C. Edwards 1970, "Oral hypoglycemic agents: Report of the Food and Drug Administration, October 30, 1970," *Diabetes* 19(2): viii–ix.

Chris Feudtner 1995, "The want of control: Ideas, innovations, and ideals in the modern management of diabetes mellitus," *Bulletin of the History of Medicine* 69: 66–90.

Stan N. Finkelstein, Stephen B. Schechtman, Edward J. Sondik, and Dana Gilbert 1981, "Clinical trials and established medical practice: Two examples," in *Biomedical Innovation,* ed. E. B. Roberts, R. I. Levy, S. N. Finkelstein, J. Moskowitz, and E. J. Sondik, (Cambridge: MIT Press).

Elizabeth Fones-Wolf 1994, *Selling Free Enterprise: The Business Assault on Labor and Liberalism* (Urbana: University of Illinois Press).

Jeremy A. Greene 2007, *Prescribing by Numbers: Drugs and the Definition of Disease* (Baltimore: Johns Hopkins).

Neal C. Hogan 2003, *Unhealed Wounds: Medical Malpractice in the Twentieth Century* (New York: LFB Scholarly Publishing).

Elliott Joslin 1959, *The Treatment of Diabetes Mellitus,* 10th edn (Philadelphia, Lea & Febiger).

Jay Katz 1984, *The Silent World of Doctor and Patient* (New York: Free Press).

Harvey C. Knowles, Jr. 1977, "An historical view of the medical–social aspects of the UGDP," *Transactions of the American Clinical and Climatological Association* 88: 150–56.

Gina Kolata 1979, "Controversy over study of diabetes drug continues for nearly a decade," *Science* 203: 990.

Louis Lasagna 1970, "1938–1968: The fda, the drug industry, the medical profession, and the public," in *Safeguarding the Public: Historical Aspects of Medicinal Drug Control,* ed. John B. Blake (Baltimore, Johns Hopkins University Press), 171–79.

Michael Lynch 2004, "Ghost writing and other matters," *Social Studies of Science* 34:147–48, 219–45.

Harry M. Marks 1997, *Progress of Experiment: Science and Therapeutic Reform in the United States: 1900–1990* (Cambridge, Cambridge University Press).

"In Boston: A Diabetes Tea Party Hits FDA" *Medical World News*, December 18, 1970, 13–14.

Curtis L. Meinert and Susan Tonascia 1986, *Clinical Trials: Design, Conduct and Analysis* (New York: Oxford University Press).

Morton Mintz 1970, "Antidiabetes pill held causing early death," *Washington Post*, May 21.

Barbara Mintzes and Catherine Hodges 1996, "The consumer movement: From single issue campaigns to long-term reform," in *Contested Ground: Public Purpose and Private Interest in the Regulation of Prescription Drugs*, ed. Peter Davis (New York: Oxford University Press), 76–91.

James M. Moss 1975, "The UGDP scandal and cover-up" *Journal of the American Medical Association* 232(8): 806–08.

Ralph Nader 1965, *Unsafe at Any Speed: The Designed in Dangers of the American Automobile*, (New York: Grossman).

David Nathan 2002, "Initial management of Glycemia in Type 2 Diabetes Mellitus," *New England Journal of Medicine*, 347(17): 1342–49.

John B. O'Sullivan and Ralph B. D'Agostino 1975, "Decisive factors in the Tolbutamide controversy." *Journal of the American Medical Association* 232(8): 825–29.

James Wright Presley 1991, "A history of Diabetes Mellitus in the United States, 1880–1990" (Ph.D. diss., University of Texas at Austin).

Robert N. Proctor and Londa Schiebinger (eds.) 2008, *Agnotology: The Making and Unmaking of Ignorance*. (Palo Alto: Stanford University Press).

David Rothman 1991, *Strangers at the Bedside, A Story of How Law and Bioethics Transformed Medical Decision-Making* (New York: Basic Books).

Harold M. Schmeck 1970a, "Doubts about oral diabetes drugs" *New York Times*, June 7.

Harold M. Schmeck 1970b, "Diabetes drug use backed by council," *New York Times*, June 15

Harold M. Schmeck 1970c, "Pills for the diabetic: Dilemma for doctors," *New York Times*, June 21.

Harold M. Schmeck 1970d, "fda cites doubt on diabetes pill: Says two medical groups share its misgivings about popular drug," *New York Times*, October 22.

Edward Tolstoi 1950, "Treatment of diabetes with the 'Free Diet' during the last ten years," in *Progress in Clinical Endocrinology* (New York: Grune & Stratton), 292–302.

Edward Tolstoi 1953, *The Practical Management of Diabetes* (Springfield, IL: Charles C. Thomas).

United States Food and Drug Administration (FDA) 1970, "FDA Statement, Friday, May 22, 1970," *Diabetes* 19(s1): 467.

The University Group Diabetes Program (UGDP) 1970, "a study of the effects of Hypoglycemic agents on vascular complications in patients with adult-onset diabetes: i. Design, methods, and baseline results," *Diabetes* 19(2): 747.

6
Thalidomide, Drug Safety Regulation, and the British Pharmaceutical Industry: The Case of Imperial Chemical Industries

Viviane Quirke

Introduction

Since World War II, the pharmaceutical industry has become one of the most heavily regulated sectors of the economy, often to the despair of company managers, who have sometimes blamed the current dearth of new drugs on the increasingly complex and costly procedures involved in complying with the regulations imposed on them. As well as patenting, these have included pricing, and – most importantly – drug safety regulations. Indeed, as Gaudillière and Hess note in their introduction to this volume, the products of the pharmaceutical industry are not like those of the car or electrical industries: they require special control and surveillance in order to protect the health of the public. It is well known that, in the wake of the thalidomide tragedy, drug safety regulation became more stringent, and that this would have a profound effect on the trajectory of drugs from their invention in the laboratory to their application in the clinic. It is also generally agreed that the Food and Drug Administration (FDA), which had helped to avoid the disaster in America, became the reference point for regulatory bodies elsewhere. However, the manner in which this form of state administration became associated with, or superimposed on, other kinds of collective management, in particular within the industry, has rarely been studied. In another volume, I have written about the standardization and codification of pharmaceutical R&D practices that was associated with the tightening of drug safety regulation in the second half of the twentieth century (Quirke, in Bonah *et al.*, 2009). In this chapter, I propose to examine the impact of drug safety regulation on the organization of

R&D, focusing in particular on the integration of biomedical disciplines within pharmaceutical firms.

In a 2006 article, John Abraham and Courtney Davis argued that the beta-blocker practolol, which had to be withdrawn from the market in 1975 because of its serious side effects (which went as far as blindness, and even death in some patients) and became "the first British drug disaster of the modern post Thalidomide regulatory period," was possible because of "the culture of reluctant regulation" that persisted in the British context, even after thalidomide (Abraham and Davis, 2006). They have written that this culture was due to the British regulatory authorities' commitment to the protection of industrial interests, and to their support for a medical profession that maintained an optimistic view of drugs and their purported benefits. However, they have largely ignored the company responsible for the disaster: Imperial Chemical Industries (ICI), which in 1993 spun off its Pharmaceutical Division to create Zeneca, which in 1999 merged with the Swedish firm Astra to produce AstraZeneca. In this chapter, I therefore propose to address this major gap by examining ICI's response to post-World War II regulatory measures, more specifically drug safety regulations, and by studying the impact of such regulations on the firm not only before and after the thalidomide tragedy, but also before and after practolol. I will argue that a concern for drug safety was present within the company, even before thalidomide. I will also show that ICI adopted organizational and other R&D practices in order to meet not only the British regulatory authorities', but also the FDA's requirements. More specifically, the need to demonstrate the safety and efficacy of drugs before marketing stimulated the adoption of physical methods as well as the integration of biomedical disciplines within ICI. But first, I begin by summarizing the development of British and American pharmaceutical regulation in the twentieth century, setting the scene for the study of the impact of drug safety legislation on ICI.

British pharmaceutical regulation (Appendix 6.1)

Among the laws that shaped the British pharmaceutical industry in the twentieth century (Appendix 6.1), patent legislation and pricing regulation aimed to promote innovation, while at the same time controlling the industry. Under the 1919 Patents Act, although protecting processes was accepted, the patenting of drugs, especially those derived from natural substances, had largely been frowned upon. The 1949 Patents Act was intended to remedy this (see Bud, 2008; Slinn, 2008; also Dutfield, 2009). Coupled with the National Research and Development Corporation (NRDC), a public body created in 1948 to facilitate the patenting of discoveries made in universities (Keith, 1981), the 1949 Act had the effect of stimulating innovation in the biomedical field. Similarly, whilst endeavoring to contain the spiraling drugs bill, a major preoccupation of the Ministry of Health under the new National Health Service (NHS, created in 1946), pricing regulation encouraged expenditure on innovative R&D in order to compete on the national and international markets (Hancher, 1990; Thomas, 1994). As its name suggests, the Voluntary Price Regulation Scheme (VPRS), which was introduced in 1957, was voluntary (i.e., firms could join the scheme if they wished). It relied on a consensus between firms and the Ministry on a "fair and reasonable" price to pay for drugs, a principle that survived until the 1978 Pharmaceutical Price Regulation Scheme (PPRS) and beyond (Slinn, 2005). However, when the VPRS was revised in 1969, in addition it required companies to inform the Ministry on their costs and profits. In its rules it also enshrined the Ministry's commitment to a strong, competitive, and innovative pharmaceutical sector, and its preference for controlling the industry by regulating its products, rather than intervening directly in the market (Anderson, 2005: 213).

Throughout the twentieth century, as new drugs were being launched on the market for the treatment of a range of diseases, a succession of acts were passed in order to prevent their sale and advertising by others than medical practitioners. The first of these was the 1917 Venereal Diseases Act. It was followed by the 1939 Cancer Act, the 1941 Pharmacy and Medicines Act (which targeted a wide range of disorders, from Bright's disease to tuberculosis), and the 1956 Therapeutic Substances Act (replacing the 1947 and 1953 Acts, and restricting the sale and advertising of antibiotics). In addition, there were laws that were specifically intended to protect the public from poor quality and dangerous substances: the 1920 and 1951 Dangerous Drugs Acts and the 1933 Pharmacy and Poisons Act.

The 1925 Therapeutic Substances Act went a step further, however. It required pharmaceutical preparations, in particular vaccines, sera and toxins, posterior pituitary glandular extracts, and arsenical drugs, to be tested and deemed safe *before* manufacturers could be granted a license for their production, although their sale and supply remained out of the state's control. This was changed by the thalidomide tragedy, which between 1959 and 1961 caused deformities in 10,000 babies, 5,000 of whom were born in the United Kingdom. As a result, the British government, which had previously focused mainly on the cost and pricing of drugs under the NHS, shifted its attention toward drug safety, and set up a new joint subcommittee on the safety of drugs, chaired by Lord Cohen. It then accepted the recommendations of the Cohen Report, and in 1964 established the Committee on Safety of Drugs (CSD, later renamed the Committee on Safety of Medicines, CSM). This was an independent advisory body composed of four subcommittees – on toxicity, clinical trials, therapeutic efficacy, and adverse reactions – whose task it was to review the safety of new drugs, monitor adverse reactions to existing drugs, and inform medical practitioners. However, like pricing regulation, at first British drug safety regulation retained a voluntary element, and relied on expert bodies for the assessment of drug safety and monitoring of adverse drug reactions. These included the Register of Unexpected Toxicity, set up by the College of General Practitioners, or the Expert Committee on Drug Toxicity, established by the Association of British Pharmaceutical Industries (ABPI) at about the same time as the CSD. Furthermore, a number of areas remained outside the remit of the new committee: the control of quality and over-the-counter sales of drugs, and the regulation of names used for new drugs and of claims made by manufacturers. These issues were addressed in a 1967 White Paper, and led to the 1968 Medicines Act, which overhauled the multitude of regulations that until then had controlled access to dangerous substances, including drugs. It required license holders to satisfy the new CSM of the quality, safety, and efficacy of their products before they could be marketed and sold, thereby establishing a coherent system of drug safety regulation through licensing (Anderson, 2005: 208–10).

Up until 1968, British drug safety legislation therefore was a piecemeal system, put in place in what retrospectively seems as a reactive, rather than proactive fashion, in response to drug discoveries and their use (and sometimes misuse) in the medical marketplace (Tansey and Reynolds, 1997; Anderson, 2005). In contrast to the other two types of legislation, the primary purpose of drug safety regulation was not to encourage innovation, but rather to protect the health of the public. As such, it has been blamed for hindering innovative R&D, most vocally by company managers, but also by scholars who have attributed to it one of the causes of the current dearth of new drugs (Wells, 1983: 17–21). Nevertheless, one of its unintended consequences has been to encourage the integration by drug companies of biological disciplines such as biochemistry, pharmacology, and toxicology, and of physical methods and instrumentation, especially high-performance chromatography, mass spectrometry, and nuclear magnetic resonance, thus transforming the organization and practice of pharmaceutical R&D (Quirke, 2009b). This transformation has benefited the British pharmaceutical industry, which has had a comparative advantage in the biological field (Corley, 2003; Thomas, 1994). Nowhere has this been more striking than in ICI's Pharmaceutical Division, which gradually assimilated a wide range of biological disciplines and techniques in order to comply with drug safety legislation, as well as compete on the medical market, more especially the American market. Because the FDA controlled access to this market, the largest market for drugs, its influence on biomedical science and the pharmaceutical industry has been immense. Moreover, because it had helped to avoid the thalidomide disaster in the United States, it became a model for regulatory agencies elsewhere, including the CSD/CSM in Britain, and the yardstick against which other regulatory systems have been measured by Abraham, Davis, and others (Abraham, 1995; Marks, 1999; Abraham and Davis, 2006; Daemmrich, 2002; Daemmrich, 2004). I will therefore say a few words about it here.

Drug safety regulation: The FDA as yardstick and model? (Appendix 6.2)

American drug regulation began later than, and was modeled on, its British counterpart, principally because of initial US dependence on drugs imported from Britain and elsewhere. Hence, in a first phase of creation and consolidation of drug regulation, the 1906 Federal Food and Drugs Act, which prohibited adulteration and misbranding, was inspired by Britain's 1875 Sale of Food and Drugs Act. It built upon the 1902 Biologics Act, which had required manufacturers of biological products to be licensed by government before they could be marketed (see Appendix 6.2). Like the British system, American drug regulation was put together in many successive stages, often in response to major drug disasters, such as the accidental deaths in children following administration of contaminated diphtheria antiserum in St. Louis, Missouri, and Camden, New Jersey, which had led to the 1902 Biologics Act; or by the fatalities caused by Elixir Sulphonamide in 1937, which precipitated the 1938 Federal Food, Drug and Cosmetic Act (Marks, 1997: 82; Swann, 2010). As in Britain, in the 1950s a major preoccupation in the United States was with the cost and pricing of drugs. Thus, Senator Estes Kefauver's well-known attacks on the FDA and the industry had at first been mainly concerned with patents, prices, and profits, before they turned to drug safety and led to the 1962 New Drug Amendments (Hutt, 2007: 22).

Even though there have been limits to the FDA's power, in particular with respect to the definition of drug safety, on the whole left up to the medical profession, or to the strategy by which the FDA has been able to implement legislation, that is, mainly been through control of drug labeling (Marks, 1997), a major distinction between British and American drug legislation has been the greater institutionalization of the American system, and the greater and more direct influence of the FDA over the market for drugs in the United States. Indeed, the 1906 Act gave the FDA legal powers, thereby establishing it as a law enforcement agency. These powers grew in 1938, when it was given the right to inspect factories, and then issue guidance to industry. As a government body responsible for the premarket approval of drugs, which became a legal requirement in 1962, the FDA has therefore long been more interventionist and proactive than many of its counterparts in other countries. However, in a second phase that began in the

1980s, partly under pressure from disease-based organizations, the pharmaceutical industry, and patient-activist movements, the FDA appeared to step back, as drug regulation was revised to stimulate innovation in the treatment of rare conditions (the 1983 Orphan Drug Act), facilitate access to generic drugs whilst extending patent coverage of branded medicines (the 1984 Drug Price Competition and Patent Term Restoration Act), and accelerate review of, as well as expand access to, experimental drugs for patients suffering from serious diseases like AIDS (1987 and 1991 revised regulations).

So how can we explain these differences between the British and American regulatory systems? Although Abraham has acknowledged that "the improvement of regulatory standards for patients has sometimes lagged behind even industrial support for it," and that major pharmaceutical firms have often been at the forefront of regulatory reform, especially in the United States, partly because this has provided them with competitive advantage, he has also argued that, in Britain more than in the United States, "corporate bias" has molded regulation, often at the expense of consumers/patients (Abraham, 1995: 83). Developing this idea further, in their 2006 joint article, Abraham and Davis have written that the practolol disaster was possible because of "the culture of reluctant regulation" created by the British regulatory authorities' commitment to industrial interests, and to their support for a medical profession with an optimistic view of drugs. However, Abraham and Davis's sources have been rather limited, and they have left the company responsible for developing the drug out of their analysis, an omission all the more regrettable if corporate bias has indeed molded regulation. Using the archives (mainly research reports and organizational files) of ICI's Pharmaceutical Division, I will therefore examine the attitude to regulation by as well as the impact of drug safety regulation on the firm that developed practolol. By describing developments that took place within the industry, sometimes in anticipation of, but more often in response to pharmaceutical legislation, I hope to offer a more balanced view, that is, one that takes into account not only Britain's regulatory culture, but also its wider "therapeutic culture" (Daemmrich, 2004: 3–5).

Britain's therapeutic culture

The Medical Research Council (MRC), a government-funded, semiautonomous body responsible for medical research, has played a crucial role in building a relationship between the state, the pharmaceutical industry, and the medical profession, hence in shaping Britain's therapeutic culture in the twentieth century. The collaboration between the MRC and British drug companies began soon after the Council's creation, during World War I. The principal objective of this collaboration was the manufacture and testing of remedies, more especially synthetic drugs, such as the antisyphilitic drug salvarsan and the local anesthetic novocaine, which until then had been imported from Germany, and were required for the war effort. It continued after the war, when the strict controls that had been imposed upon the industry in wartime were replaced by "moral control" (Liebenau, 1989: 169–70). This looser form of control consisted not only in encouraging the development of in-house research, but also in providing firms with a scientific role model, and with instructions, standards, and norms for the development of new drugs, such as the pancreatic hormone insulin (Sinding, 2002). Thus, a network of "reputable" science-based companies was formed, to which the MRC and other government agencies would turn time and time again in order to develop novel drugs, from various chemotherapeutic and biological remedies between the wars, to penicillin and synthetic antimalarials during World War II, and cortisone, anesthetics, antibiotics, and interferon afterward (Quirke, 2008: chs. 2, 3, 6).

Between the wars, the MRC also encouraged clinical research – albeit not without resistance from some members of the medical profession – by awarding grants, creating clinical research units, as well as organizing and coordinating clinical trials by means of therapeutic committees, for instance, the Therapeutic Trials Committee created in 1931 with the support of the Association of British Chemical Manufacturers (ABCM) (Booth, 1989). In the same period, together with the British Medical Association, the MRC put pressure on government to improve drug safety legislation, which led to the 1925 Therapeutic Substances Act (Appendix 6.1). Then, during World War II, once again mobilization for the war effort helped to shape and transform Britain's therapeutic culture. For the main purpose of developing penicillin, the MRC collaborated with a consortium of research-based pharmaceutical firms: the Therapeutic Research Corporation (TRC), formed in 1940, and later joined by companies like ICI, whose capabilities in large-scale chemical processes complemented those of the TRC (Liebenau, 1987; Quirke, 2008: ch. 3).

The experience of wartime Anglo-American collaborative projects, particularly of developing penicillin and subsequently losing it to American firms that patented the deep-fermentation process for manufacturing the antibiotic, contributed to the MRC's change in attitude toward the patenting of drugs and other medical discoveries. Hence, in the aftermath of the conflict, the Council turned its attention toward patents, putting its full weight behind the creation of a national trustee, the NRDC (formed in 1948), as well as the new 1949 Patents Act, which were both intended to stimulate innovation in the biomedical field (Bud, 2008; Quirke, 2008: 208–10; Quirke, 2009c). Combined with the drive to rationalize the allocation and use of resources under the NHS, this led to a reconfiguration of Britain's postwar biomedical landscape, in which the MRC continued to play a key role between the state, the medical profession, and the pharmaceutical industry, but more as an equal partner of other government agencies, such as the Ministry of Health and the NRDC, of medical associations, and of drug companies, among them ICI's new Pharmaceutical Division. Emblematic of this reconfiguration and transformation of Britain's therapeutic culture after the Second World War, the Randomized Clinical Trial (RCT), which has become the gold standard in clinical research and played an important part in the development of modern biomedicine, was promoted by the MRC as part of its new role within the new, centralized, postwar state (Timmermann and Valier, 2008).

In Britain, therefore, the experience of collaboration between the MRC, which had long relied on the goodwill and cooperation of the pharmaceutical industry for the smooth running of its therapeutic committees, and science-based drug companies, whose reputation and livelihood depended on the manufacture of safe medicines, had created a therapeutic culture that was based on mutual trust, depended to some extent on self-regulation, and may be described as "consensual." This experience forms the backdrop of histories such as ours, and accounts for the faith in the coercive power of opinion that permeated the CSD/CSM, and underpinned Britain's regulatory system (Wells, 1983: 24; Tansey and Reynolds, 1997: 111). It also helps to explain why some scholars, who have implicitly, if not always explicitly, compared the British regulatory system with the American model, have described the British bodies responsible for drug safety as relatively weak, and bowing to the wishes of pharmaceutical firms. However, this has had the effect of obscuring the extent to which drug safety legislation was internalized and became embodied in the organization and R&D practices of firms like ICI.

How does one assess the impact of drug safety regulation on industry in the era of biomedicine and of national health systems? One possibility is to study its effects on the relationship between government agencies, drug companies, and the clinic. Another is to examine its impact directly on the industry, in a way that may be summarized thus:

- Standardization, formalization and codification of R&D and drug testing practices
- Integration and assimilation of biomedical disciplines in industry
- Organization/institutionalization of links between industry and the clinic
- Organization/institutionalization of links between industry and the state

In what follows, I will describe the origins and early history of ICI's Pharmaceutical Division. Then, taking advantage of the availability of organization charts, I will describe the changing organization of ICI's R&D in the 1960s and 1970s.

The origins and early history of ICI's Pharmaceutical Division

Following the discovery that the antibacterial action of the red azo dye Prontosil resided in the colorless part of the molecule, sulphanilamide, which had long been known and therefore could not be patented, in 1936, that is, ten years after ICI's foundation, ICI's main board decided to set up a Medicinal Chemicals Section within the Dyestuffs Group at Blackley, North of Manchester (Reader, 1975; Sexton, 1962: 362). At first, this new section consisted of six Ph.D. chemists, and depended on University Medical Schools to do the biological testing of the compounds they synthesized. However, in 1937, the company started recruiting biologists, and with the outbreak of war the team's research became concerned with developing not only sulfa-drugs, but also antimalarials, which until then had been imported from Germany, such as atebrine and mepacrine (Quirke, 2005a). As well as unraveling their chemical structure, the team began to study their pharmacodynamic properties (i.e., their absorption, distribution, metabolism, and excretion), and for this acquired among the first spectrophotometers and fluorimeters to become available in Britain (Johnson *et al.*, 1984: 571–72).

At first, ICI's management objected to the time spent on such studies, then rarely carried out in the industry. However, the determination of the Section leader, the azo dye chemist Frank Rose (Suckling and Langley, 1990), the skill of the researchers, in particular Alfred Spinks, an organic chemist who in 1942 had moved from Imperial College to Blackley, and the successful outcome of their research, which produced sulphamethazine in 1940, and paludrine in 1945, ensured that the study of chemical structure–drug absorption relations would become research policy at ICI (Johnson *et al*, 1984: 571–72). At the same time, the company became involved in the collaborative scheme to develop penicillin, devising one of the first industrial-scale processes for its manufacture in a government agency factory at Trafford Park, near Manchester, and working on the chemistry of the antibiotic in the hope of finding a route to its synthesis (Kennedy, 1986: 127–29; Cook, Nov. 1947). However, whereas the experience of developing synthetic antimalarials was to have a massive impact on ICI's pharmaceutical R&D after the war, the manufacture of penicillin and other antibiotics largely remained outside its area of expertise, synthetic organic chemistry. Nevertheless, ICI's participation in these wartime developments strengthened the position of Medicinal Chemistry within Dyestuffs, and in 1944 a new Pharmaceutical Division was created.

War had also strengthened the faith in the power of science and modern medicine to improve health and treat diseases. Combined with the NHS established after the war, it stimulated companies into investing in research. At ICI, this led to the creation of a Pharmacological Section within the Pharmaceutical Division. Headed by Alfred Spinks, who had been sent to Oxford to gain a physiology degree in 1950–52, it adopted a fundamental approach (based on scientific hypotheses, as opposed to routine chemical investigations) in order to find treatments for chronic diseases, in particular hypertension, which represented a potentially vast market for drugs (Quirke, 2005b; Greene, 2007, part I). By 1953, it had become a research project, led by Spinks (AstraZeneca, hereafter AZ, CPR 3, March 24, 1953). By 1957, when the pharmaceutical laboratories moved from Blackley to a new research center situated in Alderley Edge, south of Manchester, this novel approach to drug development, in which narrow projects were established as part of a wider, and more fundamental program of research, had gained importance within ICI's Pharmaceutical Division, and hypertension became of a broader scheme to study cardiovascular diseases.

It is significant that an international conference was organized to mark the opening of the new laboratories, and that the topic chosen for the symposium was "the evaluation of drug toxicity" (Spinks and Walpole, 1958). In this way, ICI placed its Pharmaceutical Division squarely within the ethical and science-based sector of the British drug industry, not only by the ambitious projects it selected for study, but also by its concern for drug safety.

The impact of drug safety regulation on ICI: The organization of pharmaceutical R&D

ICI's Pharmaceuticals Division in 1960

Soon after the opening of Alderley Park (see Figure 6.1) an organigramme was produced (AZ DO759, May 1, 1960), the first in a series spanning the period from 1960 to 1976. It showed that there were seven directors, each one responsible for a different area of activity. The one that interests us here is the Research Department, directed by the chemist W.A. Sexton (see Figure 6.2). In 1960, the Research Department was composed of a Techno-Commercial Department, which dealt amongst other things with patents, and of three groups – the Chemistry Group, the Biology Group, and the Administrative Group – each headed by an Associate Research Manager. The Chemistry Group was managed by Frank Rose, and comprised six different sections, numbered 1 to 6. The Biology Group had as its Manager the parasitologist D.G. (Garnet) Davey, and as well as the already mentioned Pharmacology Section, run by Spinks, it included sections on Parasitology, Bacteriology and Virology, Pathology and Tissue Growth. As to the Administrative Group, it included a section responsible for monitoring the progress of clinical trials.

Plan of the main laboratory block at Alderley Park, ca 1958

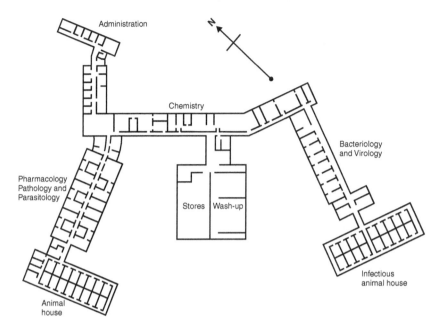

Figure 6.1 Drawing produced by Jayne Stuart, Media Workshop, Oxford Brookes University, from a ground floor plan designed by Harry S. Fairhurst & Son. With kind permission from The Fairhursts Design Group Ltd

Figure 6.2 ICI Pharmaceuticals Research Department, 1960
Source: AZ DO759: May 1, 1960.

ICI's Research Department in 1965

By 1965, that is to say post thalidomide, the number of research departments had increased from three to four, and as well as Biology, Chemistry, and Administrative (later renamed "Research Services"), now included a new Biochemistry Group, led by Spinks (AZ DO770, May 1965) (see Figure 6.3).

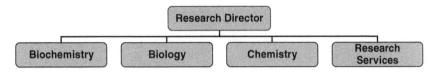

Figure 6.3   ICI Pharmaceuticals Research Department, 1965
Source: DO770: May 1965.

This reflected the growing importance of biochemical methods in generating the detailed biological information required for the assessment of drug safety and efficacy, and the increasingly prominent role played by biochemists in pharmaceutical R&D – interestingly at a time when they were competing unsuccessfully with molecular biologists in the academic arena (Kohler, 1982: ch. 12).

ICI's Research Department in 1971

By 1971, the Research Director, Garnet Davey (who in 1969 had become the first non chemist to attain such a high level on the managerial ladder), was responsible for six departments (Figure 6.4). These included Chemistry, managed by Rose, and an offshoot of Chemistry, entitled "Biological Chemistry," managed by the organic chemist Bernard Langley (AZ DO772, March 18, 1971). Between them, Chemistry and Biological Chemistry totaled ten sections. However, Rose retired soon after the organigramme was produced, and this led to the two departments being reunited under the umbrella of "Chemistry," run by Langley (AZ DO1415: S.I. 16/71, April 21, 1971).

Biology was then combined with Biochemistry, and the newly merged department was managed by the biochemist Brian Newbould, who would later succeed Davey as Research Director. It was organized not only along disciplinary lines, as it had been before, but also along therapeutic lines, including sections for the study of allergic and inflammatory conditions, thrombosis, cardiovascular diseases, anesthesia and gastrointestinal diseases, and diseases of the central nervous system (CNS), suggesting a closer alignment of pharmaceutical R&D on clinical research and a growing interconnection between the two.

In 1972, however, Biology and Biochemistry once again separated, partly so that researchers working on the chemical and biological aspects of natural products could be united under a single roof, in Biochemistry.[1] The Research Department was therefore returning to the earlier "trinity" of Chemistry/ Biochemistry/Biology, with Biochemistry acting as a bridge between Chemistry and Biology (AZ DO1415: S.I. 17/72, August 22, 1972). There was also a large Research (Administration) Department, which included a Scientific Services Group, as the former Research Services Department was now known, and absorbed the sections on Physical Methods (formerly of the

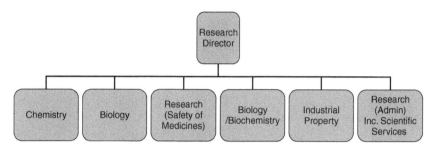

Figure 6.4   ICI Pharmaceuticals Research Department, 1971
Source: AZ DO772: March 18, 1971.

Biological Chemistry Group) and Data Services (formerly of the Industrial Property Department) (Quirke, 2004). Then, in 1979, the Physical Methods Section was spun off into a fully fledged research department, called "Physical Chemistry." This reflected the expanding use of physical instruments within ICI's Pharmaceutical Division, partly in response to the ever-increasing amounts of quantitative data needed to satisfy regulatory bodies at home and abroad (AZ CPR 109B).

ICI's Research (Safety of Medicines) Department in 1971

In addition, the growing intensity of exchanges between the laboratory and clinic and the creation of the CSM in 1969 had led first to a new Clinical Research Department being formed as an offshoot of the Medical Department, and then, in 1971, to a new Research (Safety of Medicines) Department, which worked mainly on pharmacological and toxicological topics (Figure 6.5). This was achieved by demerging the Biochemical Pharmacology Section from the Biochemistry Department and the Toxicology Section from the Biology Department, before amalgamating them into the new department (AZ DO1415: S.I. 7/71, February 8, 1971). In 1979, recognizing that there was "an increasing involvement of the Safety of Medicines Department in many aspects of ICI's pharmaceutical business on an international basis," it was decided from then on that the Safety of Medicines Department Manager would no longer report to the Research Director, but directly to the Technical Deputy Chairman above him in the corporate hierarchy (AZ DO1415: SI 3/79, Feb 19, 1979).

The quasi-organic evolution of pharmaceutical R&D depicted in ICI's successive organization charts resulted from internal factors and local contingencies, such as the movement of personnel up the corporate ladder, or in and out of the firm. However, it also mirrored the evolution of academic disciplines, and reflected the growing complexity of the drug discovery

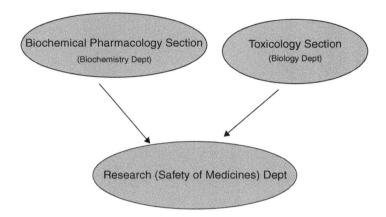

Figure 6.5   ICI Pharmaceuticals Research (Safety of Medicines) Department, 1971
Source: AZ DO1415: S.I. 7/71, February 8, 1971.

process in an increasingly competitive and tightly regulated environment. But what of the research practices in ICI's Pharmaceutical Division? I describe these next, through the history of ICI's beta-blocker project, which was ICI's most successful pharmaceutical program (both commercially and scientifically), but which as well as propranolol and atenolol produced practolol. It shows that a concern with drug safety was present and was internalized within the company, long before it became institutionalized in the Research (Safety of Medicines) Department, and even before thalidomide. Tracing the evolution of this concern through a specific research program, that of the beta-blockers, enables us to highlight the flexibility of the notion of drug safety, which changed along with the knowledge accumulated in what was then a new drug class.

The development of the beta-blockers, and the concern for drug safety before thalidomide

In 1957, the physiologist James (later Sir James) Black was appointed to Alderley Park to work on the beta-adrenergic receptors of the heart in the hope of finding a drug treatment for coronary artery disease (Barrett, 1972; Shanks, 1984; Vos, 1991; Quirke, 2006). This work fitted into the pharmacological research program that had been established five years earlier, and Black benefited from the experience of laboratory technicians who had previously worked with Spinks on hypertension, in particular Brian Horsfall, whom Black encouraged to set up an instrumentation section in his laboratories. Horsfall devised various pieces of physiological apparatus to help measure heart rate, force of contraction, and blood flow, and his contribution was later acknowledged by making him the head of his own unit, the Biological Electronic Unit of the Biology/Biochemistry Department (AZ DO772). Black was also assisted by two chemists: E.H.P. Young, who like Horsfall had worked with Spinks on hypertension, and J.H. Stephenson, who in 1960 synthesized compound ICI 38,174 (pronethalol, Alderlin). Although it was the most promising compound in terms of blocking the beta-receptors in the heart (hence the term "beta-blocker"), Black did not consider that it was ready yet for clinical trials. Indeed, in one of his reports he wrote: "It has not been shown that ICI 38,174 is a candidate for clinical trial and much work on toxicity, and particularly on pharmacological side-effects, including hypotensive action, will be necessary before its suitability for trial can be assessed" (AZ CPR 50/6: June 14, 1960). The basic principle in novel drug discovery is to demonstrate specificity of action, that is, to show that at concentration levels required for biological activity, a compound does not act on systems other than its primary target. At this early stage of the research, the hypotensive action of pronethalol was therefore considered to be an undesirable side effect, even though within five years of writing this report hypertension would become a major therapeutic target for ICI's beta-blockers.

Black set the agenda for ICI's R&D strategy in the cardiovascular field thus: "This work will be given very high priority, as will the evaluation of new analogues." By January 1962, 136 analogues and related compounds had been tested. While these tests were mainly concerned with the completion of patents, the team were still looking for a compound that would "a) be longer acting; b) have greater resistance to catecholamine 'breakthrough'; c) show less penetration of the CNS," that is, be more effective and have fewer side effects (CPR 56, 56/8: Jan. 26 1962). These were all areas in which Black felt that improvements could be made, but so far pronethalol remained the best compound to take to trial.

Small-scale clinical trials with pronethalol therefore began, and an additional set of tests could now be carried out, using biological material obtained from the trials. These tests were described under the title "biochemistry" (in addition to the already existing pharmacology and chemistry sections of the reports), and coincided with the creation of a separate Biochemistry Section (Figure 6.3). They were performed by W.A.M. Duncan, who devised a spectrofluorometric method for estimating the presence of pronethalol in blood and tissues, in order to obtain basic information about the behavior of the drug not only in animals, but also in man. This allowed him to measure the concentration of pronethalol, and study its degradation products in vivo.

Thus the biological assessment of drugs was largely left to the discretion of project leaders such as Black, and of individual researchers such as Duncan. At the preclinical stage of the drug development process, the "safety" of pronethalol corresponded to the results of the pharmacological experiments and toxicity tests, and therefore was as yet indistinguishable from laboratory phenomena. This stage also involved the synthesis of large numbers of derivatives in order to provide wider patent coverage, and to find a backup should the compound of choice fail at the last hurdle, that of the large-scale clinical trials, or be displaced on the market by competitors.

As part of the biological and chemical work involved in the development of pronethalol, which in preparation for market launch was named Alderlin (a near-anagram of Alderley Park), compound ICI 45,520 was identified, which had 10 to 20 times greater activity than Alderlin. Moreover, its LD50[2] i.v. was the same, which suggested that it had a better therapeutic ratio (i.e., a greatly increased ratio between blocking and toxic doses in man) (AZ CPR 56, 56/8: Jan 26, 1962). Compound ICI 45,520 was, in fact, going to become Alderlin's replacement, as propranolol, or Inderal.

The development of the beta-blockers, and the concern for drug safety after thalidomide

From pronethalol (Alderlin) to practolol (Eraldin)

Black was then joined by a clinical pharmacologist, R.G. Shanks, who would inherit the beta-blocker project once Black had left ICI for SmithKline & French in 1964. They began studying three prototypes of the different series of analogues of Alderlin, which had been selected for pharmacological and toxicological studies: the naphthyloxy analogue ICI 45,520, the phenoxy analogue ICI 45,763, and the benzodioxane analogue ICI 45,847, and their properties were compared with those of Alderlin. The tests included in vitro and in vivo beta-blockade, effect on resting heart rates, and half-life (a measure of the drug's persistence in the body). Toxicity tests included both acute and chronic tests, the latter showing that, whereas with ICI 38,174 thymic tumors appeared within 120 days, "the test with ICI 45,520 had been running since 18.12.62 and no tumours had appeared before we went to press" (AZ CPR 56, 56/10: Apr. 26, 1963).

Hence, when Alderlin was launched in 1963, ICI already knew that they had a backup in propranolol, and Alderlin was marketed for life-threatening conditions only. Meanwhile, the tests on propranolol were gaining in complexity, and now included additional studies into its metabolism and mechanism of action. This was partly because of the decision to prepare a CSD submission for the compound "as an exercise" prior to clinical trials (Woodbridge, 1981: 193), and partly in response to growing competition from other companies, such as Mead Johnson, who had just launched their beta-blocker sotalol, which although less active, appeared from the animal data to be less toxic than propranolol. It was suspected that the lower toxicity of sotalol was due to its lower fat solubility. Therefore, Shanks concluded that "in view of the possible reduction in therapeutic ratio of Inderal, it would seem essential to investigate fully the inter-relationship of acute toxicity, accumulation within the central nervous system (CNS) and water/lipid partition" (AZ CPR 56, 56/13: Nov 13, 1964).

In February 1966, following the early clinical investigations of propranolol, it was recognized that, although the drug had been developed as an adrenergic beta-receptor antagonist, it appeared "to possess several other properties including 1) local anesthetic action, 2) quinidine-like (anesthetic) properties, 3) a negative inotropic action (reducing heart muscle contractions) which may not be related to its beta-blocking activity, 4) a hypotensive action" (AZ CPR 99, 4B: Feb 18, 1966). This hypotensive action had been revealed in clinical trials by the clinical pharmacologist Brian Prichard at University College Hospital (Prichard Gillam, 1964), and ICI obtained clearance from the CSM to use Inderal for the treatment of hypertension in 1969, at the same time as a new Clinical Research Department was formed as an offshoot of the Medical Department. Building on the experience it had gained in preparing CSD/CSM submissions, the company, which had begun targeting the American market by then, succeeded in obtaining clearance for propranolol in hypertension from the FDA in 1973. However, following adverse publications in the *Lancet*, ICI had also begun developing concerns about the safety of propranolol, for it had been associated with some cases of heart failure. Moreover, propranolol caused bronchoconstriction; hence it could not be used in patients with asthma. Therefore the search continued for a compound that would have therapeutic advantages over propranolol, despite the evidence that patients with angina who were given the drug were improved, that incidence of side effects was small, and that mortality was not increased (AZ CPR 99, 5B: May 25, 1966).

Compound ICI 50,172 (practolol, Eraldin), which had been synthesized in 1964, offered the biologists what appeared to be an improvement over Inderal. Not only was it cardio-selective, but it had no depressant activity on the heart. Moreover, although less potent than propranolol, practolol had a much greater duration of action. However, practolol was not viewed as an outright replacement for Inderal, but rather as a useful drug for asthmatics suffering from heart disease. In the midst of debates among clinicians as to what was the optimal combination of properties for a beta-blocker, pressure therefore mounted to take practolol through to clinical study (Woodbridge, 1981: 219–20). Trials began in human volunteers early in 1968, before moving on to double-blind trials in angina, and then in hypertension a few months later.

Growing competitive activity from rival firms also stimulated ICI's researchers into increasing the number of beta-blocker analogues and expanding the pharmacological tests carried out on them. Thus, a new beta-blocker screening program now included a bronchospasm test and an additional check on selectivity, in order to save time "on following false leads" (AZ CPR99, 7B: Feb 23 1967). At the same time, the assistance of R.H. Davies of Central Management Services (later in Research Services, Research Administration Department, Figure 6.4) was sought, so that structure–activity relationships could be presented "on a more rational basis than hither-to," using Information Theory Analysis[3] to improve the efficiency and reliability of the tests (AZ CPR99, 12B: Nov 7 1968). By the time of the next report, a new section had been created to accommodate Davies' quantitative approach, the Mathematics and Statistics Group of Management Services (AZ CPR99, 13B: March 27, 1969).

The stimulus for including additional pharmacological tests and quantitative data analysis therefore came from concerns with safety, from scientific uncertainties and debates among investigative clinicians as to the best way to achieve this, from the need to demonstrate efficacy, as well as from growing competition from other firms. The need to increase and accelerate the synthesis and testing of analogues in order to meet this competition led to divergences between the pharmacological and chemical agendas and created tensions within the company. These tensions highlight the complex link that exists between the research and commercial agendas of innovative drug companies like ICI (Sismondo, 2010), as expressed in a 1972 report:

> If we are to maintain our lead in the field of beta-adrenergic blocking research, close attention will have to be paid to preserving the balance between the negative activities of defensive patenting and preparing back-up compounds, and the constructive activity of looking for novel compounds with new actions. (AZ CPR 99, 22B: March 7, 1972)

By February 1967, Shanks had left ICI, partly as a result of these tensions. He was replaced by the pharmacologist A.M. Barrett, who reorganized the beta-blocker program in a systematic way (see Table 6.1). However, Barrett too soon left ICI. He was succeeded by the clinical pharmacologist F.D. (Desmond) Fitzgerald, who moved from the Medical to the Research Department in 1969, and became Head of Pharmacology and leader of the Cardiovascular Project Team.

The beta-blockers from practolol (Eraldin) to atenolol (Tenormin)

In 1969, compound ICI 66,082 (atenolol, Tenormin) had been synthesized as part of the continuing chemical investigations to determine the relationship between biological activity and chemical structure in the beta-blockers selected to go through to the development stage. Again, the direction of the beta-blocker project was determined by the head of the biological team:

> Analysis of the mechanism of action of Inderal and ICI 50,172 is complicated by their possession of different properties. The choice of the

eventual successor to Inderal depends in part on a better understanding of the relevant contribution of these different properties. The principal objective of the screening programme has been to find compounds which would help to define the properties of the "ideal" beta-blocker more clearly. (AZ CPR 99, 13B: March 27, 1969)

The report concluded that this objective had been achieved "now that compounds with selectivity and without stimulant action are available (e.g. ICI 66,082) and compounds without anaesthetic properties or stimulant action but which are not selective have also been found (ICI 61,081)." ICI 61,081 had lower potency than Inderal, but had little local anesthetic action and less toxicity than Inderal. On the other hand, of ICI 66,082 it was said that "this compound may be very important since it is uniquely selective and non-stimulant" (AZ CPR 99, 13B: March 27, 1969). To the researchers, ICI 66,082 obviously possessed a unique combination of properties, worthy of development (see Table 6.1).

In the light of the most recent clinical results, the positions of Inderal and Eraldin were compared at a meeting of the cardiovascular team, on November 5, 1969. Eraldin was not proving to be better than Inderal in angina, although there appeared to be fewer side effects with Eraldin than with Inderal. It was concluded that the position of Inderal in clinical practice was now reasonably secure, and its use in dysrhythmias, angina pectoris, and hypertension was based on a proven efficacy. However, competition was expected from Haessle (with Aptin), Ciba (Trasicor), and Sandoz (LB 46). ICI presumed that the main claims of their competitors would be in terms of improved safety on the grounds of intrinsic sympathomimetic (stimulant) activity, which tended to decrease the negative inotropic (depressant)

Table 6.1 Properties of beta-blocking compounds

Compound	ED_{50}^{*} (μg/kg)	Selectivity	Local anesthesia	Depressant on heart	Intrinsic activity
Inderal	62	No	yes	Yes	no
Eraldin	167	Yes	no	No	yes
Aptin	80	No	yes	Yes	yes
ICI 60,847	6	Yes	no	no	yes
ICI 66,082	96	Yes	no	No	no
ICI 65,674	25	Yes	yes	Yes	yes
ICI 61,081	112	no	no	Yes	no

*Effective Dose for 50% of the tests carried out. A lower number indicates a more potent compound.

Source: AZ CPR 99, 14B: July 8, 1969.

effect on the isolated heart. For LB 46, there was the added advantage of increased potency (about 40 times greater than Inderal). There followed a detailed analysis of the relative advantages of Inderal and Eraldin, which was complicated by the lack of information concerning the true significance in a clinical setting of certain biological and pharmacological properties – such as membrane-stabilizing activity, cardio-specificity, lipid solubility. It was against this background of competitive pressure and uncertainty as to what constituted the "ideal" beta-blocker, that Eraldin was launched onto the market, in 1970.

Meanwhile, work continued on possible successors to Inderal, around which a controversy had arisen concerning the causes of heart failure and hypotension following propranolol. In an attempt to clarify this, the biological team modified the screening procedure for new beta-blocking compounds to separate their properties into various classes (AZ CPR 99, 19B: March 10, 1971):

a)   sympathomimetic selective beta-blockers;
b)   membrane-stabilizing selective beta-blockers;
c)   selective beta-blockers without additional properties; and
d)   nonselective beta-blockers without additional properties.

One of the main protagonists in the controversy, the pharmacologist Vaughan-Williams, who had first identified the quinidine-like (anesthetic) effects of propranolol, came to give a talk in the Division. His observations had been made in vitro, but he thought that these may not be relevant to the action of propranolol in man. Matters were complicated further by a recent publication by Gibson and Coltart, which suggested that both membrane-stabilizing and intrinsic properties were irrelevant, and that cardio-selectivity really was the most important factor in achieving improved safety.

By 1972, ICI had become aware that Inderal, and to some extent Eraldin, affected intraocular pressure, and might therefore be useful in the treatment of glaucoma (AZ CPR 99, 22B: March 7, 1972). However, from the research reports it is clear that the cardiovascular team did not know then of the oculomucocutaneous syndrome brought on by Eraldin, as it did not show up in the toxicity tests carried out both before and after marketing. Nevertheless, the search for other cardio-selective beta-blockers continued. The cardiovascular team therefore returned to two compounds in earlier series, ICI 66,082 and ICI 66,081. If ICI 66,081 ultimately failed on further toxicity tests, then the team would search for a non teratogenic member of the series. However, there was still uncertainty about which had the most desirable properties, both from a scientific and from a commercial point of view (see Table 6.1).

By 1973, however, evidence of the serious side effects associated with Eraldin had begun to accumulate, and the advice of Dr. K.G. Green (since 1965 Manager of the Medical Department) was sought on how to proceed. He gave clear priority to the development of a safe beta-blocker for use in hypertension, aimed at the American market, but he did not define what exactly constituted a safe beta-blocker, and favored an altogether different compound, ICI 72,222:

> We have been concerned, for the past 3 years, with the discovery of beta-blocking drugs possessing various combinations of additional pharmacological properties. It seems that we should have been seeking a safe effective hypotensive beta-blocking drug [...] It is clear that there is now considerable urgency to bring ICI 72,222 to clinical trial [...] strengthened by the possibility that Eraldin administration is associated with a drug-induced immune complex syndrome in certain rare cases. The development policy of ICI 72,222 is markedly affected by these two considerations and in my view we should try to carry out pharmacological, toxicological, and formulation studies with a special emphasis on the American market. All our work should be done with FDA regulations even more in mind than they commonly are at this stage. (AZ CPR 99, 22B: Mar 17, 1972)

However, concerns soon arose about the teratogenic potential of ICI 72,222. This made the need to find an alternative compound even more urgent, and the opportunity was taken to remind the team that the relevant criteria for selecting such a compound were "potency, selectivity, intrinsic activity, absence of membrane activity, patentability, the existence of a method for estimating it in biological fluids, and ease of large-scale synthesis" (AZ CPR 99, 24B: Nov 7, 1972). It was noted that no single compound was without some present or potential drawback. Faced with such dilemmas and uncertainties, the most urgent need of all was "to have a potent Eraldin-like compound which would meet minimal opposition from the US Drug Authorities."

The requirements of the FDA had therefore started to define what constituted drug safety and to provide a major focus for the research of the cardiovascular team, with a particular emphasis on carcinogenicity. However, in planning the two-year carcinogenicity tests required by the American Agency, discrepancies between FDA requirements and the knowledge and practices associated with the beta-blockers were noted. Indeed, because FDA guidelines suggested that the maximum tolerated dose should be used, it was necessary to select a high dose of beta-blocking drug, even though such a high dose would not be needed to obtain a therapeutic effect as the beta-blockers were generally well-tolerated by the oral route. This meant having to "fix arbitrarily a dose which is acceptable to the FDA." A dose 20 times the ED50[4] was proposed, but it seemed unlikely that the FDA would accept this as a safety factor since "intuitively, the doses looked too low. No doubt the final dose will be 200 mg/kg daily irrespective of the science" (AZ CPR 99, 24B: Nov 7, 1972). In the event, it was the pharmacologists' favored compound, ICI 66,082 (atenolol, Tenormin), that is, the closest to a clean, "ideal" beta-blocker, which was chosen for launch.

Having warned doctors that there had been evidence of serious side effects associated with it – in 1974, that is, before the CSM had issued their own warning (Abraham and Davis, 2006: 135) – ICI withdrew Eraldin in 1975. The company set up a compensation scheme, for which the claims were to outnumber those for thalidomide by about 5 to 1 (Woodbridge, 1981: 232–35). As to Tenormin, it became one of the world's biggest selling heart drugs. Within ten years, its related products generated sales worldwide of about £500 million. Nearly 40% of sales were made in the United States, 30% in Western Europe (including the United Kingdom), 14% in Japan, and 16% in the rest of the world (Holland, Sept 12, 1987: 288).

Thalidomide, drug safety regulation, and ICI

In the wake of the thalidomide tragedy, drug safety regulation became more stringent, and the FDA became the reference point not only for regulatory bodies such as the CSD/CSM, but also for firms beginning to target the American market, like ICI in the 1960s and 1970s. In this chapter, I have explored the manner in which, as a form of state administration, drug safety regulation became associated with other kinds of management, an aspect that has rarely been studied, especially in the context of the pharmaceutical industry. Using the example of ICI, I have shown how drug safety legislation was internalized, helping to shape R&D organization and practices, not only before and after the thalidomide tragedy, but also before and after practolol. This has been described by Abraham and Davis as "the first British drug disaster of the modern post Thalidomide regulatory period," possible in their view because of "the culture of reluctant regulation" that persisted in Britain. I have argued that a concern for drug safety was present in ICI, the firm responsible for developing practolol, even before thalidomide. However, I have also shown that, after thalidomide, ICI adopted organizational and other R&D practices in response to the British regulatory authorities, and – as they turned their attention toward the lucrative American market – in response to the FDA's requirements. More specifically, the requirement to demonstrate the safety and efficacy of drugs before marketing, which became a legal requirement in Britain in 1968, stimulated the integration of biomedical disciplines and the adoption of statistical methods and physical instrumentation within ICI, in an evolution that began in the mid-1950s, mirroring the development of academic disciplines and reflecting the growing complexity of the drug discovery in an increasingly competitive environment.

The impact of thalidomide and of drug safety regulations on ICI, which I have used to illustrate the development of Britain's pharmaceutical industry after World War II, can therefore be summarized thus:

• It stimulated the integration of biomedical disciplines (e.g., pharmacology, biochemistry, toxicology)
• It stimulated the adoption of statistical methods and physical instrumentation
• It led to the alignment of pharmaceutical R&D and clinical research
• It helped to institutionalize links between industry and the clinic (with the creation of ICI's Clinical Research Department in 1969, for example)
• It helped to institutionalize links between industry and regulatory authorities as an organ of the state (with the creation of ICI's Safety of Medicines Department in 1971)

However, the history of ICI's beta-blocker project, of which practolol was a product, and which this chapter has examined in some detail, also shows that drug safety was a flexible concept, a perpetually moving target that evolved along with scientific and clinical knowledge accumulated in the process of searching for successful products. The search for therapeutically and commercially successful beta-blockers was carried out in a new drug class, in which ICI's researchers were confronted by scientific and clinical uncertainties and dilemmas, as well as by tensions between the scientific and commercial agendas of the firm. Moreover, the history of ICI's beta-blocker project highlights the contested nature of drug safety, which is molded by related and sometimes rival interests, whether they are those of industrial or academic scientists, of clinical researchers, of company managers, or of regulatory agency officials.

In this context, is it therefore fair and accurate to refer to Britain as having had a "culture of reluctant regulation," as Abraham and Davis have done? Or of reactive, perhaps even bottom-up regulation, as their implicit comparison with the United States might suggest? For the purpose of this chapter, Daemmrich's concept of "therapeutic culture" has proven to be more helpful. This was a consensual therapeutic culture, which depended to some extent on self-regulation, and in which companies, state agencies, and the medical profession collaborated in building a national health industry, which became one of the pillars of the British Welfare State in the second half of the twentieth century.

Acknowledgments

I thank the editors of this volume, Jean-Paul Gaudillière and Volker Hess, for their comments on an earlier version of this chapter, which appeared as "The impact of Thalidomide on the British pharmaceutical industry:

the case of Imperial Chemical Industries," in J.-P. Gaudillière and V. Hess (eds), "Ways of Regulating: therapeutic agents between plants, shops and consulting rooms" (Max Planck Institut für Wissenschaftgeschichte, Preprint 363, Berlin, 2009: 125–41). I am also most grateful to Desmond Fitzgerald, John Patterson, and Janet Wale, for their helpful and constructive criticisms of a first draft of this chapter. Any failings and misunderstandings that remain are entirely my own. The research for this chapter was funded by the Wellcome Trust (grant numbers 059793 and 025577).

Appendix 6.1 Milestones of British pharmaceutical regulation

Date	Legislation	Brief Description
1851	Arsenic Act	restricting the sale of arsenic to pharmacists
1868	Pharmacy and Poisons Act	extended this restriction to other poisons
1872	Drug Adulteration Act	appointing public analysts to test medicines prior to sale
1875	Sale of Foods and Drugs Act	prohibiting adulteration and misbranding
1917	Venereal Act	covering arsenical and other preparations to treat venereal diseases
1919	Patents Act	enabling the patenting of drug manufacturing processes
1920	Dangerous Drugs Act	controlling the manufacture and possession of narcotics
1925	Therapeutic Substances Act	controlling by license the manufacture of vaccines, sera, and other therapeutic substances
1933	Pharmacy and Poisons Act	restricting the sale of potentially dangerous drugs through prescription
1939	Cancer Act	covering preparations to treat cancer
1941	Pharmacy and Medicines Act	regulating the manufacture and sale of sulphonamides under poisons legislation
1947	Penicillin Act	controlling the sale and supply of penicillin
1949	Patents Act	enabling the patenting of drugs as well as their processes
1951	Dangerous Drugs Act	
1953	Therapeutic Substances (Prevention of Misuse) Act	including therapeutic substances not covered by poisons legislation
1956	Therapeutic Substances Act	covering hormones and other natural substances
1957	Voluntary Price Regulation Scheme (VPRS)	
1963	Cohen Report	establishing the Committee on Safety of Drugs (CSD)
1964	Drugs (Prevention of Misuse) Act	

Continued

Appendix 6.1   Continued

Date	Legislation	Brief Description
1968	Medicines Act	Establishing the Committee for the Safety of Medicines (CSM) and requiring evidence of safety and efficacy for licenses to be granted for manufacture, sale, and supply
1978	Pharmaceutical Price Regulation Scheme (PPRS)	
1985	Introduction of Limited List	

Source: "Select chronology of pharmaceutical legislation," in: L. Richmond, J. Stevenson and A. Turton (eds), *The Pharmaceutical Industry: A Guide to Historical Records* (Aldershot: Ashgate, 2003), pp. 49–52; also Anderson, "Drug regulation and the welfare state: Government, the pharmaceutical industry and the health professions in Great Britain, 1940–1980," in V. Berridge and K. Loughlin (eds), *Medicine, the Market and the Mass Media: Producing Health in the C20th* (London: Routledge, 2005), pp. 192–217.

Appendix 6.2   Milestones of American Pharmaceutical Regulation

Date	Legislation	Brief Description
1902	Biologics Control Act	requiring manufacturers of biological products to be licensed by government before they could be marketed
1906	Food and Drugs Act	expanding the 1902 act to other types of products, and establishing the FDA as a law enforcement agency
1913	Gould Amendment	establishing labeling requirements
1930		Department of Agriculture's Chemical Division renamed FDA
1938	Federal Food, Drug and Cosmetic Act	expanding the 1906 Act and requiring new drugs to be shown to be safe before marketing, and that the FDA inspect factories
1943	*United States v. Dotterweich*	Supreme Court ruling establishing strict criminal liability in relation to the food and drug industries
1945	Penicillin Amendment	requiring FDA testing and certification of safety and effectiveness of all penicillin products
1949	Black Book	Procedures for the Appraisal of the Toxicity of Chemicals in Food: FDA's first guidance to industry
1951	Durham–Humphrey Amendment	defining types of drugs that can only sold on prescription

Continued

Appendix 6.2   Continued

Date	Legislation	Brief Description
1962	New (Kefauver–Harris) Drug Amendment	transforming former policing statues into premarket approval requirements, and requiring drug manufacturers to prove the effectiveness of their products before these can be sold
1966	Fair Packaging and Labeling Act	FDA given powers to enforce this; National Academy of Science – National Research Council contracted to evaluate the effectiveness of drugs approved on the basis of safety alone between 1938 and 1962
1981	Revised regulations for human subject protection	providing for wider representation on institutional review boards and detailing what constitutes informed consent
1983	Orphan Drug Act	enabling the FDA to promote research and marketing of drugs for rare diseases
1984	Drug Price Competition and Patent Term Restoration Act	enabling the FDA to approve applications to market generic versions of branded drugs without repeating safety and efficacy tests. Companies of branded products can apply for an additional 5 years of patent coverage
1987	Revised Investigational Drugs regulations	for expanded access to experimental drugs for patients with serious diseases, such as AIDS
1991	Revised regulations	for accelerated review of drugs for life-threatening diseases
1993	Consolidation	of adverse reaction reporting systems launched as MedWatch

Source: A. Daemmrich and J. Radin (eds), The FDA at 100: Perspectives on Risk and Regulation (Philadelphia: Chemcial Heritage Foundation, 2007), pp. ix–xvii.

Notes

1. NB: in 1961 members of the Natural Products Division, located at the Akers Laboratories near Welwyn had been moved to the Pharmaceutical Division and split between the Chemistry and Biology groups.
2. LD50 is the median lethal dose required to kill half the tested population after a specified test duration, used as a measure of the acute toxicity of a substance.
3. A mathematical and statistical method, initially applied to electrical communications, but later applied more widely to business systems and other areas concerned with information.
4. ED50 is the median effective dose required to produce a therapeutic effect in 50% of the test population.

Bibliography

Abraham, J. (1995) *Science, Politics and the Pharmaceutical Industry: Controversy and Bias in Drug Regulation*, London: UCL Press.

Abraham, J. and Davis, C. (2006) "Testing times: The emergence of the Practolol disaster and its challenge to British drug regulation in the modern period," *Social History of Medicine*, 19: 127–47.

Anderson, S. (2005) "Drug regulation and the welfare state: Government, the pharmaceutical industry and the health professions in Great Britain, 1940–1980," in V. Berridge and K. Loughlin (eds), *Medicine, the Market and the Mass Media: Producing Health in the C20th*, London: Routledge, pp. 192–217.

AstraZeneca organigrammes:

AZ DO759 organigramme (May 1, 1960).

AZ DO770 organigramme (May 1965).

AZ DO772 organigramme (March 18, 1971).

AstraZeneca organization memos:

AZ DO1415: S.I. 7/71 organization memo (February 8, 1971).

AZ DO1415: S.I. 16/71 organization memo (April 21, 1971).

AZ DO1415: S.I. 17/72 organization memo (August 22, 1972).

AZ DO1415: SI 3/79 organization memo (February 19, 1979).

ICI Ltd, Pharmaceuticals Division, Research Department Period Reports:

AZ CPR 3: A. Spinks, "Survey of new fields for pharmacological research" (March 24, 1953).

AZ CPR 109B: Physical Methods (1966–).

AZ CPR 50/6: A.F. Crowther and A. Spinks (eds) "Coronary artery disease, hypertension, rheumatism, allergy, bronchodilators, anaemia, atherosclerosis" (June 14, 1960).

AZ CPR 56, 56/8: A.F. Crowther and J.W. Black (eds), "Anticonvulsants, tranquillisers, stimulants, analgesics, anaesthetics, coronary artery disease, hypotensives" (January 26, 1962).

AZ CPR 56, 56/10: A.F. Crowther and J.W. Black (eds), "Anaesthetics, analgesics, anticonvulsants, hypotensives, and adrenergic beta-blocking agents" (April 26, 1963).

AZ CPR 56, 56/13: A.F. Crowther and R.G. Shanks (eds), "Analgesics, anaesthetics, anticonvulsants, psychotropics, β-adrenergic blocking agents, cardiotonic agents, hypotensive agents," November 13, 1964.

AZ CPR 99, 4B: R. Clarkson and R.G. Shanks (eds), "Cardiovascular Drugs" (February 18, 1966).

AZ CPR 99, 5B: R. Clarkson and R.G. Shanks (eds), "Cardiovascular Drugs" (May 25, 966).

AZ CPR 99, 7B: R. Clarkson and R.G. Shanks (eds), "Cardiovascular Drugs" (February 23, 1967).

AZ CPR 99, 12B: A.M. Barrett (ed), "Cardiovascular Drugs" (November 7, 1968).

AZ CPR 99, 13B: A.M. Barrett, R. Clarkson and R. Hull (eds), "Cardiovascular Drugs" (March 27, 1969).

AZ CPR 99, 19B: J.D. Fitzgerald (ed), "Cardiovascular Drugs" (March 10, 1971).

AZ CPR 99, 22B: F.J. Conway (ed), "Cardiovascular Drugs" (March 7, 1972).

AZ CPR 99, 24B: F.J. Conway (ed), "Cardiovascular Drugs" (November 7, 1972).

Barrett, A.M. (1972) "Design of β-blocking drugs," in A. J. Ariëns (ed), *Drug Design* New York: Academic Press, vol. 3, pp. 205–28.

Booth, C. (1989) "Clinical research," in J. Austoker and L. Bryder (eds), *Historical Perspectives on the Role of the MRC* Oxford: Oxford University Press, pp. 205–21.

Bud, R. (2008) "Upheaval in the moral economy of science? Patenting, teamwork and the World War II experience of penicillin," in J.-P. Gaudillière (ed.) special issue of *History and Technology*, 2: 173–90.

Cook, J.G. (Nov. 1947) "Penicillin at Trafford Park," *ICI Magazine*.

Corley, T.A.B. (2003) "The British pharmaceutical industry since 1851," in L. Richmond, J. Stevenson and A. Turton (eds), *The Pharmaceutical Industry: A Guide to Historical Records*, Aldershot: Ashgate, pp. 14–32.

Daemmrich, A. (2002) "A tale of two experts: Thalidomide and political engagement in the United Sates and West Germany," *Social History of Medicine*, 15: 137–58.

Daemmrich, A. (2004) *Pharmacopolitics: Drug Regulation in the United States and Germany*, Chapel Hill/London: University of Carolina Press.

Dutfield, G. (2009) *Intellectual Property Rights and the Life Science Industries: Past, Present and Future*, Singapore: World Scientific Publishing, 2nd edn.

Edgerton, D.E.H. (2006) *Warfare State: Britain, 1920–1970*, Cambridge: Cambridge University Press.

Gaudillière, J.-P. (2008) (ed), *History and Technology*, Special Issue 24.

Greene, J.A. (2007) *Prescribing by Numbers: Drugs and the Definition of Disease*, Baltimore: The Johns Hopkins University Press.

Hancher, L. (1990) *Regulating for Competition: Government, Law and the Pharmaceutical Industry in the United Kingdom and France*, Oxford: Clarendon Press.

Holland, K. (Sept. 12, 1987) "IC Pharmaceuticals," *Pharmaceutical Journal*, 286–88.

Hutt, P.B. (2007) "Turning points in FDA history," in A. Daemmrich and J. Radin (eds), *The FDA at 100: Perspectives on Risk and Regulation*, Philadelphia: Chemical Heritage Foundation, pp. 14–25.

Johnson, A.W. , Rose, F.L. and B.L. Langley (1984) "Alfred Spinks, 1917–1982," *Biographical Memoirs of Fellows of the Royal Society*, 30: 567–94.

Keith, S.T. (1981), "Inventions, patents, and commercial development from governmentally financed research in Great Britain: The origins of the National Research and Development Corporation," *Minerva*, 19: 92–122.

Kennedy, C. (1986) *ICI: The Company That Changed Our Lives*, London: Hutchinson.

Kohler, R.E. (1982) *From Medical Chemistry to Biochemistry: The Making of a Biomedical Discipline*, Cambridge: Cambridge University Press.

Liebenau, J. (1987) "The British success with penicillin," *Social Studies of Science*, 17: 69–86.

Liebenau, J. (1989) "The MRC and the pharmaceutical industry: the model of insulin," in J. Austoker and L. Bryder (eds), *Historical Perspectives on the Role of the MRC*, Oxford: Oxford University Press, pp. 163–80.

Marks, H. (1997) *The Progress of Experiment: Science and Therapeutic Reform in the United States, 1900–1990*, Cambridge/New York: Cambridge University Press.

Marks, L. (1999) "Not just a statistic: The history of USA and UK policy over thrombotic disease and the oral contraceptive pill," *Social Science and Medicine*, 49: 1139–55.

Prichard, B.N.C. and P.M.S. Gillam (1964) "Use of Propranolol (Inderal) in treatment of hypertension," *British Medical Journal*, 2 (5411): 725–27.

Quirke, V. (2004) "Bridging boundaries: The evolution of the Scientific Services Group, and the growth of pharmaceutical research at ICI ca 1955–75," European History Association annual meeting, Barcelona, September 16–18, 2004 (EBHA website: www.econ.upf.edu/ebha2004).

Quirke, V. (2005a) "War and change in the pharmaceutical industry: A comparative study of Britain and France in the twentieth century," *Entreprises et Histoire*, 36: 64–83.

Quirke, V. (2005b) "From evidence to market: Alfred Spinks' 1953 survey of fields for pharmacological research, and the origins of ICI's cardiovascular programme," in V. Berridge and K. Loughlin (eds), *Producing Health: Medicine, the Market and the Mass Media in the C20th*, London: Routledge, pp. 144–69.

Quirke, V. (2006) "Putting theory into practice: James Black, receptor theory, and the development of the beta-blockers at ICI," *Medical History*, 50: 69–92.

Quirke, V. (2008) *Collaboration in the Pharmaceutical Industry: Changing Relationships in Britain and France, ca 1935–1965*, London/New York: Routledge.

Quirke, V. (2009a) "Standardising pharmaceutical R&D in the second half of the twentieth century: ICI's Nolvadex Development Programme in historical and comparative perspective," in C. Bonah, C. Masutti, A. Rasmussen and J. Simon (eds), *Harmonizing Drugs: Standards in 20th-Century Pharmaceutical History*, Paris: Glyphe, pp. 132–45.

Quirke, V. (2009b) "The material culture(s) of the British pharmaceutical laboratory in the Golden Age of Drug Discovery," *International Journal for the History of Engineering and Technology*, 79: 298–317.

Quirke, V. (2009c), "Developing penicillin, patenting cephalosporin, and transforming medical research in Britain: The Sir William Dunn School of Pathology at Oxford, 1930s–1970s," in J.P. Gaudillière, D.J. Kevles, and H.-J. Rheinberger (eds), *Living Properties: Making Knowledge and Controlling Ownership in the History of Biology*, Max Planck Institute for the History of Science, Preprint 382, pp. 65–73.

Reader, W.J. (1975) *Imperial Chemical Industries: A History*, London: Oxford University Press, vols I and II.

Sexton, W.A. (1963) "The research laboratories of the pharmaceutical division of ICI," *Chemistry and Industry*, March 3: 372–77.

Shanks, R.G. (1984) "The discovery of beta-adrenoceptor blocking drugs," *Trends in Pharmacological Sciences*, 5: 405–09.

Sismondo, S (2010) "Linking research and marketing, a pharmaceutical innovation," in V. Quirke and J. Slinn (eds), *Perspectives on 20th-Century Pharmaceuticals*, Oxford: Peter Lang, pp. 241–56.

Sinding, C. (2002) "Making the unit of insulin: Standards, clinical work, and industry, 1920–1925," *Bulletin of the History of Medicine*, 76: 231–70.

Slinn, J. (2005), "Price Controls or Control through Prices? Regulating the cost and consumption of prescription pharmaceuticals in the UK 1948–1967," *Business History*, 47: 352–66.

Slinn, J. (2008), "Patents and the UK pharmaceutical industry between 1945 and the 1970s," in J.-P. Gaudillière (ed.) special issue of *History and Technology*, 24: 191–205.

Spinks, A. and A. Walpole (eds) (1958) *A Symposium on the Evaluation of Drug Toxicity*, London: Churchill.

Suckling, C.W. and B.L. Langley (1990) "Francis Leslie Rose," *Biographical Memoirs of fellows of the Royal Society*, 36: 491–524.

Swann, J.P. (2010) "Reducing with Dinitrophenol: Self-medication and the challenge of regulating a dangerous pharmaceutical before the US Food, Drug, and Cosmetic Act," in V. Quirke and J. Slinn (eds), *Perspectives on 20th-Century Pharmaceuticals*, Oxford: Peter Lang, pp. 285–302.

Tansey, E. M. and L. A. Reynolds (1997) (eds), "The committee on safety of drugs," *Wellcome Witnesses to Twentieth Century Medicine*, London: Wellcome Trust, vol. 1, pp. 103–32.

Thomas, L.G. (1994), "Implicit industrial policy: The triumph of Britain and the failure of France in global pharmaceuticals," *Industrial and Corporate Change*, 3: 451–89.

Timmermann, C. and Valier, H. (2008), "Clinical trials and the reorganization of medical research in post-Second World War Britain," in V. Quirke and J.-P. Gaudillière (eds), "The Era of Biomedicine," special issue of *Medical History*, 52: 493–510.

Vos, R. (1991) *Drugs Looking for Diseases: Innovative Drug Research and the Development of the Beta-Blockers and the Calcium Antagonists*, Dordrecht: Kluwer Academic.

Wells, N. (1983) *Pharmaceutical Innovation: Recent Trends, Future Prospects*, London: OHE.

Woodbridge, J.A. (1981) "Social aspects of pharmaceutical innovation: Heart disease," Ph.D. thesis, University of Aston, Birmingham.

7
What's in a Pill? On the Informational Enrichment of Anti-Cancer Drugs

Alberto Cambrosio, Peter Keating, and Andrei Mogoutov

Introduction

The May 28, 2001, cover of *Time* splashed the following headline: "There Is New Ammunition in the War Against Cancer: These Are the Bullets."* Color-coded in orange, the words *ammunition, cancer,* and *these* pointed to a tiny pile of orange pills identified in a smaller headline: "Revolutionary new pills like GLEEVEC combat cancer by targeting only the diseased cells. Is this the breakthrough we've been waiting for?" Two years later, Daniel Vasella, Chairman and CEO of Novartis, the maker of Gleevec, reiterated both the color theme and the war metaphor in the title of his book *Magic Cancer Bullet: How a Tiny Orange Pill Is Rewriting Medical History* (Vasella, 2003).[1]

The story of Gleevec and other "magic bullets" draws on a common understanding of (anticancer) drugs as specific agents, bearing immutable and, often, heroic properties. Laws that award patents for individual substances reinforce this perception. And yet, as we will see in this chapter, drugs express somewhat more transitory properties tied to the evolution of treatment regimens and diseases. For instance, the FDA initially approved Gleevec for treating patients in three different stages of chronic myeloid leukemia: blast crisis, accelerated phase, and chronic phase "after failure of interferon treatment" (Press Release, 2001). After 1999, trials implicating Gleevec "mushroomed" to include "a variety of cancers that share common molecular abnormalities."[2] In other words, not only does a drug often redefine the boundaries of the diseases that it targets (in this case "molecular abnormalities" replaced older pathological criteria) and is in turn redefined by those reconstituted entities, but the definitions of disease entities are also subject to modification as a result of the concurrent activity of other substances, as in "after failure of interferon treatment."

In this regard, drugs can be understood as research objects, rather than unproblematic tools for the management of disease, just as they can also be described as industrial products and commercial goods. In short, drugs experience multiple "modes of existence" (Simondon, 1958). How, then, do specific compounds cut through experimental, medical, industrial, commercial and legal regimes in the course of their existence? Underlying this seemingly unproblematic question is a portrait of drugs as little more than chemical compounds that maintain their identity through different drug-making and drug-using practices. Borrowing from Andrew Barry's discussion of chemical substances as "informed material," we can go beyond this uneasy mix of (soft) social constructivism and realism, and track the circulation and testing of therapeutic substances as a process of "progressive informational enrichment" (Barry, 2005; Gomart, 2002).

Barry's original formulation construes chemical substances as "constituted in their relations to informational and material environments," that is, by a *space* defined by the distance between their emerging properties and the properties of the models used in deriving those properties, and by their relation to competitors in the legal and economic environment in which they are deployed and to whose definition they contribute. Barry notes that the notion of a *chemical space* is a native category, routinely used by chemists and pharmacologists. His observation is supported by recent contributions that mobilize this notion in a metaphorical and literal sense. Metaphorically when researchers refer, for instance, to the "global mapping of pharmacological space" and the "pharmacological target space" defined by the "promiscuity" between compounds. And literally, when they use network analysis software to represent the interactions between substances as a two-dimensional map displaying clusters of related molecules (Paolini *et al.*, 2006).

The informational enrichment of anticancer drugs provides an excellent opportunity to explore the evolving status of pharmacological substances. Our purpose in this chapter is not to strike an ontological pose by merely claiming *that* drugs are informed material. Rather, we want to understand *how* the enrichment process operates and *how*, based on this understanding, it is possible to describe the practices that produce new drugs. Randomized clinical trials have become *the* obligatory passage point for the translation of a given substance into a legitimate prescription drug (Daemmrich, 2004; Marks 1997). In the case of cancer, the situation is somewhat more complex since clinical trials are more than a simple test or a regulatory epiphenomenon: they constitute a new style of practice that defines the experimental activities of a new brand of clinicians known as medical oncologists (Keating & Cambrosio, 2007; 2012). The development of anticancer drugs has, in this process, moved "from a low-budget, government-supported research effort to a high-stakes, multi-billion dollar industry" (Chabner & Roberts, 2005: 65; DiMasi & Grabowski, 2007) and displays considerable organizational and epistemic discontinuity. In brief, cancer clinical trials are the platform *from which* the pharmacological and therapeutic space of anticancer drugs arises.

Indeed, the common understanding of cancer trials as a simple regulatory device for testing the safety and efficacy of a substance against cancer misleads on two counts: first because clinical trials coproduce the diseases against which a particular regimen has been devised, and, furthermore, because clinical trials do not operate on substances per se but on *regimens* deployed within *protocols*. The aforementioned coproduction refers both to the production of new nosological categories (e.g., a given kind of cancer is shown to consist, in fact, of two distinct patho-biological entities that respond differently to similar treatments) and to the production of subdivisions within a given nosological category. For instance, one of the early chemotherapy protocols used in the treatment of leukemia, developed in the early 1960s, innovated by treating different phases of the disease as separate stages of disease. In particular, the trial protocol defined remission and relapse in a single patient as independent events that, nonetheless, afflicted the same person (Keating & Cambrosio, 2002). By replacing the natural history of a disease with its treated history, protocols like this uncovered new targets for therapeutic substances, modifying, in the process, their therapeutic space. In addition and as previously suggested, contemporary protocols often define their target patient population (or their target disease) on the basis of prior treatment with other substances as in "Tamoxifen-treated, node-negative breast cancer" or "Desatinib in Imatinib-resistant, Philadelphia Chromosome-positive leukemias" (Paik *et al.*, 2004; Talpaz, 2006).

Figure 7.1 shows the number of papers listed in the *PubMed* database that report results of clinical trials using a selection of anticancer drugs. The constant growth in the number of papers over 40 years is obviously not due to an increase in the number of clinical trials devoted exclusively to a given substance but to the fact that a large number of trials explore the effects of specific dispositions of substances known as *regimens*. The latter vary according not only to the quantities administered and the modes and schedules of administration, but also to the type and number of substances they mobilize. Two different regimens may deploy the same substance with different routes of admission (e.g., oral vs. intravenous) or dose size. Similarly, two different regimens may use different combinations of substances.

Regimens are not stand-alone devices: they are embedded in *protocols*. For instance, a protocol may stipulate that the two regimens will be tested on patients suffering from "metastatic breast carcinoma." That qualification thus diversifies the space of therapeutic experimentation. The space can be further enlarged through the addition of qualifications such as the stage of a patient's disease trajectory as well as a growing number of parameters (from the traditional anatomical–pathological parameters to the more

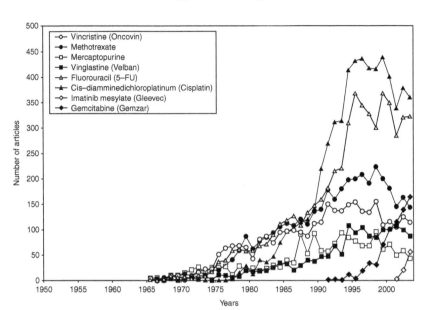

Figure 7.1   Number of articles listed in PubMed reporting the results of clinical trials with a selection of anticancer drugs. Reprinted with kind permission from P. Keating & A. Cambrosio, *Cancer on Trial: Oncology as a New Style of Practice* (Chicago: The University of Chicago Press, 2012). © 2012 The University of Chicago Press

recent molecular parameters) that define the prognostic status of a patient. These strictures continuously integrate the results of previous interventions. The comparison of two different regimens can be performed, for example, as part of a protocol that targets a first-line (standard) chemotherapy, or an adjuvant chemotherapy (i.e., after surgical removal of the tumor), or other forms of chemotherapy (before surgery, postrelapse, etc.).[3]

In their search for a magic bullet (Vasella, 2004),[4] oncologists may often foreground a given substance. They may claim, for example, that "Cisplatin has become the gold standard treatment against cervical cancer in combination with radiotherapy" (Pasetto, 2006: 67). While the importance of Cisplatin is undeniable, the foregoing quote contains a significant simplification as the same paper goes on to examine the differential qualities of an entire spectrum of Cisplatin analogues and their combination with other anticancer drugs. In other words, trialists talk substances but practice regimens. Given these variables, the number of possible regimens and of actual trials is quite large, and accounts for much of the pattern of growth displayed in Figure 7.1. This view is further confirmed by Figure 7.2, showing the number of new anticancer molecules and the number of claims for those molecules approved by the FDA. The number of molecules is quite

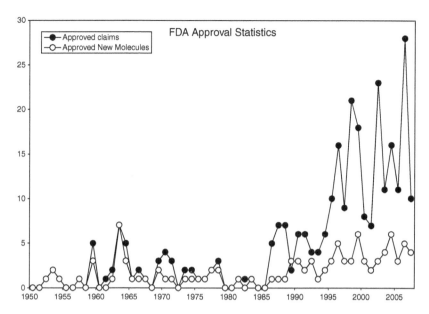

Figure 7.2 Number of new molecules and claims for antineoplastic agents approved by the FDA (Data source: FDA Web site). Reprinted with kind permission from P. Keating & A. Cambrosio, *Cancer on Trial: Oncology as a New Style of Practice* (Chicago: The University of Chicago Press, 2012). © 2012 The University of Chicago Press

small when compared to the number of trials carried out by oncologists over the last four decades. Moreover, after the mid-1980s the number of claims (diseases and disease stages for which the molecule has been approved) easily outstrips the number of new molecules. In the late 1980s, of the approximately 30 commercially available anticancer drugs, five accounted for about 70% of sales (Laker, 1989: 80). A decade later 80% of the standard treatments used only 15 out of the roughly 60 major drugs then available (Cvitkovic-Bertheault, 2000).[5] Thus, the "same" substance can be simultaneously engaged in a variety of trials, from which it will emerge with different qualities and qualifiers: its therapeutic space is thus a function of past and future trials and the regimens that those trials test.

The nature of chemotherapy and thus the structure of the therapeutic space within which anticancer agents are deployed have evolved significantly since the beginning of chemotherapy in the 1950s. We can get a rough idea of this evolution by borrowing from the visualization methods used by scientists. Figure 7.3 shows the names of the most common anticancer drugs from the over 40,000 cancer clinical trial references listed in PubMed, arranged by year of publication.[6] The map displays a neat chronological sequence. At the top and the bottom of the map we see sets of

Figure 7.3   Co-occurrence between anticancer agents (substances and categories) and years in cancer clinical trial articles listed in PubMed. Reprinted with kind permission from P. Keating & A. Cambrosio, *Cancer on Trial: Oncology as a New Style of Practice* (Chicago: The University of Chicago Press, 2012). © 2012 The University of Chicago Press

substances corresponding, respectively, to the emergence of chemotherapy and to the most recent period, characterized by the appearance of targeted therapies such as imatinib (Gleevec). The intermediary period corresponds to the staging of large phase III trials testing seemingly endless combinations of a relatively small number of substances. At the center of the map lie a few anticancer drugs (e.g., Fluorouracil, Cisplatin, Cyclophosphamide) that are connected to most years (except the initial ones) and thus appear to form the backbone, so to speak, for the majority of regimens developed since the 1980s.

While Figure 7.3 displays the general coordinates of the therapeutic space of anticancer compounds, it maintains the focus on individual substances. Figure 7.4 allows us to go further by displaying the evolving associations that one of the aforementioned key drugs, Cisplatin, established with fellow substances. Figure 7.5 does the same for the changing relations of Cisplatin

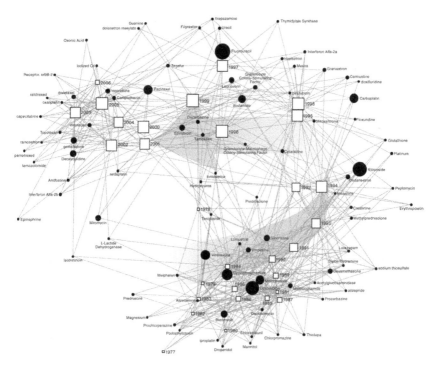

Figure 7.4 Anticancer agents associated with Cisplatin in cancer clinical trial articles listed in PubMed, 1980–2006. Reprinted with kind permission from P. Keating & A. Cambrosio, *Cancer on Trial: Oncology as a New Style of Practice* (Chicago: The University of Chicago Press, 2012). © 2012 The University of Chicago Press

with different diseases. At the center of the map we see a group of major cancer categories (e.g., lung cancer, breast cancer) against which Cisplatin has been used for many years in combination with other substances, while the most specific subcategories (e.g., metastatic adenocarcinoma of the lung, locally advanced breast cancer, etc.) targeted by regimens containing Cisplatin are displayed on the external layer of the map and are linked to the year(s) during which the results of those particular studies were published.

Relational maps sketch the outlines of the therapeutic space, but they cannot capture its more subtle dynamics. Take, for instance, the notion of a *therapeutic index* that refers to the ratio between the toxic and the therapeutic dose of a given drug and thus measures its relative safety. Researchers synthesized and tested thousands of Cisplatin analogues in the attempt to enhance the therapeutic index and other parameters such as the reduction of cross-resistance between the components of a regimen (Pasetto *et al.*, 2006: 60). Cross-resistance is obviously a relational parameter, but so, too, is the therapeutic index because the maximum tolerated dose has changed over

Figure 7.5   Diseases associated with Cisplatin in cancer clinical trial articles listed in PubMed, 1980–2006. Reprinted with kind permission from P. Keating & A. Cambrosio, *Cancer on Trial: Oncology as a New Style of Practice* (Chicago: The University of Chicago Press, 2012). © 2012 The University of Chicago Press

time due both to the simultaneous use of other substances such as growth factors and antibiotics that can rescue a substance excluded from the therapeutic space by counteracting its toxic side effects, and to the oncologists' adoption of more aggressive therapies (Cvitkovic-Bertheault, 2000: 816). Other factors that reshape the therapeutic space include the aforementioned transformation of the natural history of a given cancer into a *treated* history. Previously administered anticancer drugs modify the biological parameters of tumors, opening up different avenues of intervention for other drugs. The evolution of other treatment modalities (surgery, radiotherapy) also play a role, so that the transformation of the natural history of certain cancers can be ascribed both to the use of a particular drug combination and to the evolution of surgical techniques that redefine resection criteria (Cvitkovic-Bertheault, 2000: 815).

Substances and regimens in the 1960s and 1970s

We can further chart the evolution of the therapeutic space of anticancer agents by taking a closer look at experimental practices that generate, test, and evaluate compounds, regimens, and protocols. Let us begin in the early 1960s. In the drug procurement and testing system of the US National Cancer Institute (NCI) of that time, chemical and natural compounds moved initially through a screening system designed to allow them to express activity in a sequence of tests with various animal models (Keating & Cambrosio, 2006: 181–208). The tiny fraction of substances that survived these tests underwent further examination in clinical studies. The travel time from submission to the beginning of a clinical trial for a typical compound was four years (Zubrod *et al.*, 1966: 469) during which the fate of the compound devolved to the NCI's Decision Network Committee (DNC), entrenched within an organizational structure termed the "Linear Array." The Linear Array – an idealization of the NCI's chemotherapy program – was represented by chemotherapy program administrators in a detailed flow chart that described a series of parallel and sequential flows of knowledge and substances. The flow began with the development of new drugs and moved ineluctably toward "the determination of role of agents" in clinical trials (Zubrod *et al.*, 1966: 537; DeVita *et al.*, 1979: 200). While most of the members of the DNC were intramural staff from the NCI's Division of Cancer Treatment (DCT), the Committee also included industry and university representatives. Despite the representation of the process as a *linear* array, substances did not always follow a straight line from the beginning to the end of the evaluation process. As noted by the DCT Director, "[s]ome compounds may actually be at several stages *at the same time*."[7] At each of these (possibly simultaneous) stages, compounds became increasingly qualified and diversified according to a variety of chemical and biological parameters, until they finally entered human trials where, as shown below, they assumed multiple identities depending on the therapeutic space in which they were deployed.

As previously noted, in the case of anticancer agents the operative term is less substance than regimen. Regimens are described, at the simplest level, by the dose size, schedule, and mode of administration of individual substances, and, quite often, they consist of a combination of different substances.[8] Determining whether or not to deploy a given regimen, at which stage of a disease, that is, first occurrence, relapse, metastasis, and in conjunction with what other modalities – surgery and/or radiotherapy – involves a complex set of "trials of strength" (B. Latour, 1987).

The establishment of a number of standard chemotherapeutic substances in the 1960s created new constraints for the evaluation of substances in the 1970s. Prior to entering a regimen, substances had henceforth to undergo a primary *information* test: could they be deemed an analogue of a substance already in use? If so, clinical evaluation differed from the evaluation of other substances insofar as "the analogue must always be compared in some way to its parent *clinical* [note that it doesn't say: *chemical*] structure while the new drug stands alone and need only search for its role in competition with currently available *regimens*" (Carter, 1987: 69 (our emphasis)). In such comparisons, the analogue might show a greater or wider spectrum of activity (namely, cell-kill) or diminished acute or chronic toxicity. Yet none of these end points could normally be tested by themselves: most substances in use were already deployed in a given regimen that included other compounds so that activity, for example, was not absolute but relative to a given combination. Since every combination was different, "[t]hese new combinations then become in many ways like a new drug again requiring a phase I and phase II process which is often covered under the rubric of pilot study" (Carter, 1987: 69).

Thus, in the case of analogues the standard phase I, II, III testing sequence inaugurated in the 1950s required modification in terms of planning and assessment. For instance, while the move from a phase I study to a phase II trial usually sought "to see if a reasonable activity level can be demonstrated," if one examined an analogue in order to find a broader spectrum of activity than the parent drug, then the phase II trial would focus on "unresponsive tumors to parent structure to see if any level of activity is seen," and the subsequent phase III trial would be used "to establish a level of activity." A different and more complex strategy emerged in the case of analogues designed to diminish acute or chronic toxicity. Here, the phase II trial sought "to establish a reasonable expectation of activity and to get an early feeling for chronic toxicity in responding patients given maintenance therapy on reasonable follow-up," while the phase III trial focused on establishing "comparative activity as against comparative chronic toxicity" (Carter, 1987: 71). It goes without saying that terms such as "reasonable" or "early feel" allowed trialists to employ a certain degree of interpretative flexibility.

Obviously, these strategies idealized complex trajectories. Consider, for instance, the life of CCNU (Lomustine) in the first half of the 1970s. An analogue of BCNU (Carmustine), CCNU was synthesized as part of the program to produce analogues of successful substances. Because of this status, researchers assumed for several years that CCNU would act like its parent substance.[9] It turned out, however, that the drug's action varied according to the disease encountered. In other words, while structure indicated activity, it did not define it and so clinicians were forced to conclude in 1974 that "the active form and mechanism of action of CCNU are unknown" (Wasserman *et al.*, 1974: 131).

When the Deputy Director of the DCT, Stephen Carter, reviewed the numerous clinical trials conducted using CCNU between 1971 and 1974, he divided the trials into the following categories: (a) alone against many tumors: phase II; (b) alone against a single tumor type: phase II and III; (c) in combination against many tumors: phase I and II; and (d) in combination against a single tumor type: phase I and II. Two features of this system are particularly striking. The first is that the unit category is not the substance but the regimen. The second is that from the point of view of the evaluation of a substance, the phase I–III trajectory is subsumed under a broader classification (one/many tumors; alone/in combination) that cuts across the phase categories. The phase distinctions, in other words and somewhat counterintuitively, sometimes describe the trajectory of a compound synchronically, not diachronically.

Broad phase I–II trials of combination chemotherapy screened regimens in humans using patients with advanced disease that had failed to respond to conventional therapies. They gathered together many diseases within a single protocol. For instance, the summary table of a typical study of CCNU in combination with Adriamycin, published in 1973, listed a series of tumors (breast, melanoma, etc.) and their breakdown by the type of response (complete or partial remission, no change, progression) elicited by the combination treatment. A similar table presented an overall evaluation of CCNU as a single agent. The next step in this process assessed the comparative activity of CCNU vis-à-vis other anticancer drugs in different subspecies of a given type of cancer. Finally, came a list of "promising drug combinations" of CCNU and other agents, and of the "target tumors" against which each of these combinations should be tested (Wasserman *et al.*, 1974: 143–49). Such was the therapeutic space of a compound like CCNU in the 1970s.

The 1970s also saw the emergence of an attempt to articulate chemotherapy with radiotherapy, immunotherapy, and surgery as part of a new practice called *adjuvant therapy*, thus further enlarging the therapeutic space of anticancer drugs. The integration of the different modalities followed from the "recognition that surgery and radiotherapy [had] reached a plateau in their ability to cure solid tumors" (Carter & Soper, 1974: 1), a statement that must be understood in the context of an era when:

> the primary treatment in solid tumors without disseminated disease is surgery and/or radiotherapy. Chemotherapy is relegated to secondary or tertiary use after the local modalities fail and the disease is advanced and disseminated. Since the secondary or tertiary therapy is rarely curative in any tumor, including hematologic malignancies, it is understandable that chemotherapy in solid tumors has not been curative although it has produced tumor regression, palliative benefit and some survival gains. (Carter & Soper, 1974: 3)

In an effort to move chemotherapy to the front of the therapeutic line, chemotherapists devised a strategy that consisted of testing regimens in late-stage patients and moving the successful ones forward until they reached the front line. Specifically, the program sought to test chemotherapy regimens in patients from the advanced-late disease stage, determine the optimum drug combination in patients with secondary recurrent or advanced disseminated disease, and use a combined modality approach that would henceforth include chemotherapy in patients from the first two stages (Carter & Soper, 1974: 4; Marx, 1989).[10] An idea of the size of the project can be gathered from a "cross-reference chart of drug–tumor interactions" whose rows and columns confronted two parameters: the 29 substances that had shown activity in solid tumors and the 16 most common solid tumors. The resulting matrix provided a striking, at-a-glance visual proof that a lot of work remained to be done, for 263 (57%) of the matrix's 464 cells contained the acronym NE ("Not Evaluated") and a further 37 (8%) the symbol for "inadequately tested" (Carter & Soper, 1974: 5). A few tumors, such as lung, colon, and breast cancer, had received considerable attention, while others (e.g., pancreas, stomach, prostate) had suffered comparative neglect. Substances could thus be ranked *in relation* to the number of tumors in which they had been "adequately" or "inadequately" tested (Carter & Soper, 1974: 7).

The notions of adequate and inadequate obviously open the door to a number of interpretations of what any given trial signifies with the evolving therapeutic space of substances and their regimens. The "Cooper regimen" provides a useful example of the levels of interpretation implicated in this exercise. The story began in 1969 when Richard Cooper and his group at the Buffalo General Hospital presented the results of a five-drug antibreast cancer regimen involving a mix of oral and IV routes of administration and specific dosages following daily and weekly schedules. Cooper reported a spectacular 90% complete response rate in advanced breast cancer patients who had developed resistance to hormone therapy (Cooper, 1969). Cooper's regimen created considerable buzz and, almost immediately, six different NCI Cooperative Oncology groups set out to test variations of the regimen (Carter & Livingston, 1975: 300).

By adding and subtracting substances – for instance, L-Phenylalanine Mustard (aka L-PAM or Alkeran) – and by devising different dose schedules, researchers created a number of derivative combinations. A brief glance at the fate of these combinations and their varying rationales shows that the interpretation of trial results invariably takes place within the context of past, present, and future trials. Consider one of the six protocols derived from the Cooper regimen, ECOG (Eastern Cooperative Oncology Group) 0971.

Launched in 1971, the trial had enrolled 183 patients with metastatic breast cancer and who, in most cases, had previously failed surgery or hormonal therapy. The protocol compared two regimens: a single drug (L-PAM) regimen and a three-drug combination regimen (CMF). Some researchers suggested that the object of ECOG 0971 was yet another test of the Cooper regimen (Carter & Livingston, 1975: 300). The sponsors of the trial, the Eastern Cooperative Oncology Group, however, offered an alternative description and rationale, stating that it was "a study designed to evaluate Alkeran [L-PAM] versus a three-drug combination chemotherapy" and added that the study had "conclusively [shown] that the combination chemotherapy approach was better than the single agent" (ECOGa). Although the results remained unpublished until 1976, the fact that both arms showed some suppression of diffuse disease led to a stage II experiment, namely, the use of chemotherapeutic substances – singly and in combination – as an adjuvant to surgery (Canellos *et al.*, 1976).

Indeed, just as ECOG 0971 had set out to compare L-PAM to combination chemotherapy, ECOG and the National Surgical Adjuvant Breast Project (NSABP) subsequently set about comparing L-PAM against a placebo as adjuvant therapy following radical mastectomy (Fisher *et al.*, 1975). At first sight, such an enterprise would seem somewhat paradoxical. Why would some members of ECOG test L-PAM against a placebo while others tested the same compound against a combination of substances? If ECOG 0971 made any sense at all, then L-PAM was at least better than a placebo. The apparent paradox disappears, however, if we consider that the ECOG/NSABP tandem worked on an entirely different construal of both the disease and the therapy. In particular, they used L-PAM as adjuvant in an *early combined modality therapy*. This meant that while administered after radical mastectomy, the chemotherapy intervened prior to metastasis. Intervening shortly after surgery also meant that this use of L-PAM qualified it as an adjuvant as opposed to primary therapy. In consequence, the ECOG/NSABP protocol invoked a principle entirely foreign to 0971, namely, that "regimens with moderate effectiveness may be highly accentuated when combined with surgery and/or radiotherapy" (ECOGb).

In this regard then, the ECOG/NSABP team viewed its trial in a way that placed it in an entirely different perspective. Whatever might be accomplished in advanced breast cancer, however suggestive, did not automatically translate to adjuvant therapy. Interviewed by a reporter from *Science* in 1975 with regard to the trial, the head of the Group explained that "we are going about this in a very orderly manner. First we tested a drug versus nothing, then one drug versus two. Now, we'll look at other combinations. The point is to find the minimal treatment that will do the job with minimal toxicity" (Culliton 1976: 1030). In some respects, the ECOG/NSABP study was a success insofar as it continues to be cited as the one of the first "in principle" proofs of the utility of adjuvant chemotherapy, even though a 30-year controversy has ensued (Levine & Whelan, 2006).

L-PAM's movement and the interpretation of its movement in therapeutic space was further complicated by the fact that in 1974, that is, before the publication of the ECOG/NSABP study, a team of Italian trialists from Milan financially supported by the NCI, rather than testing L-PAM against a placebo, chose to test the CMF combination against a placebo. Results published barely a year after the publication of the ECOG/NSABP results and in the same journal (Bonadonna *et al.*, 1976) stated that the three-drug combination had dramatically reduced recurrences and were hailed as being of "monumental importance" (Holland, 1976). Picked up by such popular weeklies as *Times* and *Newsweek* under headlines such as "Spectacular Hope" and "Breakthrough," enthusiasm waned when *Science* published an article entitled "Reports of New Therapy Are Greatly Exaggerated," in which NCI officials agreed that although the study was good, it did not rise to the level of "monumental" (Culliton, 1976: 1029).

In the meantime, 14 French cancer centers had entered the fray in 1976 devising a trial to evaluate adjuvant therapy in breast cancer as a whole using patients stratified according to primary treatment (mastectomy, local resection, with or without radiotherapy) and extent of disease following treatment (number of positive lymph nodes) and then randomizing patients according to five different chemotherapeutic regimens including L-PAM alone. While the results of the trial are not pertinent to our present discussion, it is worth noting that they were sufficiently inconclusive to open the possibility for further study (Auclerc *et al.*, 1979: 100).

What is the lesson for us here? Clearly, any clinical trial admits of a number of descriptions and here we refer not simply to the statistical results but to the significance of the trial within the totality of trials deemed relevant from a variety of research and therapeutic perspectives. From a narrow substance perspective, ECOG 0971 tested L-PAM in advanced breast cancer and found it wanting as compared to CMF. This view, however, is clearly not the whole story. From the central perspective of the NCI and its clinical research/screening viewpoint, ECOG 0971 constituted a test of combination chemotherapy for solid tumors and built on previous success in hematologic malignancies. From within the Cooperative groups themselves, although 0971 applied to advanced cancer, the study was ultimately meant to pave the way for adjuvant chemotherapy and fit comfortably into the groups' ongoing effort to launch the multimodal era by integrating the different modalities into group research programs. Finally, from our point of view, 0971 is a typical case of all of the above; it is typical of a new style of practice that tests regimens within a space of substances, practices, and diseases that clearly defies reduction to chemical structures and yet is not a matter of mere "interpretation," insofar as it alters the material properties of the agents involved.

A new form of informational enrichment: The molecular turn

In the early 1990s the overwhelming evidence that a plateau had been reached in the discovery of cytotoxic drugs led researchers and clinicians to invest in the development of new classes of biological substances derived from natural sources. Shortly thereafter, however, they shifted gears once again and entered the promised land of molecularly targeted therapies (Eastman & Perez, 2006). Some confusion surrounds the notion of targeted or molecular therapy. It commonly refers to the development of novel drugs that, regardless of their chemical nature (antibodies, small molecules, etc.), have been designed to interfere with specific molecular targets within the biochemical pathways that allow cancer cells to thrive. It is also used, however, to refer to the selective use of existing cytotoxic regimens for patients whose tumor's genetic profile or signature has been determined to be particularly responsive to that particular regimen (Simon, 2006; Weinstein & Pommier, 2006). In the first case, informational enrichment results from the unexpected behavior of both substances and tumor cells in human trials that counters all "reasonable" and carefully argued expectations of efficacy (Secko, 2006; Hedgecoe & Martin, 2003). The second case yields yet another step in the informational enrichment of traditional substances, insofar as patients (or, rather, patient subgroups based on the "signature" of their tumors) are now selected for substances, and not the other way around. Promoters of targeted therapies can henceforth blame the apparent failure of otherwise promising drugs on the choice of "inappropriate patient populations" (Eastman & Perez 2006: 13).

As a consequence of the molecular turn, by 1999 sifting through natural products looking for active substances that prohibited cancer cell growth had become passé. Researchers now sifted through mechanisms and pathways.[11] Research had, to some extent, replaced screening. Molecular biology provided the tools for identifying the molecular abnormalities that made cancer cells "behave badly": these abnormalities then became targets for concerted research efforts and the increasing number of small and large biotech and pharmaceutical companies in the redefined cancer market.[12] In 2000, and in order "to promote a more molecular-oriented, knowledge-based approach to drug discovery and development" the NCI initiated a "Molecular Targets for Drug Discovery Program." The program offered funding not only to university-based investigators but also to small biotech companies for research on promising anticancer targets. In a sense, the initiative inverted the sourcing program; instead of combing through thousands of natural and synthetic compounds looking for one that showed evidence of cell-kill activity for yet unknown reasons, it was now a question of combing through thousands of potential molecules implicated in the cancer process looking for one that was sufficiently strategic to become a target.

The production of anticancer drugs has thus experienced a profound epistemic *and* economic transformation in the past 10–15 years, turning a once marginal field into a major pharmaceutical domain. Such growth has obviously affected the production and evaluation of new anticancer drugs. From the point of view of the present chapter, the interesting twist in this process has been the transformation of the sequential process of informational enrichment, and, in particular, of the institutional and organizational arrangements that define it. Clinical research networks that focused on phase II/III trials have been destabilized by the growing importance attributed to organizations specifically devoted to phase I trials. The EORTC (European Organisation for Research and Treatment of Cancer) is a case in point.[13]

Established in the early 1960s as a European network of trialists, the EORTC was modeled on US institutions and, like its US counterparts, it grew during the 1970s and the 1980s as an organization devoted principally to the conduct of large, multicenter phase III trials of combination chemotherapy regimens. Back then, because of the small number of patients needed for phase I trials, the latter were conducted mostly within individual hospital centers. In the 1970s, several EORTC members established an Early Clinical Trials cooperative group (ECTG) to focus on phase I and early phase II trials, but this group remained marginal within EORTC. The ECTG, however, soon developed unique knowledge of the statistical and procedural requirements of early trials, and its own network of clinicians, data managers, and statisticians.

In the 1980s, the EORTC established a New Drug Development Office (NDDO) that was, in large part, an organic outgrowth of the ECTG. Phase I trials continued to be largely performed within a single clinical center but NDDO representatives visited each center to inspect and monitor the conduct of the trial. The NDDO actively procured new drugs through their contacts in academic and commercial institutions. NDDO's role as an intermediary between drug producers and trialists is epitomized by its organization of the NCI–EORTC symposia on new drugs in cancer therapy. Although the NDDO conducted its activities under the official EORTC label, the new organization quickly became a "state within the state." Located in Amsterdam ("the European capital of new drugs") (Anon., 1998a) rather than Brussels (where the EORTC has its headquarters), it developed an independent and distinct legal, financial, and bioclinical profile.

This arrangement fell apart in 1998: following more than two years of negotiations and the consequent emergence of "different approaches to the philosophy of their respective activities" (Anon., 1998b), the NDDO and the EORTC parted ways, offering contrasting justifications for their actions. The EORTC framed the issue as one of commercial versus independent research, whereas the NDDO saw itself as severing its *exclusive* relation with the EORTC. In contradistinction to the EORTC's self-image, the NDDO viewed the EORTC as just one of many organizations (cooperative groups and commercial institutions) in the European cancer field, rather than the European equivalent of the NCI. As a centralized and financially powerful federal organization, the NCI reigned over a complex system of cooperative groups, research laboratories, and related cancer institutions. This was obviously not the case for the EORTC. Prior to the NDDO/EORTC separation, other clinicians, some of whom had occupied important positions in the EORTC, established SENDO, the Southern Europe New Drug Organization

(Anon. 1996). The EORTC reacted by creating a New Drug Development Program (NDDP) (Lacombe *et al.*, 2002; Meunier &. Lacombe, 2003: 14–15). Attempts to establish a European drug development network to coordinate the activities of the EORTC, the UK-based Cancer Research Campaign and SENDO never really took off (Anon., 1999). While actors in these events had clear financial motives, it would be wrong to reduce their intentions to economic self-interest. And while there is no denying that the "political economy" of cancer drugs did indeed undergo a major transformation, this transformation cannot be dissociated from the redefinition of the informational enrichment process of anticancer drugs.

From the point of view of pharmaceutical companies, large phase III trials of anticancer regimens were better left to public and academic institutions such as the cooperative groups. One may wonder, however, why commercial producers would resort to public institutions to test new, targeted molecules in phase I trials rather than carrying out the trials themselves or outsourcing them to an expanding range of private contract research organizations (CROs) known to deliver quick results (Mirowski & Van Horn, 2005). Obviously, part of the answer lies in the additional weight carried by findings published by a prestigious scientific organization such as the EORTC. But there are at least two other issues at work here. First, phase I trials require that patients be closely monitored for adverse effects and thus be performed in specialized clinical centers.[14] Second, and more importantly, while CROs do indeed deliver quick and (as far as possible) positive results, they lack the skills and expertise to turn a phase I trial from a simple test answering a predefined question into an experiment producing, sometimes, unexpected but welcome results, and even less to generate ideas for testing possible innovative uses of a new drug.

To sum up, in recent years, as new anticancer drugs undergo a sequence of trials leading to their informational enrichment, they simultaneously transit through redefined organizational arrangements that are both a cause and a consequence of the present emphasis on targeted therapies. Because of the somewhat secretive environment surrounding the early development stage of new commercial molecules and because of the flexibility of small organizations (as contrasted with the large phase III network organizations),[15] dedicated institutions conduct the initial informational enrichment of anticancer drugs.[16] Meanwhile, the focus of the more traditional cooperative groups remains the subsequent phases of informational enrichment, involving combination regimens and large multicenter phase III trials, albeit redefined by the use of molecular biomarkers (Bogaerts *et al.*, 2006). Final approval of a substance and the design of appropriate regimens still require phase III trials.

Several subtler changes are also worth noting. Recall that during the 1970s and 1980s, once a drug was found to have an anticancer effect, laboratories looked for analogues hypothesizing that those with similar chemical structures would express the same antitumor activity and perhaps provide a better therapeutic index. Phase III trials subsequently tinkered with combinations of a limited number of substances and analogues hoping to produce small incremental victories in cancer treatments. Presently, what counts is less the chemical structure of the drug than the target in the tumor cell pathway. For instance, once a consensus formed around the identification of angiogenesis (tumor-induced blood vessel growth) as a suitable target for new drugs (Denekamp, 1993; Bicknell, 1994; Baillie, Winslet & Bradley, 1995), companies around the world nearly simultaneously initiated discovery research programs for the development of angiogenesis-inhibiting agents. This, as already noted, has contributed to refocusing clinical research on phase I/II trials as the platform from which promising new substances emerge. These new substances are then fed into the phase III machine that, having lost at least some of its autonomy, is consequently less isolated from the phase I/II process. The proliferation of targeted agents aiming at similar pathways or receptors also raises the intriguing possibility that the oncology drug market will shift from a cumulative system whereby new products are added to standard – of – care combinations, to one where new products act as direct substitutes for existing products (Santoni & Foster, 2007).

The success of products like Gleevec in early clinical trials in the late 1990s was interpreted as a signal that the targeted approach had finally left the laboratory and was about to produce a major restructuring of chemotherapy and the entire cancer drug discovery process. This has been the case in many respects. In 2003, 88% of phase III trials involved studies of molecularly targeted therapies as opposed to conventional therapies (Roberts *et al.*, 2003). The consequences of molecular targeting for the conduct of clinical trials have been explored in a number of review articles, opinion pieces, and plenary speeches from the podiums of numerous conferences in the field of clinical cancer research. The general opinion is that the kind of information that will enrich substances in the course of clinical trials is now different and that "the clinical evaluation of many novel agents will require us to engage in new thinking, making use of some of the elements of existentialist thought with respect to assigning purpose and meaning to the design, measurement and interpretation of data" (Eisenhauer, 1998: 1047).[17]

The calls for change draw on a number of the novel qualities of targeted agents. While traditional agents kill cells and reduce tumor burden, many of the targeted agents interfere with molecular mechanisms, keeping the tumor under control. While the promissory notes accompanying the development of the new agents anticipate a future where cancer will become a *chronic* disease, a more immediate consequence is that new informational enrichment paths are needed. New study designs call for the use of "time to tumor progression" rather than tumor reduction. Similarly, toxicity is no longer a necessarily appropriate means of measuring dose size, and more complex means of calculating a dose–response relationship must be found (Korn *et al.*, 2001). Moreover, given the mechanism of action of the new agents, there may simply not be enough time to evaluate all the possible combinations of patients and drugs according to the old system that required thousands of patients, hundreds of millions of dollars, and a decade of research per drug (Schilsky, 2002). Unfortunately, trials testing new variables have to solve several issues the least of which is that trials of this type are the most expensive. Thus, a review of the literature in 2004 showed that most phase I studies continued to use traditional end points (Parulekar & Eisenhauer, 2004). Finally, the definition and scope of, say, certain phase I trials have also been transformed. While it remains true that the goal of phase I is not to test for efficacy, and while it is indeed possible that the target of a drug will turn out to be different from the one initially described, it makes clinical sense to determine the recommended dose of a substance defined by a given target on patients whose tumors have the appropriate profile, rather than on generic cancer patients. This strategy can change the goal of the first trials of a new drug into one that not only evaluates toxicity but also generates preliminary evidence of efficacy, thus turning phase I into phase I/II trials.[18]

As with any major innovation, the consequences and proper use of the new agents have become the subject of speculation, debate, and controversy. To begin with, the advent of targeted therapy has led to a call for an acceleration of the entire drug development process. Here, molecular targeting connects with a recurring theme. In effect, the FDA had already opened up a speedier route to drug approval in the early 1990s in part in response to AIDS activism (Epstein, 1996). Rather than wait for a full-scale phase III trial for approval, the FDA decided to allow drugs submissions to use surrogate end points in the approval process. This meant that criteria other than the standard criteria first formulated in 1985 could form the basis for FDA approval (Johnson & Temple, 1985).[19] Since the FDA rule change, a significant proportion of anticancer agents have traveled the accelerated route (Johnsohn *et al.*, 2003). More recently, the FDA issued a Guidance document that promoted the notion of "phase 0" trials as a bridge between the initial isolation of a compound and its use in humans.[20] Defined as "nontherapeutic studies that assess whether a drug has had an impact on its intended target" (Bates, 2008), phase 0 is thus a novel regulatory entity that signals the advent of targeted therapies. The FDA Guidance did not set out precise rules on how to proceed with the novel arrangement. Instead, in a sort of Kuhnian turn, it offered three examples of what a phase 0 trial might look like and invited investigators to experiment with the "paradigm" outlined by the three exemplars.

While the opening of an alternate route for anticancer drugs may have accelerated movement along the production line, it did not, once again, obviate the need for phase III trials. First and foremost, accelerated approval leads to only temporary approval. Drug sponsors remain obligated to complete phase III trials in order to obtain final approval following accelerated approval. Second, the majority of drugs seeking approval will not produce definitive results in phase II trials (Roberts *et al.*, 2003). In order to save time, those sponsoring clinical trials sometimes curtail the phase II trial and enter directly into a phase III study in order to obtain FDA approval. This relatively popular strategy has also received tacit approval by the FDA. Finally, all the issues raised by combinations in the 1970s have not been banished by a magic bullet. With a few exceptions, blocking a single pathway does not cure cancer even in mice (Hu *et al.*, 2006). While for some trialists the solution to this conundrum lies in a combination of the old and the new, for instance, of receptor inhibitors and conventional chemotherapy agents, or even of combinations of the novel substances with old modalities such as radiotherapy (Ryan & Chabner, 2000), molecularly oriented pharmacologists see this as a proof that "translation of new therapies into the clinic is an iterative process, not unidirectional," that is, a back-and-forth movement between the laboratory and the clinical trial setting (Eastman & Perez, 2006: 13).

Conclusion

This chapter drew on the notion of informational enrichment in order to redefine the ontological status of anticancer substances as informed material. As a consequence, we avoided the Scylla and Charybdis of naive realism and social constructivism. Beyond the simplistic understanding of substances as characterized by inherent properties, albeit subjected to different interpretations, we offered a series of concrete examples of *how* substances changed their relational identity in the course of clinical trials. To argue today, for instance, that the anticancer agent Cisplatin is the same substance discovered in 1965 by Rosenberg *et al.* is, under our description, largely an anachronism since Cisplatin as we know it today is a significantly different substance from the Cisplatin that emerged from Rosenberg's laboratory more than 40 years ago. While, on some levels, the practice of clinical cancer research has remained essentially intact through the sea change of concepts and techniques running from the chemotherapeutic approaches of the 1960s and 1970s to the present-day era of targeted therapies, on other levels there has been a fundamental reordering of oncology research. In particular, the relationship between biology and the clinic has undergone drastic evolution. We investigated a key aspect of this transformation by following substances as they undergo parallel informational enrichment sequences. Informational enrichment practices are in constant flux and they involve the continual rearrangement of the complex set of elements – epistemic, institutional, organizational, financial, material – that constitute the cancer clinical trial networks.

Notes

* Research for this chapter was made possible by grants from the Canadian Institutes for Health Research (CIHR MOP-64372), the Fonds Québécois de la Recherche sur la Société et la Culture (FQRSC ER-95786), and the Social Sciences and Humanities Research Council of Canada (SSHRC 410-2002-1453). We would like to thank the participants in the Drug Trajectories V Workshop (Max-Planck Institute for the History of Science, 2006), and in particular Harry Marks, for their comments and suggestions.

1. Orange is featured in Gleevec advertisements: an old couple sits at a table outside what looks like a posh secondary residence; they are smiling, her head turned shyly away while her hand rests on his shoulder; on the table an orange gift package stands out in the otherwise black & white picture: "Time Is a Gift" explains an orange headline on top of the picture. The color of the pills has led to legal confrontations: see, e.g., *Shire v. Barr*, No. 02–3647 (3d Cir. May 23, 2003), and is also credited with therapeutic effects (de Craen et al., 1996).
2. National Cancer Institute. The Nation's Investment in Cancer Research: A Plan and Budget Proposal for Fiscal Year 2003 (Bethesda, MD, NIH Publication No. 02–4373), p. 31.
3. For a definition of the different kinds of chemotherapy see http://www.chemocare.com/whatis/important_chemotherapy_terms.asp, date accessed December 2006.
4. For recent examples of the continuing fortune of the magic bullet metaphor linked to the "turn of the century" work on Salvarsan by Paul Ehrlich (popularized by the 1940 movie *Dr. Ehrlich's Magic Bullet*) see Vasella (2003) and Mann (2004).
5. There are now hundreds of compounds waiting in the pipeline: see Box 2 in Chabner & Roberts, 2005, p. 70.
6. This and following maps (Figures 7.4 and 7.5) were created using the software package ReseauLu X2 (http://www.aguidel.com). A statistical algorithm was used to select only the more specific links. For a description of the methodology used to produce the maps see Cambrosio et al. (2006).
7. Memo from Director, DCT, NCI to Members, Decision Network Committee, "The Definition of the Role of the Decision Network Committee," October 10, 1978, Abraham Goldin Papers, Acc. 1997–021, National Library of Medicine (our emphasis).
8. A regimen is often known by an acronym (e.g., CMF) derived from the initial letters of the substances it deploys (e.g., Cyclophosphamide, Methotrexate, and Fluorouracil).
9. Individual anticancer drugs were classified according to their mechanism of action: alkylating agents, antimetabolites, mitotic inhibitors, and so on.
10. An interesting complication, in this respect, is that as soon as a given regimen became the recommended standard therapy, it was unethical to perform placebo-controlled trials to assess variants of or alternatives to that regimen.
11. National Cooperative Drug Discovery Groups for Cancer, RFA, http://grants.nih.gov/grants/guide/rfa-files/RFA-CA-99–010.html.
12. Interview with Dr. Elisabeth Eisenhauer, National Cancer Institute of Canada, Clinical Trials Group, Kingston, Ontario, August 30, 2006.
13. In addition to the literature cited, the next few paragraphs are based on information obtained from interviews carried out in 2005–06 with leading European oncologists directly involved in the reported events.

14. Interview with Dr. Franco Cavalli, Istituto Oncologico della Svizzera Italiana, Bellinzona, December 22, 2006; interview with Dr. Herbert M. Pinedo, Free University Medical Center, Amsterdam, December 7, 2006.
15. Interview with Dr. Jean-Claude Horiot, Centre Georges-François Leclerc de Lutte contre le Cancer, Dijon, May 16, 2005.
16. While we contrasted the role of academic networks such as SENDO with the role of commercial CROs, a number of commercial organizations have evolved a recognized expertise in the field. NDDO was acquired in 2005 by a CRO (INC).
17. The author defined existentialism as "a philosophical theory which emphasizes the existence of an individual as a free agent in determining his or her own development, purpose and meaning."
18. Interview with Dr. Eisenhauer.
19. Prior to this change, the FDA approved oncology drugs solely on the basis of tumor response rate.
20. http://www.fda.gov/cder/guidance/7086fnl.pdf, date accessed February 2009.

Bibliography

Anon. (1996) "A New Drug Office for Southern Europe," *Annals of Oncology*, 7, 983–84.
Anon. (1998a) "New Optimism at the New Drugs Conference," *Annals of Oncology*, 9, 789.
Anon. (1998b) "NDDO Splits from EORTC," *Annals of Oncology*, 9, 1145–6.
Anon. (1999) "Creation of a European Drug Development Network," *Annals of Oncology*, 10, 365.
G. Auclerc *et al.* (1979) "Analysis of a Co-operative Study of Adjuvant Chemotherapy in Breast Cancer," *Cancer Treatment Reviews*, 6 (Suppl.), 97–100.
C. T. Baillie, M. C. Winslet, and N. J. Bradley (1995) "Tumour Vasculature: A Potential Therapeutic Target," *British Journal of Cancer*, 72, 257–67.
A. Barry (2005) "Pharmaceutical Matters: The Invention of Informed Material," *Theory, Culture & Society*, 22, 51–69.
S. E. Bates (2008) "From the Editor," *Clinical Cancer Research*, 14, 365.
R. Bicknell (1994) "Vascular Targeting and the Inhibition of Angiogenesis," *Annals of Oncology*, 5 (Suppl. 4), 45–50.
J. Bogaerts *et al.* (2006) "Gene Signature Evaluation as a Prognostic Tool: Challenges in the Design of the MINDACT Trial," *Nature Clinical Practice Oncology*, 3, 540–51.
G. Bonadonna *et al.* (1976) "Combination Chemotherapy as an Adjuvant Treatment in Operable Breast Cancer," *New England Journal of Medicine*, 294, 405–10.
A. Cambrosio *et al.* (2006) "Mapping the Emergence and Development of Translational Cancer Research," *European Journal of Cancer*, 42, 3140–48.
G. P. Canellos *et al.* (1976) "Chemotherapy for Metastatic Breast Carcinoma. Prospective Comparison of Multi-Drug Therapy with L-Phenylalanine Mustard," *Cancer*, 38, 1882–86.
S. K. Carter (1987) "The Clinical Evaluation of Analogues. i. The Overall Problem," *Cancer Chemotherapy and Pharmacology*, 1, 69–72.
S. K. Carter and W. T. Soper (1974) "Integration of Chemotherapy into Combined Modality Treatment of Solid Tumors. 1. The Overall Strategy," *Cancer Treatment Reviews*, 1, 1–13.
St. K. Carter and R. B. Livingston (1975) "Cyclophosphamide in Solid Tumors," *Cancer Treatment Reviews*, 2, 295–322.

B. A. Chabner and T. G. Roberts (2005) "Chemotherapy and the War on Cancer," *Nature Reviews Cancer*, 5, 65–72.

R.G. Cooper (1969) "Combination Chemotherapy in Hormone Resistant Breast Cancer," *Proceedings of the American Association for Cancer Research*, 10, 15 (abstract).

A. J. M. de Craen *et al.* (1996) "Effect of Colour of Drugs: Systematic Review of Perceived Effect of Drugs and of their Effectiveness," *British Medical Journal*, 313, 1624–26.

B. J. Culliton (1976) "Breast Cancer: Reports of New Therapy are Greatly Exaggerated," *Science*, 191, 1029–31.

F. Cvitkovic-Bertheault (2000) "Quarante ans de progrès en chimiothérapie," *Pathologie Biologie*, 48, 812–18.

A. A. Daemmrich (2004) *Pharmacopolitics: Drug Regulation in the United States and Germany* (Chapel Hill: University of North Carolina Press).

J. Denekamp (1993) "Review Article: Angiogenesis, Neovascular Proliferation and Vascular Pathophysiology as Targets for Cancer Therapy," *British Journal of Radiology*, 66, 181–96.

V. T. DeVita *et al.* (1979) "The Drug Development and Clinical Trials Program of the Division of Cancer Treatment, National Cancer Institute," *Cancer Clinical Trials*, 2, 195–216.

J. A DiMasi and H. G. Grabowski (2007) "Economics of New Oncology Drug Development," *Journal of Clinical Oncology*, 25, 209–16.

Eastern Cooperative Oncology Group (ECOGa) *Group Progress Report, May 1971–May 1974*, p. 10A.

Eastern Cooperative Oncology Group (ECOGb) *Group Progress Report, May 1974–May 1975*, p. 3.

Eastman and R. P. Perez (2006) "New Targets and Challenges in the Molecular Therapeutics of Cancer," *British Journal of Clinical Pharmacology*, 62, 5–14.

E. A. Eisenhauer (1998) "Phase I and II Trials of Novel Anti-Cancer Agents: Endpoints, Efficacy and existentialism," *Annals of Oncology*, 9, 1047–52, 1047.

S. Epstein (1996) *Impure Science: AIDS, Activism,and the Politics of Knowledge* (Berkeley: University of California Press).

B. Fisher *et al.* (1975) "L-Phenylalanine Mustard (L-PAM) in Management of Primary Breast Cancer: Report of Early Findings," *New England Journal of Medicine*, 292, 117–22.

E. Gomart (2002) "Methadone: Six Effects in Search of a Substance," *Social Studies of Science*, 32, 93–135.

A. M. Hedgecoe and P. Martin (2003) "The Drugs Don't Work: Expectations and the Shaping of Pharmacogenetics," *Social Studies of Science*, 33, 327–64.

J. F. Holland (1976) "Major Advance in Breast Cancer Chemotherapy," *New England Journal of Medicine*, 294, 440.

Y. Hu *et al.* (2006) "Targeting Multiple Kinase Pathways in Leukemic Progenitors and Stem Cells Is Essential for Improved Treatment of Ph+ Leukemia in Mice," *PNAS*, 103, 16870–75.

J. R. Johnson and R. Temple (1985) "Drug Administration Requirements for Approval of Anticancer Drugs," *Cancer Treatment Reports*, 69, 1155–59.

J. R. Johnson, G. Williams and R. Pazdur (2003) "End Points and United States Food and Drug Administration Approval of Oncology Drugs," *Journal of Clinical Oncology*, 21: 1404–11.

P. Keating and A. Cambrosio (2002) "From Screening to Clinical Research: The Cure of Leukemia and the Early Development of the Cooperative Oncology Groups, 1955–1966," *Bulletin of the History of Medicine*, 76, 299–334.

P. Keating and A. Cambrosio (2006) "Le criblage des médicaments: les modèles animaux en recherche thérapeutique au *National Cancer Institute* (1955–2000)" in G. Gachelin (ed.) *Les organismes modèles dans la recherche médicale* (Paris: Presses Universitaires de France), pp. 181–208.

P. Keating and A. Cambrosio (2007) "Cancer Clinical Trials: The Emergence and Development of a New Style of Practice," *Bulletin of the History of Medicine*, 81, 197–223.

P. Keating and A. Cambrosio (2012) *Cancer on Trial: Oncology as a New Style of Practice* (Chicago: University of Chicago Press).

E. L. Korn *et al.* (2001) "Clinical Trial Designs for Cytostatic Agents: Are New Approaches Needed?" *Journal of Clinical Oncology*, 19, 265–72.

D. Lacombe *et al.* (2002) "The EORTC and Drug Development," *European Journal of Cancer*, 38, S19–S23.

T. L. Laker (1989) "Assessing the Market for Cytotoxic Drugs," *Medical Marketing and Media*, 24(5), 77–82, 80.

B. Latour (1987) *Science in Action: How to Follow Scientists and Engineers Through Society* (Cambridge, MA: Harvard University Press).

M. N. Levine and T. Whelan (2006) "Adjuvant Chemotherapy for Breast Cancer. 30 Years Later," *New England Journal of Medicine*, 355, 1920–22.

J. Mann (2004) *Life Saving Drugs. The Elusive Magic Bullet* (Cambridge, UK: The Royal Society of Chemistry).

H. M. Marks (1997) *The Progress of Experiments: Science and Therapeutic Reform in the United States, 1900–1990* (Cambridge, UK: Cambridge University Press).

J. L. Marx (1989) "Drug Availability is an Issue for Cancer Patients, Too," *Science*, 245, 346–47.

F. Meunier and D. Lacombe (2003) "The EORTC and its Pan-European New Drug Development for Cancer," *Business Briefing: Future Drug Discovery*, pp. 14–15; http://www.touchbriefings.com/pdf/16/Meunier.pdf, date accessed March 2007.

P. Mirowski and R. Van Horn (2005) "The Contract Research Organization and the Commercialization of Scientific Research," *Social Studies of Science*, 35, 503–48.

National Cancer Institute. *The Nation's Investment in Cancer Research: A Plan and Budget Proposal for Fiscal Year 2003* (Bethesda, MD, NIH Publication No. 02–4373), p. 31.

S. Paik *et al.* (2004) "A Multigene Assay to Predict Recurrence of Tamoxifen-Treated, Node-Negative Breast Cancer," *New England Journal of Medicine*, 351, 2817–26.

G. V. Paolini *et al.* (2006) "Global Mapping of Pharmacological Space," *Nature Biotechnology*, 24, 805–15.

W. R. Parulekar and E. A. Eisenhauer (2004) "Phase I trial Design for Solid Tumor Studies of Targeted, Non-Cytotoxic Agents: Theory and Practice," *Journal of the National Cancer Institute*, 96, 990–97.

L. M. Pasetto *et al.* (2006) "The Development of Platinum Compounds and Their Possible Combination," *Critical Reviews in Oncology Hematology*, 60, 59–75, 67.

T. G. Roberts, T. J. Lynch and Bruce A. Chabner (2003) "The Phase III Trial in the Era of Targeted Therapy: Unraveling the 'Go or No Go' Decision," *Journal of Clinical Oncology*, 21, 3683–95.

B. Rosenberg, L. Van Camp and T. Krigas (1965) "Inhibition of Cell Division in *Escherichia coli* by Electrolysis Products from a Platinum Electrode," *Nature*, 205, 698–99.

P. D. Ryan and B. A. Chabner (2000) "On Receptor Inhibitors and Chemotherapy," *Clinical Cancer Research*, 6, 4607–09.

P. Santoni and T. Foster (2007) "Zero Sum Game" *Oncology Business Review*, available at http://www.imshealth.com/imshealth/Global/Content/Document/Value-based%20Medicine%20TL/zero.pdf, date accessed November 2009

R. L. Schilsky (2002) "End Points in Cancer Clinical Trials and the Drug Approval Process," *Clinical Cancer Research*, 8, 935–38.

D. Secko (2006) "A Target for Iressa. The Fall and Rise (and Fall) of a Pharmacogenetics Poster Child," *The Scientist*, 20(4), 67–68.

R. Simon (2006) "Development and Evaluation of Therapeutically Relevant Predictive Classifiers Using Gene Expression Profiling," *Journal of the National Cancer Institute*, 98, 1169–71.

G. Simondon (1958) *Du mode d'existence des objets techniques* (Paris: Aubier).

M. Talpaz *et al.* (2006) "Dasatinib in Imatinib-Resistant Philadelphia Chromosome-Positive Leukemias," *New England Journal of Medicine*, 354, 2531–41.

D. Vasella (2003) *Magic Cancer Bullet. How a Tiny Orange Pill is Rewriting Medical History* (New York: HarperBusiness).

T. H. Wasserman, M. Slavik, and S. K. Carter (1974) "Review of CCNU in Clinical Cancer Therapy," *Cancer Treatment Reviews*, 1, 131–51.

N. Weinstein and Y. Pommier (2006) "Connecting Genes, Drugs and Diseases," *Nature Biotechnology*, 24, 1365–66.

G.C. Zubrod *et al.* (1966) "The Chemotherapy Program of the National Cancer Institute: History Analysis and Plan," *Cancer Chemotherapy Reports*, 50, 349–540.

8

Treating Health Risks or Putting Healthy Women at Risk: Controversies around Chemoprevention of Breast Cancer

Ilana Löwy

Tamoxifen and raloxifene: Problematic preventive drugs

On April 16, 2006, the National Surgical Adjuvant Breast and Bowel Project (NSABP) of the National Cancer Institute (NCI) called a press conference in order to announce the results of the STAR (Study of Tamoxifen and Raloxifene) project. This randomized clinical trial compared the capacity of two drugs, tamoxifen (Novaldex® AstraZeneca) and raloxifene (Evista®, Eli Lilly) to prevent breast cancer in high-risk women. The STAR project was a direct continuation of an earlier NSABP trial, the Breast Cancer Prevention Trial (BCPT), which tested the capacity of tamoxifen to prevent breast cancer against placebo. Tamoxifen, an estrogen-receptor inhibitor, belongs to the family of SERMs (selective estrogen-receptor modulators), which are molecules that modify the effects of estrogen on cells. Tamoxifen was shown to lower the rate of recurrence in women diagnosed with estrogen-receptor-positive breast malignancies. It was found to be as efficient as standard cytotoxic chemotherapy in keeping breast cancer patients in remission, while avoiding most of the harsh side effects of chemotherapy. From the 1980s on, it therefore became a highly popular anticancer drug. It was not clear, however, whether this molecule could also prevent breast cancer in healthy women. The BCPT attempted to answer this question.

According to its organizers, the BCPT demonstrated that women with a higher-than-average risk to develop breast cancer halved their risk through the uptake of tamoxifen (Fisher *et al.*, 1998). The BCPT also indicated that tamoxifen induced a wide array of undesirable side effects, some of them unpleasant (hot flashes, digestive problems) and others, such as endometrial cancer and cardiovascular accidents, potentially life-threatening. While such side effects were seen as acceptable in women with cancer, it was less clear whether it was worthwhile to induce iatrogenic effects in healthy women in the hope of reducing their cancer risk. The answer to this interrogation was not identical in the United States and in Europe. The majority of US experts concluded that the advantages of tamoxifen outweighed its drawbacks, while the majority of European specialists reached the opposite conclusion: they considered that the disadvantages of the use of tamoxifen were greater than its expected benefits and that the results of clinical trials did not justify a large-scale use of the drug by healthy women (Veronesi *et al.*, 1998). In parallel, selected patients' associations criticized the promotion of chemoprevention of cancer by the NCI and the pharmaceutical industry.

In the late 1990s, NCI experts recognized that tamoxifen was far from being an ideal preventive drug. At that time they had, however, an alternative solution. They hoped that raloxifene, a SERM employed to treat osteoporosis, would be at least as efficient as tamoxifen in reducing cancer risk without the latter's shortcomings. The name of the clinical trial that compared these two molecules – STAR – probably mirrored these hopes. Alas, the STAR trial did not live up to its name. It demonstrated that, far from being innocuous, raloxifene produced its own cluster of side effects. Consequently, rather than referring to a *better* preventive drug, the organizers of the STAR trial decided to present raloxifene as a *different* one. The new trial, they explained, revealed the existence of two preventive drugs with distinct biological profiles. It therefore increased the scope of women's choices, allowing them to find a personalized solution for the management of breast cancer risk ("the treatment that is right for *you*"). The preventive use of SERMs is supported in the United States by the majority of cancer experts. By contrast, so far (early 2007), the majority of US women have rejected the specialists' advice. My chapter reports on the short and stormy controversy over the preventive use of tamoxifen and raloxifene. It is interested in the evaluation and the regulation of drugs intended to keep healthy people from becoming ill and in the role of users in this process.

What is "breast cancer risk"?

In the early twentieth century, the majority of women diagnosed with breast cancer had an advanced malignancy and suffered from distressing symptoms such as chronic pain and ulcerating wounds. The success of "do no delay" campaigns led to an increase in the number of women who consulted for a painless lump in their breast and who agreed to undergo extensive, mutilating surgery if the lump was declared malignant (Aronowitz, 2001). For many women, breast cancer diagnosis was therefore more a diagnosis of a potential danger than of an already existing pathology. Women who usually felt perfectly well were offered a trade-off: the acceptance of an immediate disease and of a mutilation in order to prevent a worse disease and a painful death. The advent of chemotherapy aggravated the iatrogenic effects of cancer therapy, while the generalization of mammography – a technique that detects minimal breast lesions that cannot be felt by the patient – accentuated the dissociation of breast cancer diagnosis from the advent of clinical signs of the disease.

The logical extension of this trend were attempts to act before breast cells fully expressed their malignant potential. Two in situ (i.e., noninvasive) malignancies, ductal carcinoma in situ (DCIS) and lobular carcinoma in situ (LCIS), played a key role in the blurring of boundaries between breast cancer and breast cancer risk. Before the advent of mammography, these lesions were nearly always diagnosed accidentally, usually when a breast biopsy was made for a different reason. They were accordingly viewed as rare conditions. With the generalization of mammography, doctors increasingly performed biopsies to clarify the meaning of the frequently observed suspicious radiological images Consequently, the number of diagnoses of in situ cancers also increased. Until the 1980s, LCIS and DCIS were seen as direct precursors of breast malignancies and treated, as a rule, by mastectomy. In the 1980s oncologists reviewed the epidemiological data and decided to reclassify LCIS as a risk marker rather than the direct precursor of an invasive tumor. Women diagnosed with this condition were accordingly classified as being at high risk of breast malignancy (Fisher, 1996).

In parallel, from the 1950s on, epidemiological studies confirmed the previously suspected link between family history and breast cancer: daughters and sisters of women with breast cancer were shown to have a higher probability of developing this disease. Epidemiological studies displayed other variables that increase breast cancer risk: age (the chances of developing breast malignancy increase with age), age at menarche (the first menstrual period) and at menopause (the last period) (early menarche and late menopause are risk factors, because a long period of ovarian activity is associated with an increased cancer risk), age at the birth of first child and number of children (childless women and those who have children late are at greater risk of breast cancer), lifestyle elements (nutrition, alcohol uptake, weight, practice of sports, breastfeeding), and race/ethnic origins (breast cancer is more frequent among white women and is relatively rare among Asian ones).

In the late 1980s and the 1990s, statisticians constructed models that evaluated the chances of a given woman to develop breast cancer. The most influential among these models was developed by the NCI statistician Mitchell Gail and his colleagues (Gail *et al.*, 1989). It focused on variables shown to have the strongest effect on breast cancer risk: age, the number of first-degree relatives with the disease, age at the birth of the first child, ethnic origins, age at menarche, and "morphological risk," that is, the presence of abnormal and potentially premalignant changes in the breast such as LCIS and atypical hyperplasia (AH, another form of proliferative lesions of the breast). The last variable is, however, constructed in a peculiar manner. The Gail model takes into consideration not only positive results of breast biopsies but also the very fact of having such a biopsy, even if its result was normal. The rationale for viewing a biopsy as a risk factor, regardless of its results, was the observation that women who undergo breast biopsies (often because radiologists observed suspicious findings on a mammogram) have a higher than average chance of developing breast cancer, indicating that some negative results of biopsies are false negatives. On the other hand, the number of biopsies in a given population is directly proportional to the number of performed mammograms. In the United States, experts recommend an annual mammography between age 40 and 75; in the United Kingdom, a triannual mammography between age 50 and 70; and in France, a biannual mammography between ages 50 and 70. Women who undergo more frequent mammographic screenings – such as US women – have higher chances of being diagnosed with proliferative breast lesions, which puts them in the high-risk category, but also of acquiring an independent risk element – a past biopsy.

The Gail model was widely disseminated by the NCI. Doctors were invited to evaluate the breast cancer risk of their patients with the help of the "risk assessment tool" (a computer program) distributed by the NCI. At the same time, with the propagation of the Internet, women gained independent access to Web sites that allowed them to calculate their own risk. Every responsible woman, suggested the promoters of the BCPT and the STAR trials, should undergo such assessment and be aware of her risk (Vogel 1998: Vogel, 1999). The Gail model was also elected as a tool for recruitment of women "at risk," first for the BCPT and then for the STAR trial. In both trials, the entry point was a 1.66% (or, in a simplified version 1.7 %) risk of developing breast cancer in the next five years.[1]

How does such risk evaluation work in practice? The official Web site of the STAR project offers a link to two "self-risk assessment tools," one promoted by the NSABP and the other by the NCI. Both risk assessment tools follow the Gail model with slight variations. As a "Caucasian" woman with no specific risk elements such as personal or family history of the disease, I attempted to calculate my breast cancer risk in November 2006 by using these two tools.

The tool developed by the NSABP starts with the declaration: "We have determined that women who are at high risk are those who have at least a 1.66% chance of developing breast cancer." It explains that high-risk women should calculate their risk and, if appropriate, take "NCI approved drugs" to reduce their risk.[2]

My results on this Web site were:

Your projected chance of developing breast cancer over the next 5 years is **1.43%**. Your chance of developing breast cancer by the age of 80 is **7%**.

In comparison, the average risk of a woman of your age with no risk factors is **0.84%** over 5 years and is **4%** by age 80.

The verdict was: "This is not considered high risk, but it's important to remember that many women get breast cancer who are not at high risk. Your risk could also change because of a change in your age, medical history, or family history. If this occurs, you should recalculate your risk and take appropriate steps for a 'good breast health.'"

The NCI risk-calculating tool does not include information about chemoprevention, and did not provide a specific number (1.66%) that defines "high risk."[3]

My results with this tool were:

5 Year Risk

* This woman **1.4%**

* Average woman of same age: **1.6%**

Lifetime Risk

* This woman (to age 90): **8.7%**
  • Average woman (to age 90): **9.8%**
  • The NCI site explains that the estimated risk for developing invasive breast cancer is compared to the average risk of a woman of the same age and race/ethnicity from a general US population. The text also adds that this result means that the woman's chance of *not* getting breast cancer during the next 5 years is 98.6. [4]

Both tools delivered similar estimates of my chances of developing breast cancer, but each provided a different framing for these results. The NSABP Web site presented my risk as being definitively above average and dangerously close to the fateful 1.66% "high-risk" number. The NCI tool was more reassuring: it presented my cancer risk as slightly below average, and provided the information that I have 98.6% chances of avoiding breast cancer in the next five years.[5]

The Breast Cancer Prevention Trial controversy

The BCPT started in 1992. It reflected the strong commitment of the NASBP to chemotherapy of breast cancer. The National Surgical Adjuvant Breast and Bowel Project (NASBP) was set up by doctors who opposed "surgical dogma" in cancer therapy. The project was an offshoot of the NCI's Cancer Chemotherapy National Service Center (CCNSC), dedicated to the organization of clinical trials of drug therapies for cancer (Zubord *et al.*, 1984, Löwy, 1996: 39–63). From 1957 on, the CCNSC sponsored a small program called "surgical adjuvant therapy project," which assessed the efficacy of the combination of surgery and chemotherapy. In 1967, Bernard Fisher was appointed chairman of the Surgical Adjuvant Chemotherapy Breast Project. Fisher was an outspoken advocate of the view that breast cancer was a systemic disease.

Accordingly, he strongly opposed reliance on surgery alone and advocated treatments able to deal with hidden micrometastases: radiotherapy, chemotherapy, and immunotherapy (Lerner, 2001). Fisher, like other promoters of cancer chemotherapy, was thoroughly committed to the clinical trial ethos. Under his leadership, the Surgical Adjuvant Chemotherapy Breast Project organized large-scale, randomized clinical trials of cancer therapies. The first and best-known among them demonstrated that a conservative treatment of breast cancer (lumpectomy and radiotherapy) was as efficient as radical mastectomy. The NACBP program then evaluated the efficacy of numerous anticancer drugs, tamoxifen among them.

In the 1980s, a clinical trial conducted by this program (Protocol B-14) displayed the efficacy of tamoxifen in the treatment of estrogen-receptor-positive invasive breast cancer. In the early 1990s, another clinical trial (Protocol B-24) indicated that women with DCIS who received tamoxifen after their surgery had a lower incidence of invasive breast cancer than a control group, the first indication of the preventive potential of the drug.[6] The proposal made the following year to test whether tamoxifen could also prevent the development of breast cancer in women classified as "at risk" of breast malignancy – Protocol P-1 or BCPT – was a logical extension of Protocol B-24. The trajectory of tamoxifen within the NSABP – from the prevention of the recurrence of invasive breast cancer to the prevention of cancer in women with in situ malignancies, to attempts to lower the incidence of cancer in women classified as being at "high risk" of breast cancer using the Gail model – mirrored the increasing permeability of the boundaries between breast cancer, precancerous conditions, and cancer risk.

The BCPT protocol included 13,388 women who had a 1.66% chance or more of developing breast cancer in the next five years and compared tamoxifen with a placebo. The trial was interrupted 14 months before its scheduled end because, according to its promoters, the results had already indicated that tamoxifen reduced cancer risk, and it became immoral to continue to treat women in the control group with sugar pills. In a press conference held on April 30, 1998, Bernard Fisher and his colleagues announced that tamoxifen halved (51%) the incidence of breast cancer among high-risk women (Fisher *et al.*, 1998). The effect on the development of in situ cancer was similar: an approximately 50% incidence of in situ carcinoma in the tamoxifen group. Public releases of BCPT data were systematically presented as absolute numbers (a 50% risk reduction) and not as natural frequencies. When expressed as such frequencies, the results were less impressive: 89 women out of 6,694 included in the experimental group developed cancer, 86 women probably avoided cancer thanks to tamoxifen (there were 175 incidences of cancer in the control group), and 6,605 took tamoxifen with no benefit to themselves. Tamoxifen induced in parallel an increase (again, a small one in absolute numbers) in life-threatening incidents such as deep vein thrombosis, pulmonary embolisms, and endometrial cancer. Moreover, it frequently induced non-life-threatening complications. In the United States, these complications were presented as acceptable side effects of a substance found to be remarkably efficient in reducing breast cancer risk.

The FDA shared the enthusiasm of the BCPT organizers and approved the use of tamoxifen (Nolvadex®) as a preventive therapy in healthy women barely one month after the publication of the trial's results.[7] Some breast cancer activists were less enthusiastic. They protested against the FDA's rapid endorsement of preventive uses of tamoxifen, arguing that focus on individual chemoprevention deflected attention from the environmental causes of carcinogenesis. Critics of the preventive use of tamoxifen also sustained that the claim that tamoxifen prevented breast cancer was grounded on insufficient scientific proof. It was unclear whether tamoxifen truly reduced the incidence of cancer or merely delayed the onset of malignancy, and there was not enough data on the long-term effects of this molecule. Finally, according to some activists, the NCI invited women to exchange one set of risks for another (Brenner, 1999a). However, in 1998, the voice of activists could hardly be heard in the chorus of positive assessments of BCPT, and it did not seem very likely that it would be able to neutralize the powerful publicity campaign launched by AstraZeneca to promote Novaldex® for preventive use.

Contesting tamoxifen: Professional and lay reactions

In 1998, AstraZeneca launched an important, direct-to-consumer advertisement campaign for the preventive use of Novaldex® (tamoxifen). Billboards and magazines displayed an image of a young, good-looking woman wearing a black lace bra with the caption, "If you care about breast cancer, care more about being 1.7 than a 36B [36B is a bra size]." The text went on to explain: "Know your breast risk assessment number... Knowing your number gives you power, and knowing about Novaldex® should give you hope. Novaldex® has now been proven to significantly reduce the incidence of breast cancer in women at high risk... Call your doctor for your Breast Cancer Risk Assessment test."[8] The Novaldex® publicity provoked angry reactions from some activists. Leaders of Breast Cancer Action (BCA), the National Women's Health Network (NWHN), Boston Women's Health Book collective, Massachusetts Breast Cancer and DES Action strongly protested against this direct-to-consumer publicity campaign. In addition, BCA produced a counteradvertisement that criticized the AstraZeneca arguments (Cody, 1999; Press and Burke, 2000).

Social scientists who studied this controversy pointed to the sophistication of the arguments used by both sides. The AstraZeneca publicity used the empowerment language developed by feminists ("Knowing your number gives you power") in order to persuade women that their product would allow them to control their destiny. On their side, the activists pointed to the lack of consistency and inaccurate information in the

Novaldex® advertisement, such as in its presentation of a reduction in a five-year incidence of breast cancer in relative terms ("50% risk reduction") and of the side effects of the molecule in absolute terms ("less than 1% of the women suffered from severe side effects of tamoxifen") (Klawiter, 2002). The anthropologist Linda Hogle compared women's reactions to the AstraZeneca advertisement with their reactions to the BCA's counteradvertisement and found that women viewed the BCA's "mud-racking" announcement as less honest and more manipulative than the AstraZeneca "its your choice" language (Hogle, 2001). It is plausible that many women found the AstraZeneca advertisement persuasive. It should not be forgotten, however, that activists were not addressing the drug consumers exclusively. One of their main targets was official regulatory bodies. Activists wrote to the FDA to protest against AstraZeneca's advertisements, and at the same time used their public campaigns as a way to put additional pressure on government agencies. The pressure turned out to be effective. Several times, the FDA division of drug marketing, advertising, and communication contested AstraZeneca's use of BCPT results in its commercial, and it is possible that the BCA and NWHN campaign also favored informal pressure on AstraZeneca to change its form of advertising tamoxifen (Klawiter 2002: 333–35).

The protesters had an additional ally – the tamoxifen molecule itself. AstraZeneca's advertisement was indeed convincing – in theory. In practice, it is not easy to persuade healthy people to take a pill every day, all the less so that the pill is associated with cancer chemotherapy and induces unpleasant and bothersome secondary effects. Tamoxifen is usually better tolerated than standard cytotoxic chemotherapy, but this does not mean that a molecule that mimics distressing menopausal symptoms is a problem-free drug. A French woman treated with tamoxifen for a local recurrence of an invasive breast cancer describes her feelings:

> To diminish the risk of recurrence I need to take a daily hormonal treatment... This is a dilemma: how can I swallow the pill when I know that it will immediately induce hot flashes, insomnia, stress and the disappearance of my libido? How can I accept survival by harming myself, by ageing, by destroying all the remains of youth? I bought the box of medication and have put it on my night table, but I do not know when I'll be able to force myself to take it every day. (Ruber, 2006: 153–54)

It can be assumed that if some cancer patients find it hard to accept tamoxifen treatment, this may be even truer for healthy women who are told that they have a higher-than-average risk to develop a malignancy. Indeed, circa 2001, it became increasingly clear that "high-risk" women did not become enthusiastic users of the new preventive treatment. Women who enrolled in BCPT were highly motivated and therefore willing to disregard the problematic image of tamoxifen and its side effects. Other women did not share this attitude. General practitioners were reluctant to recommend tamoxifen, and women were reluctant to take it. In one widely discussed study, only 5% of women to whom their doctors recommended tamoxifen decided in favor of chemotherapy in spite of the fact that in this study physicians were instructed to systematically discuss chemoprevention with eligible patients (Port *et al.*, 2001). When doctors were not given any specific instructions, less that 2% of the eligible "high-risk" women decided to take tamoxifen (Taylor and Taguchi, 2005). Professional associations recommended that doctors should discuss with their clients the advantages and disadvantages of chemoprevention, but medical practitioners often elected to stress the disadvantages of this approach, and women elected to refuse SERMs (US Preventive Task Force, 2002). In addition, the critical attitude toward the chemoprevention of breast cancer outside of the United States might have affected some North American doctors as well (Goel, 1998; Powell, 2002).

Advocates of tamoxifen developed two strategies to counter the lack of enthusiasm for a preventive use of this drug. The first was to redefine the target population for chemoprevention of breast cancer. NCI statisticians, led by Gail, developed a new algorithm that made it possible to single out a subgroup of "at-risk" women for whom the benefits of this type of prevention clearly outweighed its drawbacks: younger women with higher-than-average risk of breast cancer. These women have a low risk of stroke, endometrial cancer, pulmonary embolism, and thrombosis, and therefore lower chances of suffering from severe complications due to tamoxifen treatment (Gail *et al.*, 1999). The new criteria greatly reduced the size of the potential target group for breast cancer chemoprevention, but this group still included more than 2 million women in the United States (Freeman *et al.*, 2003). Alas, the women targeted by the revised criteria were still unwilling to undergo preventive therapy, perhaps because younger, premenopausal women may be specially reluctant to take a drug that makes them infertile and induces menopausal symptoms (Port *et al.*, 2001, Taylor and Taguchi, 2005). Hence the second persuasion strategy: to make women more aware of their risk – or, to put it otherwise, to increase their level of cancer fear.

Advocates of chemoprevention had found that the best way to persuade women to accept preventive uptake of tamoxifen was to transform their statistical risk into an "embodied risk," that is, to convince them that they already had premalignant changes in their breast, and that these precancerous lesions needed to be kept in check by appropriate medication. A diagnosis of LCIS or AH, researchers had found, was the best predictor of acceptance of preventive chemotherapy (Tchu *et al.*, 2004). The logical conclusion was that women at risk should undergo more breast biopsies:

The current study finding regarding the importance of atypia in the decision to undertake tamoxifen chemoprevention provides a strong rationale for epithelial sampling in the high-risk woman who is uncertain regarding tamoxifen use. (Tchu *et al.*, p. 1805)

Similarly, the promoters of the STAR trial argued that a display of "embodied risk" was an especially efficient way to persuade women to accept chemoprevention with SERMs. High-risk women, they explained, may consider the finding of cellular atypia to be a more reliable measure of risk than other factors that contribute to a high Gail score. The data from both BCTP and STAR suggested that a diagnosis of atypia strongly influenced a woman's decision to enter a randomized trial to study breast cancer risk reduction. Such a diagnosis affected women's perception of the "reality" and the magnitude of their risk and therefore their willingness to undergo preventive treatment (Vogel *et al*, 2002).

Tamoxifen, advocates of chemoprevention stressed, was a highly useful drug that, unfortunately, failed to reach its targets. As a *Time* magazine article put it in 2006:

> What if doctors had a pill that prevents cancer and nobody wanted to take it? That has pretty much been the situation with tamoxifen, an estrogen-like drug that was proved in 1998 to cut in half the chance of developing breast cancer if taken for five years by women with increased risk of the malignancy. (Gorman, 2006)

One possible solution for the low enthusiasm for chemoprevention was to increase the number of women diagnosed with an "embodied risk" of breast cancer.[9] An alternative solution was to find another, better-accepted drug.

The STAR trial: Hopeful researchers and wary women

In 1999, at the time of the launching of the STAR project, one of its main promoters, Craig Jordan, expressed the hope that raloxifene would be a precursor of a new trend in pharmacology (Jordan, Gapstur and Morrow, 2001). "The emerging data on the multiple effects of raloxifene have generated great excitement that this might be the first multifunctional medicine" (Jordan and Morrow, 1999: 331). The statement "first multifunctional medicine" is puzzling, specially coming from a professor of pharmacology. The majority of drugs have multiple physiological effects, as do nearly all biologically active substances. Jordan's article, we can assume, is not alluding to the multiple functions of a molecule, but rather to its multiple marketing targets. A drug that was initially developed to reduce bone loss can be prescribed at the same time for the prevention of other conditions as well.

The model for multilevel activity, Jordan explained, was hormone replacement therapy (HRT). HRT had initially been employed to alleviate menopausal symptoms, but later, doctors found that it reduced the risk of cardiovascular disease and osteoporosis. Accordingly, HRT was recommended to all post-menopausal women. A treatment that prevents breast cancer should also be targeted toward all women, not only those with known risk factors, because the latter are only a fraction of all the women who develop breast cancer. While the efficacy of raloxifene to diminish the incidence of breast cancer was yet to be proven, Jordan concluded, "It is clear that a new era of preventive therapies has arrived" (Jordan and Morrow, 2001: 332). With the publication of results of the WHI trial of HRT in the United States (2002) and of the "Million women trial of post-menopausal hormones" in the United Kingdom (2003), HRT, once seen as highly beneficial therapy, was reclassified as problematic treatment that increased the risk of cancer and of cardiovascular pathologies (Writing Group for the WHI, 2002; Pettiti, 2002; Million Women Study, 2003). The demise of HRT as a miracle cure for age-related pathologies in women did not put an end to Jordan's high hopes for SERMs. "The recent negative assessment of HRT," he argued in 2006, "far from being a discouraging event, opens more possibilities for drugs of the SERM type. These multifunctional drugs are here to stay" (Jordan, 2006: 5013).

The STAR project was directed Dr. Victor Vogel, the director of the comprehensive breast program at the University of Pittsburgh and an enthusiastic supporter of medical management of breast cancer risk. In a 2000 review of breast cancer prevention, Dr. Vogel explained that:

> No lifestyle modifications have yet been proved to prevent or definitively lower the risk of breast cancer... women whose personal risk is high may consider reducing risk by pharmacological or surgical means... Breast risk assessment and appropriate counselling are becoming standard components of breast cancer screening and overall health maintenance. (Vogel, 2000: 156)

In parallel to his task as the director of the STAR trial, Vogel was a consultant both for AstraZeneca (the producers of Nolvadex®, the patent of which expired in early 2006, just before the publication of the STAR results) and for Elli Lilly (the producers of Evista®). The STAR trial mobilized 19,737 women.[10] The STAR organizers made a special effort to recruit minority women. BCPT had been criticized for its low percentage – 4% – of nonwhite women. STAR fared somewhat better: 6.6% of the participants were classified as "non-whites," among which 2.5% African Americans, 2.0% Hispanics, and 2.1% "other." All the women were postmenopausal because raloxifene had not been authorized by the FDA for use in premenopausal women. Of the participants, 9.1% were diagnosed with LCIS – a very high percentage with regard to the prevalence of this lesion in the general population. The NCI provided two-thirds of the funding for the trial, while the drug manufactures AstraZeneca and Eli Lilly provided one-third of the funding and the tested drugs.[11]

The press conference that announced the results of STAR was held three months before the official publications of the trial results in the *JAMA*.[12] A *Lancet Oncology* editorial argued that this was unacceptable procedure. The editorial also criticized the presentation of the trial data using relative values (50% risk reduction) and not absolute numbers, a presentation that "is unlikely to help women to make informed decisions, and cynics may argue that there were other motivating forces behind this publicity" (Editorial, 2006a). The first reactions in the United States were, in contrast, mostly enthusiastic. The accounts were based on the NCI press release that stated that women on raloxifene developed less iatrogenic complications than those on tamoxifen, but it omitted to mention that these differences were not statistically meaningful.[13] The NSABP director Dr. Norman Wolmark declared in the press conference:

> Today, we can tell you that for postmenopausal women at increased risk of breast cancer, raloxifene is just as effective, without some of the serious side effects known to occur with tamoxifen.[14]

It is not surprising that headings of articles published immediately after the press conference proclaimed: "Exciting News from NSABP"; "New Breast Cancer Drug Safer Than Tamoxifen"; "Raloxifene Preferable to Tamoxifen for Breast Cancer Prevention"; "Raloxifene: Less Risky Breast Cancer Prevention"; "New Breast Cancer Drug Has Fewer Side Effects." These articles quoted experts such as the acting director of NCI, Dr. John Niederuber, who declared that "STAR is a significant step in breast-cancer prevention," or Dr. Lawrence Wickham, one of the main STAR investigators, who explained that raloxifene was an attractive preventive option for postmenopausal women because "it has fewer adverse effects and is already available for the treatment of osteoporosis." Dr. Leslie Ford, associate director for clinical research at the NCI, explained that now women could determine with the help of their physicians what drug was best for them and then take appropriate steps to lower their breast cancer risk.[15] Dr. Judy Garber, an oncogeneticist from the Dana Farber Institute, Boston, stated that the STAR study was an important step forward because it showed that both tamoxifen and raloxifene were remarkably effective in reducing breast cancer risk:

> We hope people will begin to think of breast-cancer prevention in the same way that they think of taking a statin drug to prevent heart disease.[16]

The possibility of "doing something" to reduce breast cancer threat is indeed an important selling argument for chemoprevention. In the past 25 years, partly thanks to activists' efforts, breast cancer has become highly visible in the media and in public discourse. One of the main messages promoted by activists is the rapid extension of "breast cancer epidemics": 1 out of 12, 1 out of 10, and now 1 out of 8 women will be diagnosed with breast cancer. Many women feel like Paula Johnson, a participant in the STAR project, interviewed on the Vanderbilt Medical Center Web site:

> Every year the odds get a little bit worse. Everybody I know has someone who has to deal with it. You don't have to look too far to find someone.[17]

Advocates of chemoprevention hoped that the combination of the widespread fear of breast cancer, a better public image of raloxifene, and the expectation that this molecule would induce less side effects would persuade women who were reluctant to use tamoxifen for chemoprevention to adopt raloxifene instead.[18]

Other experts were more cautious. They pointed to the absence of statistically meaningful differences in side effects induced by the two drugs and to the fact that raloxifene, unlike tamoxifen, did not reduce the number of in situ cancers (although the meaning of prevention of these lesions is unclear). The media coverage of the data from the STAR trial suggested that raloxifene was a "winner," but some specialists explained that the only valid conclusion from this trial was that the two tested SERMs were equally good – or equally bad. Dr. Len Lichtenfeld, the deputy chief medical officer of the American Cancer Society, explained that the STAR results were much less clear-cut that it had been hoped.[19] Susan Love's popular site on breast cancer treatment stated that the final result of the STAR trial was: "We now have two drugs that are about the same. Both reduce breast cancer risk and fractures. And both have side effects."[20] The *JAMA* paper that published the official results of the STAR trial confirmed this view. It concluded that the only safe result of the STAR trial was that the two tested SERMs had a similar risk/benefit ratio (Vogel *et al.*, 2006). They also have a similar record of induction of iatrogenic effects. Women treated with tamoxifen complained more often of gynecological problems, vasomotor symptoms, leg cramps, and bladder control problems, while those in the raloxifene group complained more frequently of muscle and bone pain, painful intercourse, and weight gain (Land *et al.*, 2006). More than half of the STAR participants reported such bothersome symptoms. It is not surprising that many women elected to abandon the trial before the allocated five years.

Predictably, activists reacted negatively to what they saw as a new evidence of the tendency of NCI to focus on pharmaceutical compounds and neglect social issues. Writing in the BCA newsletter in 1999, Barbara Brenner explained that the goal of both the BCPT and STAR trials was to fulfill the "dream of the federal cancer establishment to find a simple, profitable way to prevent breast cancer." The new trial, Brenner argued, had all the shortcomings of the preceding one as well as one more: the absence of a placebo group (Brenner, 1999b). Similarly, the October 2003 National

Breast Cancer Coalition's (NBCC's) "Position Statement on Tamoxifen and Raloxifene Use in Healthy Women" stressed that the STAR trial did not have a "placebo arm," included too few African American women, did not have a sufficiently long follow-up period, and did not answer the question of whether SERMs prevent breast cancer or merely delay its onset.[21] The NWHN's official statement affirmed that at this point chemoprevention of breast cancer seemed to benefit only a small subset of women with specially high risk of breast cancer and urged other women and their health providers "not to get caught up in the hype and wait for the facts."[22]

The president of the NBCC, Fran Visco, explained that the NCI's presentations of the results of the STAR mistakenly *led* women to believe that SERMs prevented cancer, while in fact this was not prevention at all, but "risk reduction, short-term, for a subset of women we cannot identify" (quoted by Grady, 2006). The NBCC's official declaration repeated the warning issued after the publication of the BCPT results: the long-term consequences of healthy women taking SERMs daily are unknown, and women are invited to "make a choice" without sufficient information on the consequences of this choice.[23] The New Hampshire Breast Cancer Coalition newsletter concluded its analysis of the results of the STAR trial with the suggestion: "When you hear the words 'prevention,' 'breakthrough' or 'victory' in breast cancer, take a deep breath and then look carefully behind the headlines" (NHBC, 2006). Cindy Pearson from the NWHN, Judy Norsigian from the Boston Women's Health Book Collective, and Barbara Brenner from the BCA described the results of the STAR trial as "disease substitution, not disease prevention" (Wilkinson, 2006).

Environmental activists explained on their side that an exclusive focus on chemoprevention and the neglect of other important determinants of carcinogenesis is bad science and bad news for women (Martin, 2006). A review in the BCA newsletter revealed close ties between organizers of the STAR trial and the pharmaceutical industry: three out of the four main investigators of the trial worked as consultants to AstraZeneca, Eli Lilly, or both. The report discussed the 2005 guilty verdict of Eli Lilly for illegally promoting raloxifene as a "breast cancer preventative."[24] The STAR trial, added the article, violated the majority of rules of good clinical practices set by the NCI itself: raloxifene was not tested against a placebo, there was no provision for a long-term follow-up of the effects of the tested drugs, and the results were first disclosed through a press conference (Zones, 2006).

Even the usually strongly pro-NCI American Cancer Society issued a cautious declaration in November 2006 on the preventive uses of tamoxifen and raloxifene. The American Cancer Society (ACS) stated that SERMs were shown to prevent breast cancer but also enumerated their side effects and explained that randomized clinical trials did not show that women who took SERMs lived longer than women in the placebo control group. The ACS Web site recommends the preventive use of SERMs only for women with an especially high probability of developing breast cancer (e.g., carriers of BRCA mutations), a view quite distant from the earlier hopes of transforming raloxifene into a "multifunctional drug" addressed to all postmenopausal women.[25]

Tamoxifen and raloxifene were shown to reduce the risk of hormone receptor-positive breast cancer but had no effect on overall mortality of this disease, a result that does not encourage the preventive use of substances that produce many unpleasant side effects, and some very serious ones (Cutuli *et al.*, 2009). The preventive uptake of these substances indeed remained low. In 2000, approximately 0.2% of US women aged 40 to 79 years without a personal history of breast cancer took tamoxifen for chemoprevention, while five years later, 0.08% of women with no personal history of breast cancer took this drug (Waters *et al.*, 2010). SERM's advocates proposed the problem with the low preventive uptake of this substances is the fact that doctors and pharmaceutical companies make profit from the cure of cancer, not from its prevention, and are therefore reluctant to recommend a prophylactic means (Veronesi *et al.*, 2007). A competing explanation may point out to the strengthening of association between the routine uptake of female sex hormones and breast cancer. The decrease of incidence of invasive breast cancer in the early twenty-first century was attributed to the fall in consumption of menopausal hormones after the publication of WHI results (MacNeil, 2007; Kondro, 2007; Cordiz, 2007, Krieger, 2008). Women and health professionals might have had good reasons to be suspicious of substances that modulate hormone receptors.

The lack of success of SERMs as a preventive therapy probably affected efforts to develop akin treatments. A new NCI program to uncover other molecules able to prevent breast cancer, the STELLAR (Study to Evaluate Letrozole) of (Femara®Novartis) and Raloxifene (Evista ®, Eli Lilly), approved in December 2006, was halted in March the following year by NCI's director, John Niedehuber (Marshall, 2007). A 2009 review of recent developments in chemoprevention of breast cancer summed up the status of this domain: "a vision not yet realized" (Blaha *et al.*, 2009).

Drugs for healthy people: The user's role

The difficulty of persuading women and doctors to use tamoxifen to prevent breast cancer may be contrasted with the impressive success of mammography. For many years, mammography was presented by the great majority of cancer experts, as well as by cancer activists, as a highly efficient way to lessen the burden of breast cancer, despite the absence of convincing proof that this approach saved lives or increased life expectancy, or that its advantages exceeded its drawbacks (Schwartz and Woloshin, 2002). The NCI's 1994 recommendation to start mammography screening at the age of 50 (instead of 40, as proposed by the ACS and associations of radiologists) led to mass protest, a political storm and, in the end, to the retraction of the NCI's original proposal (Fletcher 1997). Gradually, selected groups of activists (among them the BCA and the NBCC) developed a more critical stance toward mammography.[26] The fate of "preventive SERMs" was different. From the very beginning, these drugs were a contested technology. The sophisticated and apparently successful commercial strategy of AstraZeneca failed to convince women to use Novaldex® to prevent breast cancer, and it seems – as things stand today – that Eli Lilly was not any more successful with Evista®. The present text suggests three possible explanations for the lack of enthusiasm for the chemoprevention of breast cancer: the intrinsic properties of SERMs; the changing attitude toward female sex hormones as drugs; and the role of activists.

Tamoxifen was handicapped by the negative image of cancer chemotherapy, by reports about potentially serious complications of the therapy, and by the fact that this molecule frequently induced bothersome side effects. Raloxifene's side effects may similarly reduce the quality of life of its users. HRT, another problematic treatment, was initially successful partly because it was efficiently promoted by its producers and supported by important segments of the medical profession, but probably partly also because women were pleased with its immediate effects. HRT eliminates the distressing symptoms of menopause, and women who suffered from such symptoms and who took HRT rapidly felt better. By contrast, women who took tamoxifen or raloxifene often felt worse. Many developed debilitating menopausal symptoms such as hot flashes, night sweats, and insomnia, and premenopausal women who took tamoxifen (raloxifene is not administered before the onset of menopause) might have also suffered from the psychological effects of having "shut off" their ovaries and become infertile.

The recent fate of cancer chemoprevention, the present text also suggests, was in all probability affected by the "WHI effect." The reluctance to prescribe and use tamoxifen to prevent breast cancer preceded the publication of the WHI results, but the decision to conduct this clinical trial might have already reflected the status of menopausal hormones as a controversial drug. Activists of the women's health movement and organizations such as the NWHN played a central role in persuading the National Institute of Health (NIH) to reexamine the risk/benefit ratio of HRT (Krieger *et al.*, 2005; Löwy, 2005, Löwy and Gaudillière, 2006; Löwy, 2009). The same organizations also played an important part in the rise of a cautious attitude toward the chemoprevention of breast cancer. The oncogeneticist Judy Garber suggested that the STAR trial reaffirmed "the role that patients can play in advancing progress against cancer."[27] Similarly, NCI expert Leslie Ford explained that tamoxifen and raloxifene were "vital options for women who are at increased risk of breast cancer and want to take action."[28] Such affirmations should be taken seriously. One can suggest that women active in organizations such as the NBCC, the BCA, or the NWHN indeed "took action." They warned women to be wary of experts' hype, revealed the financial interests that connected the drug manufactures to the organizers of the trial, and put pressure on government regulatory agencies.

These activities probably played an important role in the rise of a cautious approach to preventive uses of tamoxifen and raloxifene.

Officially the main mechanism of regulation of chemoprevention of cancer in the United States – the main site of development of this treatment – was an administrative one. A governmental agency, the NCI was the main mover behind clinical trials of this approach. In fact, however, the boundaries between industrial and administrative regulations were porous, a development illustrated by the fact that four among the five first authors of the scientific paper that summed up the results of STAR clinical trial disclosed that they act as consultants for manufacturers of SERMs (Vogel *et al.*, 2006). Producers of these substances occasionally had antagonist relationships with regulatory agencies, but usually they relied on a close collaboration with key opinion leaders in these agencies to validate preventive SERMs through clinical trials, and then to favor their use. One can therefore describe the regulation of preventive SERMs as an "administrative–industrial" regulatory pattern.

The administrative–industrial regulation of chemoprevention of breast cancer aimed at promoting a widespread diffusion of this treatment. This did not happen, however. The preventive use of SERMs was held back by several types of user-driven regulation: (a) women's spontaneous resistance to a drug which decreased their quality of life here and now, while promising reduction of their chances to develop breast cancer sometimes in the future; (b) a direct criticism of the preventive use of tamoxifen and raloxifene by Women's Health Movement activists, and (c) similarities between objections to preventive use of SERM and objections to generalization of HRT, vindicated by WHI clinical trials. The troubling parallels with HRT strengthened doctors' reluctance to prescribe preventive SERMs and women's reluctance to use this treatment. The combination of women's spontaneous resistance to the chemoprevention of breast cancer, the generalization of a critique of the use of female sex hormones as preventive drugs, and the intervention of activists became a potent mix. Activists and lay constituencies who voiced their skepticism regarding "preventive SERMs" seem to have fulfilled the role advocated by Robert Aronowitz, which is:

> to balance the message of powerful and appealing coalitions with a compelling message about caution, iatrogenic risk, and medical hubris. (Aronowitz, 2006: 163).

Notes

1. Note that 1.7% is the risk of a 60-year-old woman with no additional risk factor or that of a younger woman with one or more risk factors. On the history of the Gail model, see Foscet, 2004.
2. The tool can be found at http://www.breastcancerprevention.org/breastcancer-risks.html.

3. The tool can be found at http://www.cancer.gov/bcrisktool/.
4. I calculated my breast cancer risk with identical data, but adding two *negative* breast biopsies my risk increased in the NASB tool to 2.03% within five years, and a lifetime risk of 9%, in the NCI tool to a 2% five-year risk, and an 11.5% lifetime risk, a non negligible increase.
5. The NCI site may still be criticized for expressing data as a percentage and not natural frequencies. The latter, with the statement "If one takes 1000 healthy women with your profile, 14 of these women will develop cancer in the next five years and 986 will remain cancer free,," is perceived as less threatening. Gigerenzer, 2002.
6. Experts agree that DCIS progress to fully invasive cancer, but the conditions of this progression and its prevalence are uncertain: current evaluations are between 10% and 60% of progression to malignancy (Burstein *et al.* 2004). The demonstration of the efficacy of tamoxifen in reducing progression to cancer in women with DCIS in trials conducted by the NSABP did not lead to the acceptance of tamoxifen as routine therapy for this condition.
7. FDA approval was issued on October 29, 1998, thus shortly after the publication of the results of the BCPT study (September 15, 1998). This was a positive response to a "supplemental new drug application" issued by Zeneca Pharmaceuticals shortly after the press conference that announced preliminary results of the BCPT trial. For a fine-grained analysis of debates on the preventive use of tamoxifen in the United States see Klawiter, 2002.
8. The advertisement also adds: "Novaldex® decreases but does not eliminate the risk of breast cancer and did not show an increase in survival." The last point is important, since absence of survival gain is seldom mentioned in accounts of preventive strategies. See on this point, Welch, 2004.
9. Intensified mammographic surveillance of women at risk and a higher number of biopsies quasi-automatically increase the number of women diagnosed with proliferative conditions of the breast, since the number of such diagnoses is directly proportional to the number of breast biopsies (Welch, 2004).
10. The number of women recruited for this trial was slightly below the planned 21,000 because the average risk of the participants was higher than expected, making it possible to obtain statistically significant results with a smaller number of participants.
11. http://www.cancer.gov/star (accessed August 2010).
12. The press conference was held on April 16, 2006, and the article was published in the *JAMA Express*, June 5, 2006 (and then in the June 21 issue of the *JAMA*). In the BCPT trial too, preliminary results were announced during a press conference on April 30, 1998, whereas the official results were published in the September 16 issue of the *Journal of the National Cancer Institute*.
13. The information that these differences are not statistically meaningful was included only in the text of the NCI's teleconference.
14. http://www.cancer.gov/newscenter/pressreleases/STARresultsApr172006 (accessed August 2010).
15. http://www.cancer.gov/newscenter/pressreleases/STARresultsApr172006 (accessed August 2010)
16. http://www.dfci.harvard.edu/res/research/unveiling-of-star-trial-results-brings-kudos-to-participants.html (accessed August 2010)
17. http://sitemason.vanderbilt.edu/newspub/crmQtG?id=25968 (accessed August 2010)

18. http://www.healthtalk.com/breastcancer/programmes/2_606/page02.cfm (accessed August 2010)
19. http://www.cancer.org/docroot/MED/content/MED_2_1x_Researchers_Release_ Results_of_Breast_Cancer_Prevention_Trial_STAR.asp?sitearea=MED&viewmod e=print& (accessed August 2010)
20. http://www.susanlovemd.com/printview.asp?PID=42&SID=130&CID=391& L2=1&L3 (accessed August 2010)
21. http://www.natlbcc.org/bin/index.asp?strid=435&depid=9&btnid=1 (accessed August 2010)
22. http://www.nwhn.org/alerts/alerts_details.php?aid=63 (accessed August 2010)
23. http://www.stopbreastcancer.org/bin/index.asp?strid=810&btnid=&depid=20 (accessed August 2010)
24. Eli Lilly started promotion in 1999; at that time, AstraZeneca filed a suit against Lilly, and the latter firm was forced to pay 36 million dollars in fines as part of its settlement with the government (thus more than it was paying for its share in the STAR trial). Mokhiber, 2005; Zones, 2006. Details on the AstraZeneca verdict can be found at http://www.usdoj.gov/opa/pr/2005/December/05_civ_685.html (accessed August 2010)
25. http://www.cancer.org/docroot/CRI/content/CRI_2_6X_Tamoxifen_and_ Raloxifene _Questions_and_Answers_5.asp?sitearea=(2010a)http://www.cancer. org/docroot/CRI/conCRI_2_6X_Tamoxifen_and_Raloxifene_Questions_and_ Answers_5.asp?sitearea=10a)
26. Activists might have been slow to question mammography because breast cancer activism in the United States was built around the idea of "breast cancer epidemics,," that is, the sharp increase in incidence of breast cancer in the 1970s and 1980s. This increase is seen today by many experts mainly as an artifact of the introduction of mammographic screening (Schwartz and Wolochin, 2002; Welch, 2004).
27. http://www.dfci.harvard.edu/res/research/unveiling-of-star-tri- al-results-brings-kudos-to-participants.html (accessed August 2010)
28. http://www.cancer.gov/newscenter/pressreleases/STARresultsApr172006 (accessed August 2010)

Bibliography

Aronowitz R. (1998) *Making Sense of Illness,* Cambridge: Cambridge University Press.
Aronowitz A. (2001) "Do not delay: Breast cancer and time, 1900–1970," *Milbank Quarterly,* 79, 355–86.
Aronowitz R. (2006) "Situating health risks: An opportunity for disease prevention policy," *in* C. Rosenberg, R. Stevens and L. R. Burns (eds.), *History and Health Policy in the United States: Putting the Past Back In,* New Brunswick, NJ: Rutgers University Press, pp. 153–65.
Blaha P., P. Dubsky, F. Fitzal, *et al.* (2009) "Breast cancer chemoprevention – a vision not yet realized," *European Journal of Cancer Care,* 18(5), 438–46.
Brenner B. (1999a) "From the executive director: Rolling the dice," *Newsletter* of the Breast Cancer Action, #53, April/May.
Brenner B.(1999b), "STARs in their eyes," *BCA Newsletter* #55, August/September.
Burstein H.J., K. Polyak, J. S. Wong, *et al.* (2004) "Ductal Carcinoma In Situ of the breast," *New England Journal of Medicine,* 350, 1430–41.
Cody N. (1999) "New Zeneca Ad campaign misleading to women," *Motion Magazine,* May 5.

Cordiz G.A. (2007), "Decline in breast cancer incidence due to removal of promoter: Combination estrogen plus progestin," *Breast Cancer Research*, 9, 108–10.

Cutuli B., A. Lesur, M. Namer *et al.* (2009) "Breast cancer chemoprevention: Rational, trials results and future," *Bulletin du Cancer*, 96(5), 519–30.

Editorial (2006a) "A STARring role for raloxifene?" *Lancet Oncology*, 7, 443.

Editorial (2006b) "Raising our voices," *National Breast Cancer Coalition Newsletter*, Vol. 14, N°. 1, Spring/Summer 2006.

Fisher B., J.P. Constantine, D. L. Wickerham *et al.*, (1998) "Tamoxifen for Prevention of Breast Cancer: Report of the National Surgical Adjuvant Breast and Bowel Project P-1 Study," *Journal of the National Cancer Institute*, 90(18), 1371–88.

Fisher E.R. (1996) "Pathobiological considerations relating to the treatment of intraductal carcinoma (ductal carcinoma in situ) of the breast," *CA – Cancer Journal for Clinicians*, 47(1), 52–64.

Fletcher S. (1997) "Whither scientific deliberation in health policy recommendations? Alice in the wonderland of breast-cancer screening," *New England Journal of Medicine*, 336(16), 1180–83.

Foscet J. (2004) "Construction of 'high-risk women': The development and standardisation of a breast cancer risk assessment tool," *Science, Technology and Human Values*, 29(3), 291–313.

Freedman A.N., B. I. Graubard, S. R. Rao *et al.* (2003) "Estimates of the number of US women who could benefit from tamoxifen for breast cancer chemoprevention," *Journal of the National Cancer Institute*, 95(7), 526–32.

Gail, M.H., L.A. Brinton, D.P. Byar *et al.* (1989) "Projecting individual probabilities of developing breast cancer for white females who are being examined annually," *Journal of the National Cancer Institute*, 81(24), 1879–86.

Gail M., J. Costantino, J. Bryant *et al.* (1999) "Weighing the risks and benefits of Tamoxifen treatment for preventing breast cancer," *Journal of the National Cancer Institute*, 91(21), 1829–46.

Gigerenzer G. (2002) *Calculated Risks: How to Know When Numbers Deceive You*, New York: Simon and Schuster.

Goel V. (1998) "Tamoxifen and breast cancer prevention: What should you tell your patients?" *Canadian Medical Association Journal*, 158(2), 1615–16.

Gorman, C. (2006), "A better option?" *Time Magazine*, April 24.

Grady D. (2006), "Sorting out pills to reduce breast cancer risk," *New York Times*, May 9.

Hampton T. (2007) "Breast cancer prevention trial in limbo," *JAMA*, 297, 1968–69.

Hogle L. (2001) "Chemoprevention for healthy women: Harbinger of things to come?" Health, 5(3), 311–33.

Jordan V.C. and M. Morrow (1999) "Raloxifene as a multifunctional medicine?" *British Medical Journal*, 319(7206), 331–32.

Jordan V.C., S. Gapstur and M. Morrow (2001) "Selective estrogen receptor modulation and reduction in risk of breast cancer, Osteoporosis and Coronary Heart Disease," *Journal of the National Cancer Institute*, 83(19), 1449–57.

Jordan V.C. (2006) "The science of selective estrogen receptor modulators: Concept to clinical practice," *Clinical Cancer Research*, 12(17), 5010–13.

Klawiter M. (2002) "Risk, prevention and the breast cancer continuum: The NCI, the FDA, health activism and the pharmaceutical industry," *History and Technology*, 18(4): 309–53.

Kondro W. (2007) "Decline in breast cancer since HRT study," *Canadian Medical Association Journal*, 176(2), 160–61.

Krieger N., Löwy I. and the group "Female sex hormones and cancer" (2005), "Hormone replacement therapy, cancer, controversies and women's health: Historical, epidemiological, biological clinical and advocacy perspectives," *Journal of Epidemiology and Community Health*, 59, 740–48.

Krieger N. (2008) "Hormone therapy and the rise and perhaps fall of US breast cancer incidence rates," *International Journal of Epidemiology*, 37(3), 627–37.

Land S.R., D. L. Wickerham, J. P. Costantino *et al*. (2006) "Patient-reported symptoms and quality of life during treatment with tamoxifen or raloxifene for breast cancer prevention," *Journal of the American Medical Association* 295(23), 2742–51.

Lerner B. (2001) *Breast Cancer Wars: Hope, Fear and Pursuit of a Cure in Twentieth Century America*, NY: Oxford University Press,.

Li C.I., K. E. Malone, B. B. Saltzman and J. R. Daling (2006) "Risk of invasive breast carcinoma among women diagnosed with ductal carcinoma in situ and lobular carcinoma in situ, 1988–2001," *Cancer*, 106(10), 2104–12.

Löwy I. (1996) *Between Bench and Bedside: Science, Healing and Interelukin-2 in a Cancer Ward*, Cambridge, MA: Harvard University Press, pp. 39–63.

Löwy I. (2005), "Le féminisme a-t-il change la recherche biomédicale? Le Women's Health Movement et les transformations de la médecine aux États-Unis," *Travail, Genre et Sociétés*, 14, 89–108.

Löwy, I. (2009) *Preventive Strikes: Women, Precancer and Prophylactic Surgery*, Baltimore: Johns Hopkins University Press, 2009, pp. 184–89.

Löwy I. and Gaudillière J.P., (2006) "Médicalisation de la ménopause, mouvements pour la santé des femmes et controverses sur les thérapies hormonales," *Nouvelles Questions Féministes*, 25(2), 48–65.

Martin A.R. (2000) "Activist perspective: The dark side of the STAR trial", *The Ribbon, Newsletter of the Program on Breast Cancer and Environmental Risk Factors*, Cornell University, Vol. 5 N°. 4. http://envirocancer.cornell.edu/Newsletter/articles/v5darkstar.cfm (accessed August 2010).

Marshall E. (2007) "Budget pressure puts high-profile study in doubt," *Science*, 315, 1477.

McNeil C. (2007) "Breast cancer decline mirrors fall in hormone use, spurs both debate and research," *Journal of the National Cancer Institute*, 99(4), 226–67.

Million Women Study Collaborators (2003) "Breast cancer and hormone-replacement therapy in the Million Women Study," *The Lancet*, 362, 419–27.

Mokhiber R (2005). "Lilly's second disappointment," *Multinational Monitor*, 26(11–12), November/December, pp. 49–50.

NHBC (2006), "Raising our voices," *NHBC Newsletter*, 14(1), Spring/Summer.

Page D.L. (2004) "Breast lesions, pathology and cancer risk," *Breast Journal*, 10, Suppl. 1, S3–S4.

Petitti D.B. (2002) "Hormone replacement therapy for prevention: More evidence, more pessimism", *Journal of the American Medical Association*, 288, 99–101.

Port E.R., L. L. Montgomery, A. S. Heerdt and P. I. Borgen (2001) "Patient reluctance toward tamoxifen use for breast cancer primary prevention," *Annals of Surgical Oncology*, 8(7), 580–85.

Powell T.J. (2002) "Breast cancer prevention," *Oncologist*, 2002, 7: 60–64.

Powles T. J., R. Eeles, S. Ashley, *et al*. (1998) "Interim analysis of the incidence of breast cancer in the Royal Mardsen Hospital Tamoxifen Randomised Chemoprevention Trial", *Lancet*, 352, 98–101.

Press N. and W. Burke (2000) "If you care about women's health, perhaps you should care about the risks of direct marketing of tamoxifen to consumers," *Effective Clinical Practice*, July/August, 202–04.

Ruber F. (2006) *La Vie est là, simplement*, Paris: Albin Michel, pp. 153–04.

Schwartz L. and Woloshin S. (2002) "News media coverage of screening mammography for women in their 40s and tamoxifen for primary prevention of breast cancer," *Journal of the American Medical Association*, 287(23), 3136–42.

Taylor R. and K. Taguchi (2005), "Tamoxifen for breast cancer prevention: Low uptake by high risk women after an evaluation of a breast lump," *Annals of Family Medicine*, 3(3), 242–45.

Tchou J., N. Hou, A. Rademaker, *et al* (2004) "Acceptance of tamoxifen chemoprevention by physicians and women at risk," *Cancer*, 100(9), 1800–06.

US Preventive Task Force (2002) "Chemoprevention of breast cancer: Recommendation and rationale," *Annals of Internal Medicine*, 137(1): 56–58.

Veronesi U., P. Maisonneuve, A. Costa, *et al.*, (1998) "Prevention of breast cancer with tamoxifen: Preliminary findings from the Italian randomised trial among hysterctomised women," *Lancet*, 352, 93–97.

Veronesi U, P. Maisonneuve and A. Decensi (2007) "Editorial: Tamoxifen: An enduring star," *Journal of the National Cancer Institute*, 99, 258–59.

Vogel V.(1998) "Breast cancer prevention: a review of current evidence." *CA001ECancer Journal*, 50,156001E170.

Vogel V (1999) "Tools for evaluating a patient's 5001 year and lifetime probabilities." *Postgraduate Medicine*, 105(6), 782–86.

Vogel V.G. (2000) "Breast cancer prevention: a review of current evidence," *CA Cancer Journal for Clinicians*, 50(3), 156–70.

Vogel V.G., J. P. Costantino, D. L. Wickham and W. M. Cronin (2002) "Re: Tamoxifen for prevention of breast cancer," *Journal of the National Cancer Institute*, 94(19), 1504.

Vogel V.G., J. P. Costantino, D. L.Wickerham *et al.* (2006) "Effects of Tamoxifen vs Raloxifene on the risk of developing invasive breast cancer and other disease outcomes: The NSABP Study of Tamoxifen and Raloxifene (STAR) P-2 Trial," *Journal of the American Medical Association* , 295(23), 2727–41.

Waters EA, K.A. Cronin, B.I. Graubard *et al.*, (2010) "Prevalence of tamoxifen use for breast cancer chemoprevention among US women," *Cancer Epidemiology, Biomarkers and Prevention* 19(2), 443–46.

Welch H.G. (2004), *Should I be Tested for Cancer? Maybe Not and Here's Why*, Berkeley: The University of California Press.

Wilkinson K. (2006). "Lies about designer Estrogens and breast cancer," *Curve Magazine*, http://www.curvemag.com/Detailed/426.html (accessed August 2010).

Winchester, D.P. J. M. Jeske and R. A. Goldschmidt (2000) "The diagnosis and management of ductal carcinoma in-situ of the breast," *CA – Cancer Journal for Clinicians*, 50(3), 184–200.

Writing Group for the Women' s Health Initiative Investigators (2002) "Risk and benefits of estrogen plus progestin in healthy postmenopausal women: Principal results from the women's health initiative randomized controlled trial," *Journal of the American Medical Association*, 288, 321–33.

Zones J.S. (2006) "Pushing Raloxifene – Reports of the STAR trial," *BCA Newsletter*, #92, August/September.

Zubrod G. C., S.A. Schepartz and S.K. Carter (1977) "Historical background of the National Cancer Institute's drug development thrust," *National Cancer Institute Monographs*, 45, 7–11.

Zubrod G.C. (1984) "Origins and development of chemotherapy research at the National Cancer Institute," *Cancer Treatment Reports*, 68(1), 9–19.

9
AZT and Drug Regulatory Reform in the Late 20th-Century US

Donna A. Messner

Introduction

In discussing the development of regulations for accelerated approval of thera-peutic drugs in the United States, one cannot give a full account of the story without reference to the rise of the AIDS epidemic and the efforts of AIDS activists. Their persistent calls for access to experimental drugs and changes in clinical protocol design for drug evaluation were highly influential in regula-tory reforms implemented in the late 1980s and 1990s. On this aspect of the events of the period, the work of Epstein (1995; 1996; 1997) is authoritative. (For another view of the period, see Carpenter 2010.) However, as important as this work is, it would be a mistake to assume that activism was the only force to shape events in this period, or that prior to AIDS there was no premarket access to investigational drugs or significant efforts to expedite development and approval of important drugs. Patient lobbying for the right to choose their own therapy predated AIDS by at least a decade, although certainly not with the same stridency or radicalism as the AIDS activists (Markle and Peterson, 1980; Carpenter, 2010). Additionally, the US Food and Drug Administration (FDA), the agency responsible for assuring the safety and effectiveness[1] of new drugs before they can be marketed, had been informally granting seriously ill patients access to experimental drugs on a premarket basis for decades (US Senate, 1997).[2] Indeed, inspection of the various streams of influence colliding during this period will reveal the pivotal role in shaping a posture toward drug approval played by a specific instance of drug development and approval: the first antiretroviral drug developed to fight AIDS, AZT (azidothymidine).

In the next section, I will supply a brief history of the drug's clinical evaluation and describe the new rules written by the FDA to expedite drug approval designed explicitly on the AZT model. In the third section, I provide historical context to show the roots of the approach taken to AZT. In the fourth and fifth sections, I will argue that the experience with AZT lent support for certain practices to become established in practice and codified in regulation and legislation – even one practice that the FDA had consistently opposed. I will develop this argument with respect to the types of evidence accepted for approval of drugs intended to treat life-threatening diseases: in the fourth section, the topic will be clinical study end points acceptable for evidence of effectiveness; in the fifth section, the subject will be single-study clinical trials used as the basis for drug approval. I will conclude by arguing not only that the experience with AZT was influential throughout this period, but that it likely helped to shape the AIDS activism movement itself.

A remarkable story in drug development

In 1984, after a consensus had been achieved on the identity of the causal agent for AIDS, Dr. Samuel Broder, chief of the clinical oncology program at the National Institutes of Health (NIH), teamed with the drug company Burroughs Wellcome to screen a series of chemical compounds in storage at Burroughs for activity against AIDS. They quickly discovered that one compound, AZT, showed promise. Phase I testing – the first introduction of the drug into human subjects – began in June 1985. Nineteen patients were enrolled in the study, all of whom were very ill when testing began. In response to AZT, all the patients experienced small but statistically significant increases in CD4 cell counts, or helper T cells (Yarchoan, 1998). Since these important immune system cells were destroyed by the virus, an increase in the number of cells per cubic centimeter of blood was taken to be a sign that the drug was interfering with the action of the virus; that is, CD4 counts were taken to be a valid "surrogate marker," a laboratory measure thought to be predictive of clinical benefit for the patient.

Researchers proceeded rapidly to a placebo-controlled phase II study with 282 patients enrolled at 12 medical centers across the United States. At this time, phase II trials were considered relatively small intermediate studies to *begin* defining effectiveness and to enhance the understanding of toxicity gleaned in phase I. So the large, multicentered, placebo-controlled design of this phase II study were features more typical of a phase III study, the sort of large, statistically valid trial conventionally required for drug approval in the three-phase system of drug evaluation that evolved over many years.[3]

After four months, nine patients on placebo had died while the AZT cohort saw only one death. The data review took place on September 19, 1986. The study was discontinued the next day and the drug was made available to AIDS patients almost immediately, not only to the patients on the study, but to seriously ill patients nationwide. Such premarket access of very ill patients to investigational new drugs (INDs) – a procedure called "treatment IND" – had been practiced informally by the FDA in various ways on a limited basis for at least a decade. At the time of this premarket distribution of AZT, the practice was still considered informal, although proposed rules had been written to formalize it (FDA, 1983).

The Anti-Infective Drugs Advisory Committee (ADAC) met four months later, in January 1987, to consider a new drug application (NDA) based on this single study. The Committee was faced with clear deficits of information on drug toxicity and long-term effects. There was no information at all regarding whether less ill or asymptomatic patients would respond to the drug, nor what the effects of longer-term administration might be. Even so, the Committee realized that upon marketing approval, personal physicians would begin to prescribe this therapy to *all* their patients having AIDS, including those less ill patients for whom information was lacking. Hence, among other risk-related knowledge deficits, clearly approving the drug for seriously ill patients created an indirect risk to less ill patients who were likely to receive the drug without knowing the associated risk or benefit. The Committee consciously weighed those uncertainties against the potential benefits of approving the drug and voted for approval of the drug with the understanding that the sponsor would conduct additional studies on less seriously ill patients to fill the information gaps as quickly as possible (FDA, 1987a).

As Epstein (1996) notes, AZT "proceeded from in vitro studies to full approval in just two years and had been approved without a Phase III study"(199). This was an exceptional event because of its rapidity. It was also unusual in its compression or abbreviation of the conventional three-phase drug evaluation process, as well as its approval on the basis of a single study rather than on multiple studies. The latter decision was particularly notable because it contravened the traditional interpretation of the "substantial evidence" clause. This requirement, written into the 1962 Kefauver–Harris amendments to the 1938 *Food, Drug and Cosmetic Act,* calls for effectiveness to be proven for all new drugs using "evidence consisting of adequate and well-controlled investigations, including clinical investigations" by scientific experts qualified to conclude that the drug has the purported effect under the conditions of use prescribed.[4] The FDA's 1970 regulatory interpretation of that standard called for a study employing a randomized control group for statistical comparison to a treatment group (FDA, 1970), essentially a randomized, controlled trial or RCT. Significantly for this discussion, the traditional interpretation of the substantial evidence clause was that the statute called for "investigations" (plural) and that at least two such investigations would be required as a form of scientific replication (FDA, 1988: 41521). While the FDA has always exercised case-by-case judgment, the resulting system was one in which the FDA strongly encouraged drug sponsors to provide at least two RCTs as proof of effectiveness, which themselves ideally represented the final phase of three-phase clinical drug development.

At the January 1987 Anti-Infective Drugs Advisory Committee meeting to consider AZT approval, FDA's Dr. Ellen Cooper characterized the experience with AZT as "a remarkable story in drug development" (FDA, 1987a:

8). Years later, this event is still characterized by people like the National Cancer Institue's (NCI) Broder as a success story (Broder, 1997). Indeed, AZT became the chief exemplar for future action, even when observers disagreed on the action to be taken. For instance, in the 1987 congressional hearings on the OMB's role in revising proposed rules for treatment IND (US House, 1987), one observer objected to allowing patients access to investigational drugs until the clinical studies were complete, and cited AZT as the model to follow in this regard; meanwhile, another observer testifying at the hearings upheld AZT as a successful example of allowing access to investigational drugs *before* clinical trials are completed (Messner, 2008).

Accordingly, when Vice President Bush (as chairman of President Reagan's Task Force on Regulatory Relief[5]) called on the FDA to develop new procedures for expediting new therapies for life-threatening diseases (FDA, 1988: 41516), the FDA turned to the experience with AZT as its model for rule making. In the preamble to the "Subpart E" rule the Agency wrote, "These procedures are modeled after the highly successful development, evaluation, and approval of zidovudine, the first drug approved to treat the AIDS virus" (p. 41517). Subpart E was to apply to new chemical or biological products "that are being studied for their safety and effectiveness in treating life-threatening or severely debilitating illnesses" (p. 41517).[6] The main provisions of the rule are that, first, the drug sponsor should confer with the FDA at the end of phase I to consult on a design for the phase II study suitable for producing the kinds of evidence normally derived from a phase III study (thus eliminating the need for a phase III study). This provision represents a change of traditional practice, in that FDA previously held such conferences *after* phase II testing was complete, in preparation for phase III testing. Second, when Subpart E drugs show "early evidence" in phase II testing of being "promising" for life-threatening or severely debilitating illnesses, the FDA will work with the sponsor to develop an appropriate therapeutic regimen under the treatment IND program. Third, once the phase II study is complete, the FDA will use a risk–benefit analysis to "consider whether the benefits of the drug outweigh the known and potential risks of the drug" (p. 41529). This risk–benefit analysis will determine whether the drug should be approved on the basis of phase II data. Finally, following market approval, the "FDA may seek agreement from the sponsor to conduct certain postmarketing studies (phase 4)" (p. 41517) since approval will have been often been granted "on the basis of limited, but sufficient, clinical trials" (p. 41521). The experience with AZT thus became the blueprint for new regulation to accelerate drug development and approval.

The roots of expedited approval

While explicitly based on the procedures used for AZT, a foreshadowing of many of these provisions can be seen in the previous decade. Prior to

AZT there had been many examples of the FDA using its discretionary power to expedite approval of drugs perceived to be important. Indeed, the FDA used some of these cases as evidence to support the validity of the interim Subpart E rule (FDA, 1988: 41519). There had likewise been previous examples of the FDA stretching the traditional interpretation of substantial evidence (p. 41521). More than that, there had been several previous attempts to legislate the kinds of changes the FDA was now writing into regulation – legislative proposals designed to counteract the lengthening of the drug development and approval process arising from statutory and postthalidomide regulatory tightening of the 1960s (Hilts, 2003; Messner, 2008; Temin, 1980).[7]

In May 1977, a Review Panel convened by the Department of Health, Education and Welfare (HEW) (often called the Dorsen panel, after its chairman Norman Dorsen) concluded a two-year study and published four volumes of reports (HEW, 1977a; 1977b; 1977c; 1977d) refuting many FDA critics and making recommendations to strengthen and improve the FDA's drug review process, including recommendations to accelerate approval of "breakthrough" drugs and to give the FDA the authority to require postmarket studies. In cooperation with the FDA, members of Congress subsequently attempted to incorporate many of the Panel's recommendations into law.[8] Among the most substantive of these proposals was Senate bill S. 1831, called the *1977 Amendments to the Federal Food, Drug, and Cosmetic Act*, which sought to create a form of expedited drug approval and to require postmarket study. The following year, the *Drug Regulation Reform Act of 1978* (S.2755) proposed "provisional approval" for drugs intended for "life-threatening or severely debilitating or disabling disease or injury." Under this proposal, a lesser standard of evidence, "significant evidence" would be used as the basis of provisional approval.[9] These and other proposals recognized that quicker drug development required decision making regarding drug effectiveness to be made earlier in the clinical trials process. S.2755 explicitly acknowledged that such decision making required an *information deficit*; it created a standard to define that deficit. It accordingly created safeguards: automatic revocation of provisionally approved drugs after three years unless additional clinical evidence was presented.

Unfortunately, the more proposals for change in a given bill, the more opportunities for objections to proposals; these comprehensive legislative efforts failed to become law. So the FDA convened its own Task Force and published its own, more modest recommendations for regulatory reform in a 1979 report (FDA, 1979). This document was followed by a series of regulatory proposals in the early 1980s to create streamlined application procedures, stressing flexibility in earlier phases of drug approval, and offering end-of-phase II conferences to all drug sponsors (FDA 1979; 1982; 1983; 1987b). At this time, the Agency also developed proposed rules for codifying treatment IND (FDA, 1979; 1983).

Notably, while the FDA had long used a concept of risk–benefit assessment for decision making on drugs, the version of it prescribed in Subpart E represents a formal shift in concept. Historically, the FDA recognized that any drug might cause an adverse effect, depending on the patient and the circumstances of administration. As George Larrick, Commissioner of Food and Drugs, stated in a testimony before a 1964 Congressional subcommittee: "There is no such thing as absolute safety in drugs. There are some drugs that are less liable to cause harmful reaction than others, but people die every year from drugs generally regarded as innocuous." Accordingly, said Larrick, quoting a decade-old statement from Dr. Torald Sollman, "the administration of potent drugs involves a 'calculated risk' where the presumptive benefit is balanced against the possibility of toxic effects" (US House, 1964: 147). Hence, even before proof of effectiveness was required by statute, a "presumptive benefit" had to be taken into account to assess the relative safety of a drug. It was this concept of risk–benefit that had been used for many years; namely, that in assessing the safety of drugs, the acceptable level of toxicity was proportional to the perceived therapeutic importance of the drug.

In the 1988 Subpart E rule, however, the risk–benefit approach is modeled on the procedure used by the ADAC to evaluate AZT: when a decision must be made on the basis of less information than would normally be desired for approval, the risks of both drug toxicity and information deficit should be weighed against the anticipated benefits of approval. As stated in the Subpart E publication, the "FDA will consider whether the benefits of the drug outweigh the known and potential risks of the drug *and the need to answer remaining questions about risks and benefits of the drug,* taking into consideration the severity of the disease and the absence of satisfactory alternative therapy" (FDA, 1988: 41520, emphasis added). While clearly full and perfect information had never been available for any instance of drug approval, this rule reflects a willingness to push back the comfort level of decision making into a zone where there is a conscious need for more information, much like the "provisional approval" proposed in 1978. However, whereas S.2755 explicitly changed the definition of "evidence" to accomplish this goal, Subpart E implicitly revises the traditional conception of "risk." Moreover, Subpart E lacks the safeguards included in S.2755 to assure that additional studies would be done. Certainly in the past the FDA sometimes used a soft version of this principle informally (with the power only to "request" postmarket study, lacking any force of regulation to require it); here it finally becomes a codified principle.

AZT became a successful example of drug approval and a template for future action *not* because the regulatory approach to the situation was fundamentally novel, but because a historically contingent confluence of events conspired to make the time ripe for the codification of concepts already in circulation (and to some extent, informally in use). In the 1980s, the most strident FDA critics from the previous decade found support for their views in the deregulationist Reagan administration, which was bolstered in turn by the scourge of AIDS and the rise of AIDS activism (Edgar and Rothman 1990).[10] Any voices of opposition to proposed changes would have difficulty getting traction in this context. Perhaps as important for their acceptance, the new regulations were a stripped-down version of the previous attempted reforms – not a consciously designed new system with safeguards, but a toolbox of actions.

The successful experience with AZT not only added to the knowledge base for future drug development, but constituted the first bracing step toward further regulatory reform. Subpart E reproduced the approach taken with AZT, and pointed toward future directions in four notable ways: (1) in its movement of key meetings and decisions to earlier points in the drug approval process;[11] (2) in its codification of a possible requirement for supplemental postmarket data collection; (3) in its movement toward reducing the burden of evidence required for approval (with this lack of evidence comprising part of the "risk" of the drug);[12] and (4) in its use of a single study as the basis for a drug approval. These themes are evident in drug approval decisions and rule making throughout the 1990s and beyond.

AZT and experience with CD4 counts

Another obvious way to accelerate clinical studies is to alter the criteria by which a study can be considered complete – to change the end points. Clinical end points are direct measures of patient benefit in which the clinical manifestations of a disease are tracked in response to treatment. Clinical end points are generally reliable for assessing any salutary effects of an investigational therapy. However, tracking clinical status can take time, especially if the investigational therapy has some beneficial effect. Moreover, when studying life-threatening disease, one of the most obvious and useful clinical end points is time until death. For many in the AIDS community, the idea of taking "body counts" in clinical studies was morally repugnant (Epstein, 1996). Hence, for earlier decision making in clinical trials, the use of laboratory measures or other markers of patient condition thought to be predictive of clinical benefit, so-called surrogate end points, were needed.

Unfortunately, AIDS researchers in the late 1980s had not been able to establish the validity of various laboratory measures then available as possible end points for study.[13] Worse yet, surrogate end points could be misleading. Especially in trials of cardiovascular drugs, there had been cases in which surrogate end points pointed to positive patient benefit when, it was later realized, the drugs in question actually increased patient mortality.[14] This lesson was not lost on members of the ADAC, who met in September 1990 to discuss the criteria to be used for design of pivotal studies to support NDAs. Dr. Paul Meier noted with apparent alarm that "[s]urrogate endpoints are, by their nature, treacherous" (FDA, 1990c: 93). Nevertheless, the ADAC members were also physicians and researchers eager for measures other than survival by which to judge the effectiveness of therapies against AIDS.

A consensus began to build that there was little choice but to work with theoretically promising, if uncertain, surrogate end points in AIDS clinical drug studies. The first FDA acceptance of CD4 leukocytes as a surrogate end point in AIDS clinical trials took place in March 1990, when AZT was approved for pediatric use (FDA, 1990b). To avoid the politically awkward circumstance of collecting mortality data on children, the sponsor used CD4 cell counts as end points and supplemented this information with survival probability calculations from the available data. This approval took place just a few months *before* the ADAC meeting in which surrogate end points were described as "treacherous." However, by this time a good deal more data had been accumulated on AZT in adults, including clinical end point data, helping to bolster confidence in the validity of the data presented on pediatric use.

Still, it was clear that AZT was no miracle cure. It was quite toxic, and there was evidence that the virus became resistant to AZT after prolonged therapy (FDA, 1990c). The obvious answer was to get more therapeutic options quickly. However, testing of a nucleoside analogue "cousin" to AZT, dideoxycitidine (ddC), had faltered in 1988 due to severe toxicity (Ibid.). A new study was not initiated until 1989. With ddC delayed for toxicity problems, another nucleoside analogue, dideoxyinosine (ddI), emerged as the focus for desperate AIDS patients at this time.

The application for ddI came before the ADAC on July 18, 1991 (FDA, 1991). The drug sponsor, Bristol Myers Squibb (BMS), presented a sprawling patchwork of data from a total of 12 different studies: eight treatment group studies, most of which were either phase I trials or expanded access "trials" (treatment IND groups from which some response data was collected by personal physicians), and four trials of patients conducted from 1986 to 1989 to be used as historical controls – a practice generally considered fraught with hazard, difficult to interpret, and prone to bias (FDA, 1990c). Effectiveness-related response rates were based on some combination of changes in CD4 cell counts, p24 antigen, increases in patient weight, and other measures. However, CD4 counts constituted the bulk of the data collected on ddI in adults, and formed the core of attempted data analyses in the drug application. Variations in protocol between the studies made comparisons difficult: just within the core group of four studies, there were 24 dose levels, 3 different schedules of administration, 2 routes of administration, and 4 different drug formulations, among other variables. The problem was not that the individual studies presented in the NDA were flawed. As the FDA's medical reviewer, Dr. Rachael Berman, noted, each study met its stated goal. But the studies were now being used for purposes other than their original design (FDA, 1991: 245). According to Berman, "What is very unusual about this NDA, if you look at the treatment group, is that there are no randomized, concurrently-controlled, trials contained in the NDA" (pp. 220–21).

The FDA's approach to this data was also extraordinary. When the FDA was dissatisfied with the sponsor's inconclusive analysis, which found no significant dose–response relationship for ddI, they did their own analysis using a different technique in an effort to "separate signal from noise" (FDA, 1991: 261–77). When the FDA found the results inconclusive even after this additional analysis, they requested interim CD4 data from another, ongoing study of ddI. The FDA's institutional investment in this particular drug was literally embodied in the highly unusual presence of FDA Commissioner Dr. David Kessler, who was in attendance for the entire two days of meetings. In an opening statement, Kessler noted that this "new drug application pushes the envelope of drug development and review," and claimed that the "intense and innovative approach to drug development in evidence here today is the paradigm of the future" (pp. 14, 16). Notably, Kessler was a member of the Working Group on the Drug Approval Process under the purview of the new "Council of Competitiveness," which was the new incarnation of the Task Force for Regulatory Relief now under the direction of Vice President Dan Quayle.

In the end the ADAC was split on the vote, narrowly recommending approval of ddI on the basis of an unproven surrogate end point. As I describe elsewhere (Messner, 2008), the FDA waited until October to approve the drug, very likely to review interim data from another, ongoing study. Clearly, impetus to accept the clinical studies came from many directions, including political pressure. However, an argument can also be made that the use of CD4 counts as an indicator of drug action had been developed and used from the first clinical testing of AZT (Yarchoan, 1998), and clinical data amassed since that time provided some support for the notion that CD4 counts were clinically meaningful, at least for AZT and plausibly also for its chemical analogues. Indeed, in a 1997 interview, Samuel Broder remarked:

> AZT is important both for what it did clinically – there is no question about it – but also for the principle that it established, AZT laid the groundwork for defining surrogate endpoints in other studies, for illustrating that anti-viral agents could work in patients, and for providing a template for moving quickly from a laboratory observation to a proof-of-concept clinical study, and from their [*sic*] to a randomized prospective clinical trial. (Broder, 1997: 16)

In any event, the approach used for ddI moved well beyond the regulatory parameters established by Subpart E, even as it upheld some of the precedents created by it. In the approach to ddI, key decision points continue shifting earlier in the drug development process (i.e., deciding for approval on the basis largely of phase I data).[15] Correspondingly, the logic of risk and benefit articulated in Subpart E is aggressively applied here. However, Subpart E sought to create a phase II trial that mitigated the need for phase III, so the data deficit envisioned by that rule was comparatively limited. Also, Subpart E assumed that clinical end points would form the basis for approval.

Significantly, a year *after* the ddI approval and two years subsequent to the approval of AZT for pediatric AIDS, the FDA published a new set of rules, the Subpart H rules for accelerated approval (FDA, 1992a; 1992b), which authorized the use of unproven surrogate end points to support an NDA for drugs intended to treat serious or life-threatening diseases. While there would have to be a plausible scientific justification for believing an end point clinically meaningful, the clinical proof of that surrogate marker's validity, or confirmation of the drug's clinical benefit, could be pursued in postmarket confirmatory trials. Importantly, whereas the pursuit of post-marketing studies to fill information gaps was *possible* under Subpart E, to be requested at the FDA's discretion, here postmarketing studies are *mandatory*. Subpart H is not an alternative to Subpart E, but in effect a regulatory overlay. In this way, the promulgation of Subpart H represents not only a continuation of the trend started with Subpart E toward postponing selected data gathering until postmarketing, but a *compounding* of it.[16]

AZT and FDA Modernization Act

Many of the practices described above were further enshrined when Congress passed the *Food and Drug Administration Modernization Act* (FDAMA) of 1997 (PL 105–115). Among other things, the law established a new form of expedited approval, "Fast Track," incorporating key features of the Subpart E and H accelerated approval regulations.[17] The Fast Track provisions of the FDAMA apply to drugs "intended for the treatment of a serious or life-threatening condition" and demonstrating "the potential to address unmet medical needs" (Sec. 112). Under the FDAMA, a drug sponsor can apply for "Fast Track" designation coincident with, or any time following, submission of the IND application (i.e., from the beginning of phase I testing). If a drug qualifies for Fast Track designation, then the FDA must "take such actions as are appropriate to expedite the development and review of the application for approval of such product" (Ibid.). Such action includes granting the FDA specific authority to approve new drugs upon determination "that the product has an effect on a clinical endpoint or on a surrogate endpoint that is reasonably likely to predict clinical benefit" (Ibid.). Hence, FDAMA incorporates the early consultation and abbreviation of the traditional clinical trial system characteristic of Subpart E with the Subpart H rule's use of surrogate end points.[18]

Additionally, the FDAMA codifies a practice that had been informally used by the FDA as a rare exception: granting of drug approval on the basis of only one clinical study. As we have seen, although the FDA had approved AZT on the basis of a single phase II study, the Agency did not wish to codify the practice. The FDA noted in the interim publication of the Subpart E rules that the statutory requirement for drug applications to be supported by adequate and well-controlled clinical investigations "has long been interpreted to mean that the effectiveness of a drug should be supported by *more than one* well-controlled clinical trial and carried out by independent investigators" (FDA, 1988: 41521, emphasis added). Moreover, this requirement is considered to be "consistent with the general scientific demand for replicability to ensure reliability of study results." According to the Agency, there have been "a few unusual circumstances in which a particularly persuasive multi-center study has been accepted in support of a claim of increased survival" – such as the already-described case of AZT approval. However, the FDA cautioned drug sponsors to plan for study replication because such "persuasively dramatic results are rare"; the results of "two entirely independent studies will generally be required" for approval (p. 41521). More than that, according to Dr. Ellen Cooper, former FDA Division Director, the original AZT trial could be considered a "study within a study" (Cooper 2006). After the trial was discontinued and the placebo patient group were given AZT, follow-up data on that group of former placebo patients could be analyzed separately from the original study, effectively using the former placebo group as their own control. In this way, the former placebo group provided additional evidence for the effectiveness of AZT.[19] This subtlety was lost on many observers, however. Given the very visible example of AZT, which in all other ways was being held out as an example to follow, the precedent for single-study trials had been established in many minds. According to Cooper, "in general there was this sense that one controlled study would be enough" (Cooper, 2006). This perception was pervasive enough that in a September 1990 meeting of the Antiviral Drugs Advisory Committee, Cooper was moved to declare that "sponsors should stop planning to rely on a single Phase II study to demonstrate efficacy and then scramble to retrospectively turn Phase I or expanded access data into adequate and well-controlled studies" to fabricate additional data, as needed (FDA, 1990b: 28). FDA reiterated this position in its publication of the Final Rule for the Subpart H regulations (FDA, 1992b: 58948).

Nevertheless, the FDAMA modified the existing statutory language such that, at its discretion, the FDA may accept "data from one adequate and well-controlled clinical investigation and confirmatory evidence (obtained prior to or after such investigation)" as evidence to support a NDA if it deems such evidence scientifically sufficient to establish effectiveness.[20] Moreover, and importantly, the FDAMA inserts this language into the section of the Food, Drug, and Cosmetic Act pertaining to *all* new drugs, not the newly created "Fast Track" section of the Act. In other words, this provision of FDAMA modifies the definition of "substantial evidence" generally, and is not limited to products intended to treat life-threatening diseases.

The FDAMA seems to have put the FDA in the awkward position of discouraging single-study trials even while advising on how to design them. The following year the FDA produced a guidance document, *Providing Evidence of Clinical Effectiveness for Human Drug and Biological Products*, which, on the one hand, stresses the importance of study replication to bolster the validity of clinical study findings, yet, on the other hand, describes how one could design a single-study protocol to support an NDA (FDA, 1998). An informal practice that the FDA did *not* wish to codify, nevertheless, ultimately became law, and something on which they would be obliged to provide guidance. The durability and ultimate codification of this practice is traceable to the highly visible precedent set by AZT. Accordingly, the FDAMA's alteration of the substantial evidence requirement extends the original influence of AZT well beyond its original conceptual domain of "drugs intended to treat life-threatening diseases" into the generalized domain of "new drugs."

Conclusions: a pivotal event

According to Dr. Samuel Broder, prior to the development of AZT, many researchers in the nascent antiviral community at the time took the attitude, "A cure or nothing. Give me 20 years and I'll give you a cure." However, said Broder, with AZT came "a recognition that one could make certain partial advances with a single agent" and that recognition opened the door to "bring in things like didanosine, or the other products. One could bring in non-nucleoside inhibitors. One could be more confident that viral protease inhibitors would work and were worth pursuing" (Broder, 1997: 10). For Broder, the first step with AZT had far-reaching effects, conceptually, scientifically, and methodologically.

The example of AZT also had far-reaching effects in the regulatory and legal realms. The widely acclaimed AZT drug development and approval became a resource for future action, forming the explicit model for the Subpart E rules. It also became an important resource for researchers and regulators in their early willingness to use surrogate end points as a basis for AIDS drug decision making. Given its visibility as a success story, it even became a model for some conceptual elements that the FDA would have preferred to suppress, especially single-study trials as the basis for approval. In the process, key decision points moved progressively earlier in the drug development process. Such shifting was evident in decisions for when to hold FDA/sponsor meetings, when to discontinue trials, when to allow expanded access, and when to approve drugs. While these general trends were consistent with historical efforts to reform drug development and approval procedures, the specific, contingent application to AZT and AIDS resulted in an extrapolative effect on many previous reforms. The trajectory of AZT's influence extends well beyond its initial approval in 1987, eventually reaching into the writing of the FDAMA and, through that, to all new drugs.

I would suggest that while many historically contingent factors must be taken into account to describe the circumstances that led to the decisions and events described (including previous and ongoing reform efforts, political climate, the disease itself, and the rise of AIDS activism), this description of conditions is, by itself, insufficient to explain the events of this period. Without AZT, everything changes. If nucleoside analogues had been missed in the early preclinical screening for antiviral activity, leading to prolonged preclinical testing with no promising clinical trials and no visible hope of help for AIDS patients, how different might have been the next decade, even with all the other factors remaining the same? Without a successful example of rapid drug development and approval in the face of AIDS, what regulatory and legislative rules for accelerated approval would have been written? What need would there have been to write them?

More than that, what kind of activist movement would have arisen? The activism movement, as deftly described by Epstein (1996), found its voice and energy mainly in demanding access to experimental drugs and demanding modifications to the traditional approach to clinical trials, both for reasons of ethics and for earlier marketing of important drugs. It is notable that the calls to get "drugs into bodies" escalated in 1987 and 1988 (p. 222), at the apex of visibility and enthusiasm for AZT. Not unique to AZT but also notable is that objections to placebo controls and "body counts" in clinical trials escalated in the aftermath of the first successful AZT trial, which used a placebo control and survival end point. Hence, it is not unreasonable to suggest that the particular approach taken to AZT, perhaps magnified in light of its success, helped to shape the activist agenda. Without visibly efficacious drugs like AZT in clinical testing, there is no focal point for demands for access to therapies, and no center of gravity for such a movement to cohere. Militant activists may have pushed for greater funding, a different research agenda, or modifications to how research is done. However, it seems likely that at least part of the inspiration for their agenda came from the example before their eyes, and that the lack of such an example would necessarily modify the agenda, and very possibly attenuate their stridency, which subsequently might have affected future advocacy movements.[21] In turn, without a fully energized AIDS movement to bolster the deregulatory policies of the Reagan administration and muscularly push reform proposals, the impulse toward regulatory and statutory change would be that much weaker. In this light, AIDS activism appears in part an *effect* of aggressive drug development, as well as a cause of it.

Throughout this period, FDA regulators can be seen as remarkably adaptable to new circumstances. In the 1980s, most obviously, but in the 1970s as well, the FDA consistently used "discretion" to make "exceptions" as deemed appropriate, while seeking to change the rules to match their practice when the practice seemed useful. As we have seen, Congress ultimately followed suit and changed the law, in part, to match reforms already instituted by the FDA in rule and practice. While this practice-before-rule principle may not be surprising to those of us in the social study of science, it is the *opposite* of the linear model of legislative and regulatory action typically assumed in law and economics.[22] However, in a dynamic and changing therapeutic landscape, formal rules cannot adequately guide day-to-day practice. More importantly, this account suggests that rule making is constitutively related to rule "breaking"; each instance of deviation (or "exception") creates a menu of technically proscribed actions available for appropriation as "precedent" and incorporation into new rules. While such exceptions were not necessarily unusual for the FDA, the visibility of AZT in the ripeness of the historical moment facilitated that process of appropriation and incorporation. As a result, the experience with AZT provided one important focal point around which the rising AIDS activism movement could cohere and established new directional vectors for subsequent regulatory and legislative decision making.

Notes

1. In current usage, a distinction is often made between "efficacy" and "effectiveness." The former term is taken to mean the observed effect of an intervention under experimental conditions designed to maximize the likelihood of seeing any effect. The latter term signifies the net benefits and harms experienced by patients under real-world conditions of use of an intervention (see, e.g., Institute of Medicine, 2009, p. 31). Through the period I describe in this chapter, however, the FDA made no such distinction. Indeed, the foundational laws and regulations cited here usually used the word "effectiveness" in describing evidence standards, even though many contemporary observers would describe FDA standards as defining efficacy. For the purposes of this discussion, I have generally adopted the language used in the documents I cite.
2. According to this Senate report, the practice began after the passage of the 1938 *Food, Drug and Cosmetic Act*. Elsewhere I have postulated that the frequency of the practice must have increased after the legislative tightening of 1962 (Messner, 2008).
3. And the system is still evolving. Strictly speaking, the FDA does not dictate the phases of drug development, but it publishes opinions and guidelines. For FDA's view at the time of AZT approval, see FDA, 1988, p. 41518.
4. See US Code Title 21, Chapter 9, Section 355 (d)(7).
5. The Task Force was part of a strategy by the Reagan Administration to control the rule-writing process for all agencies, largely through the intervention of the Office of Management and Budget (OMB) (Morrison, 1986).
6. In later rules, the language "life-threatening or severely debilitating" was modified to "serious and life-threatening" with refinements of the definition of eligible disease types.

7. See drug approval statistics in Gelijns (1990) and Grabowski (1982). Peltzman (1973) provided one of the early, influential criticisms of the 1962 amendments based on the numbers of new chemical entities (NCEs) introduced before and after the amendments. Wardell (1973) introduced the idea of "drug lag" – delay in introduction of new drugs compared to other industrialized nations. See Temin's (1980) criticism of these and other perspectives.

8. In a personal interview (Temple, 2005), Dr. Temple indicated that he worked on legislation in the late 1970s that would have expedited approval of certain drugs by modifying the standards of evidence. This legislation was likely the *Drug Regulation Reform Act of 1979* (Senate bill S.1975).

9. For a more detailed discussion see Messner, 2008.

10. See also Rothman and Edgar, 1992.

11. Although not addressed in this chapter for the sake of space, a parallel shifting is apparent in the decision points for when to grant premarket treatment access to investigational drugs. See Messner, 2008.

12. In the preface to Subpart E (FDA, 1988), the FDA stressed that the issue was not how many phases of study have been completed, but whether convincing data had been generated from the studies done. Hence, the FDA might deny that there is any reduction in the burden of proof required for Subpart E approval. However, the advisory committee that approved AZT (FDA, 1987a) was consciously aware of information gaps that would have been filled by a phase III study. Furthermore, the trend toward less evidence will become increasingly apparent, particularly in the approval of the AIDS drug ddI, for which there was a great quantity of data, but, as we will see, undeniably poor quality.

13. See the September 18, 1990, ADAC meeting on designing pivotal studies for drugs to treat AIDS (FDA, 1990c), where the lack of meaningful surrogate end points for AIDS is discussed at some length. See especially comments by Dr. Daniel Hoth (p. 69) and Dr. Fred Valentine (71). For an insightful account of the debates over the validity of the CD4 cell count marker, see Epstein (1997).

14. For one notorious example, see Anonymous, 1991.

15. The decision to implement treatment IND for this drug was also made on the basis of phase I data (FDA, 1990a).

16. In apparent recognition of this fact, Subpart H contains provisions for accelerated removal from the market of drugs that fail confirmatory trials.

17. The FDAMA is a wide-ranging document providing for a variety of reforms in drug development, review, and approval. However, this discussion will be limited to Fast Track and FDAMA's relationship to other, already existing procedures for accelerated approval.

18. Indeed, Congress takes support of surrogate end points a step further in the FDAMA and requires the FDA to "establish a program to encourage the development of surrogate endpoints that are reasonably likely to predict clinical benefit for serious or life-threatening conditions for which there exist significant unmet medical needs" (FDAMA, Sec. 112).

19. Characteristically, not everyone involved interpreted the situation in these terms. With respect to the data collected in the aftermath of AZT study discontinuation, one ADAC member observed, "It is not a study anymore; it is a continuous monitoring of patients" (FDA, 1987: 178).

20. FDAMA Sec.115[a].

21. It is notable that cancer patient advocates and others subsequently adopted some of the demands and tactics of the AIDS activists. See Epstein, 1996: 348.

22. For a critical review of major works in regulatory theory, see Croley (1998).

Bibliography

Anonymous (1991) "CAST deaths revealed," *Lancet*, 337 (April 20), 969.

Broder, S. (1997) Interview by V. Harden and C. Hannaway, Bethesda, MD, Feb 2, http://aidshistory.nih.gov/transcripts /transcripts/Broder97.pdf

Carpenter, Daniel (2010) *Reputation and Power, Organizational Image and Pharmaceutical Regulation at the FDA* (Princeton and Oxford: Princeton University Press).

Cooper, E. (2006) Interview by D.A. Messner (digital recording, telephone interview), August 10 and 14.

Croley, S.P. (1998), "Theories of regulation: Incorporating the administrative process," *Columbia Law Review*, 98(1), 1–168.

Edgar, H and D. J. Rothman (1990) "New rules for new drugs: The challenge of aids to the regulatory process," *Millbank Quarterly*, 68, sup. 1, 111–42.

Epstein, S. (1995) "The construction of lay expertise: aids activism and the forging of credibility in the reform of clinical trials," *Science, Technology, & Human Values*, 20(4), 408–37.

—— (1996) *Impure Science: AIDS, Activism, and the Politics of Knowledge* (Berkeley, Los Angeles, and London: University of California Press).

—— (1997) "Activism, drug regulation, and the politics of therapeutic evaluation in the aids era: a case study of ddc and the 'surrogate markers' debate", *Social Studies of Science*, 27(5), 691–726.

Food and Drug Administration (FDA) (1970) "Hearing regulations and regulations describing adequate and well-controlled clinical investigations," *Federal Register*, 35(May 8), 7250–53.

—— (1979) *Investigational and New Drug Regulation Revisions: Concept Document* (October), FDA Docket 79N-0388.

—— (1982) "New drug and antibiotic regulations: Proposed rule," *Federal Register*, 47 (October 19), 46622–66.

—— (1983) "New drug, antibiotic, and Biologic drug product regulations: Proposed rule," *Federal Register*, 48 (June 9), 26720–49.

—— (1987a) Anti-Infective Drugs Advisory Committee Meeting, January 16, Bethesda, MD [Transcript: New Drug Application for AZT].

—— (1987b) "New drug, antibiotic, and Biologic drug product regulations: Final rule," *Federal Register*, 52 (March 19), 8798–847.

—— (1988) "Investigational new drug, antibiotic, and Biologic drug product regulations: procedures for drugs intended to treat life-threatening and severely-debilitating illnesses. Interim rule," *Federal Register*, 53 (October 21), 41516–24.

—— (1990a) Antiviral Drugs Advisory Committee Meeting, Volume II, January 30, Bethesda, MD [Transcript: Update on Expanded Access: the ddI Experience].

—— (1990b) Antiviral Drugs Advisory Committee Meeting, March 30, Gaithersburg, MD [Transcript: AZT for Paediatric AIDS].

—— (1990c) Antiviral Drugs Advisory Committee Meeting, Volume II, September 18, Bethesda, MD [Transcript: Design of Pivotal Efficacy Study to Support NDA Approval of New Antiretroviral Agents and Immunomodulators for the Treatment of Individuals with HIV Infection].

—— (1991) Antiviral Drugs Advisory Committee Meeting, Volume I, July 18, Bethesda, MD [Transcript: New Drug Application for ddI].

—— (1992a) "New drug, antibiotic, and Biologic drug product regulations; accelerated approval. Proposed rule," *Federal Register*, 57 (April 15), 13234–42.

—— (1992b) "New drug, antibiotic, and Biologic drug product regulations; accelerated approval. Final rule," *Federal Register*, 57 (December 11), 58942–60.

244 *Donna A. Messner*

—— (1998) *Guidance for Industry: Providing Clinical Evidence of Effectiveness for Human Drug and Biological Products* (May 1998).

Gelijns, A.C. (1990) "Appendix a: Comparing the development of drugs, devices, and clinical procedures," in A.C. Gelijns (ed.), *Modern Methods of Clinical Investigation* (Washington DC: National Academies Press) http://www.nap.edu/catalog/1550.html

Grabowski, H. (1982) "Public policy and innovation: The case of pharmaceuticals," *Technovation*, 1(3), 157–89.

Health, Education and Welfare, Department of (HEW) (1977a) *Interim Reports, Volume I: Review Panel on New Drug Regulation* (November 15) (Washington, DC).

—— (1977b) *Interim Reports, Volume II: Review Panel on New Drug Regulation* (February 28) (Washington, DC).

—— (1977c) *Interim Reports, Volume III: Review Panel on New Drug Regulation* (April 25) (Washington, DC).

—— (1977d) *Final Report: Review Panel on New Drug Regulation* (May 31) (Washington, DC).

Hilts, P.J. (2003) *Protecting America's Health: The FDA, Business, and One Hundred Years of Regulation* (New York: Alfred A. Knopf).

Institute of Medicine, Committee on Comparative Effectiveness Research Prioritization (2009). "What is comparative effectiveness research?" *Initial National Priorities for Comparative Effectiveness Research* (Washington, DC: The National Academies Press).

Markle, G.E. and J.C. Petersen (1980) *Politics, Science, and Cancer: The Laetrile Phenomenon* (AAAS Selected Symposium 46) (Washington, DC: American Associated for the Advancement of Science).

Messner, D.A. (2008) *Fast Track: The Practice of Drug Development and Regulatory Innovation in the Late Twentieth Century US* (Thesis, University of Edinburgh).

Morrison, A.B. (1986) "omb interference with agency rulemaking: The wrong way to write a regulation," *Harvard Law Review*, 99 (March), 1059–74.

Peltzman, S. (1973) "An evaluation of consumer protection legislation: The 1962 drug amendments," *Journal of Political Economy*, 81, 1046–91.

Rothman, D.J. and H. Edgar (1992) "Scientific rigor and medical realities," in E. Fee and D.M Fox (eds), *AIDS: The Making of A Chronic Disease* (Berkeley, Los Angeles, Oxford: University of California Press), 194–206.

Temin, P. (1980) *Taking Your Medicine: Drug Regulation in the United States* (Cambridge, MA, and London: Harvard University Press).

Temple, R. (2005) Interview by D.A. Messner (tape recording, Rockville, MD), May 31.

Wardell, W.M. (1973) "Introduction of new therapeutic drugs in the United States and Great Britain: An international comparison," *Clinical Pharmacy and Therapeutics* 14, 773–90.

US House (1964) *Hearing before the Subcommittee on Government Operations*, 88th Congress, 2nd session, March 24 and 25, April 8, and June 3.

—— (1987) *Hearing before the Human Resources and Intergovernmental Relations Subcommittee of the Committee on Government Operations, FDA Proposals to Ease Restrictions on the Use and Sale of Experimental Drugs*, April 29 (Washington, DC: US Government Printing Office).

US Senate (1997) *Food and Drug Administration Performance and Accountability Act of 1997*, Senate Report 105–43, July 1 (Washington, DC: US Government Printing Office).

Yarchoan, R. (1998) Interview by V. Harden and C. Hannaway, April 30, Bethesda, MD. http://aidshistory.nih.gov/transcripts /transcripts/Yarchoan98.pdf

10

Professional, Industrial, and Juridical Regulation of Drugs: The 1953 Stalinon Case and Pharmaceutical Reform in Postwar France

Christian Bonah

Introduction

Stalinon, 1953: evidence lies beyond the obvious.

Obviously, the drug Stalinon, freshly marketed in France in 1953, had nothing to do with an old Soviet leader and agitator who by that time was almost 75. There is in any case no published or unpublished account claiming such a connection, which today's (and 1950s?) readers or listeners might be quick to establish. There is in fact scientific evidence underpinning the name.

Obviously, Stalinon was a safe, controlled, and certified drug given that it had cleared the severe, formal French regulation channels of official state-approved drug registration (called the "Visa"). In fact, when the "Stalinon case" came to trial in a Paris penal court four years after Stalinon was introduced into market, there was not a single voice to claim that the specialty had complied with any of the three prerequisites for official drug approval: novelty (innovation), therapeutic interest, and nontoxicity.

Obviously, drug regulation is designed as state intervention to protect citizens and public health against what has been stigmatized by insufficient scientific trials and corporate cupidity when there has been a drug scandal. In a socio-historical perspective, authors, including Harry Marks (Marks, 1997; Marks in this volume), Arthur Daemmrich (Daemmrich, 2004; Daemmrich, 2007), and John Abraham (Abraham, 1995; Abraham & Davis, 2006; Abraham & Davis, 2010), have argued on subjects including sulfonamides (1930s), thalidomide (1960s), and Vioxx (2004) that the regulation of drugs in the legal and institutional meaning is usually designed as the state's intervention in the therapeutics market following a major drug-related scandal. Nonetheless, although, as suggested by many articles in this volume, the remedial of a scandal through the introduction of state regulation or the reform of existing regulation is expected to avoid future accidents – as reflected in the victims' and politicians' "never again" argument – neither state regulations nor their continuous reform has ever put a stop to the classical drug-affair scenario: after long years of research, a potent potential drug clears the regulatory requirements; industrial promotion boosts its medical use; sometime later, adverse effects are observed; the mass media cover what becomes an affair of social injustice – a certified drug causes harm, and the harmful drug effect as identified after the drug has been released appears to have "existed" before its approval; the victims go to the courts while the media, politicians, and society at large raise questions about drug safety and control (see for a slightly different version of this scenario, Marks in this volume). From sulfonamides in the 1930s to the very recent French case around Benfluorex (Mediator), drug scandals, far from disappearing, seem to remain permanent fixtures of modern industrial drug development, distribution, promotion, and consumption, and ways of regulating, be they professional, industrial, administrative, public, or juridical, have not made them disappear.

The medicine Stalinon F was developed between 1952 and 1953 by the pharmacist Georges Feuillet. The 1953 marketing authorization (*Visa*) granted by the Central Service for Pharmacy, SCP (*Service Central de Pharmacie*), a department of the Ministry of Public Health, was followed in 1954 by the death of 102 patients and 117 severe injuries with lasting disabilities. Once initial inquiries had indicated that severe adverse reactions and deaths were probably related to the medicine Stalinon, the pharmacist in charge and the directors of the processing firm were taken to court and convicted five years later. Beyond the individual responsibility of the producer, the victims and their association directly challenged the National Laboratory for the Control of Medicaments (*Laboratoire National de Contrôle des Médicaments*, LNCM), which had conducted the toxicology tests, and the SCP, which had granted the necessary authorization in the first place. The Stalinon case triggered a major parliamentary debate on the reorganization of the control and regulation of drugs in France, which lasted from 1954 until the end of the Stalinon appeal trial in 1958. Laboratories working for the Ministry of Public Health were reorganized in 1955 under the auspices of the National Laboratory for Public Health (*Laboratoire National de Santé Publique*, LNSP), and a law of February 4, 1959, established tightened controls on market authorization and (re)established specific patent rights for drugs in France.

My intentions in this chapter are twofold. First, by using the event of the French Stalinon affair between 1953 and 1960 as a snapshot (despite its duration), the present contribution will analyze and test how aims, forms of evidence, actors, and regulatory tools can be described as practiced in the French context, both in light of the Stalinon case and in reference to the ideal, typically postulated ways of regulating characterized and developed by Jean-Paul Gaudillière and Volker Hess in the present volume (Gaudillière & Hess, Introduction). Considered as an instant, a specific moment in the history of drug invention, production, and regulation, my account is not historical and chronological aiming to underscore insufficient remedial action in a struggle to advance biomedicine and its regulation as an "interactive history of invention and convention" (Cambrosio, 2010), notwithstanding that the Stalinon affair and the associated criminal trial led to major changes in public-health law in France: the 1959 reform of administrative drug regulation and the 1959/60 law legalizing patenting of therapeutics for the first time since 1844, albeit with a special provision granting the state marching rights in case of public-health emergency. On the basis of the Stalinon affair, I will seek shifts and alignments, discretion and alliances in regulatory practices, using as a reference the ways-of-regulating model. The model will structure the analysis. At the same time, my specific analysis of the moment's snapshot will challenge the heuristic model in action.

Second, my intention is to introduce law, jurisdiction, and the courts as an additional body and social world ruling on litigations around therapeutic agents. The twentieth century was regularly marred by court cases on a therapeutic agent. The concept of "judicialization" has brought forth, since the 1980s, the idea that court litigation has increasingly become a way of "regulating social questions through jurisdiction" (Thouvenin, 2007: 462–68). Not specific to the world of medicine and drugs, complaints about "judicialization" are part of the discourse of politicians, managers of huge corporations, and physicians in which they claim that social controversies are immediately taken to court, that, accordingly, lawsuits are supposedly growing in numbers, and that court intervention is an inappropriate way of solving growing conflicts of a social nature that would best be dealt with elsewhere. As such, the discourse about "judicialization" is, according to Dominique Thouvenin, part of a strategy to defend the professional interests of the groups involved rather than an obvious and established fact. The fundamental question addressed by this contribution is whether court litigation prior to the 1980s or since then represents a fifth, juridical, way of regulating, or rather an ultimate regulation of regulation. This chapter will argue that throughout the twentieth century, lawsuits were bodies of control and enforcement of all of the four ways of regulating initially described by Volker Hess and Jean-Paul Gaudillière. Courtrooms should thus be considered as complementary and ultimate bodies, the regulation of regulation. At the same time, I will argue that since the 1980s the "judicialization" issue can be read and understood as the emergence of a fifth juridical way of regulating located not above the other four ways of regulating but at their same level. This suggests the hypotheses that in a time of public controversies and in arenas where players needed to "act in a world of uncertainty" (Callon *et al.*, 2001), courtrooms and their procedures were transformed and invested with a new role, a fifth, juridical way of regulating. Courtroom activities and lawsuits since the second half of the twentieth century may thus be read not only as "judicialization" – meaning, here, filing lawsuits more quickly and more frequently – but as the genesis of a double courtroom identity: a fifth way of regulating and a regulation of regulation. These arguments recast the issue of "judicialization" as not merely an (undue) extension of court intervention in social conflicts, but rather as the genesis of a new way of regulating. Glen O. Robinson argued in 1982 that "accident liability rules serve three commonly recognized purposes: compensation of accident victims, fairness, and deterrence of accident-causing behavior" (Robinson, 1982). The first two purposes identified in the 1980s can thus be considered as enforcement mechanisms of a regulation of regulation, while the third purpose, deterrence, can clearly be associated with new patterns and goals of regulation.

Elements for a history of Stalinon F accidents – A timeline

During the latter part of 1952 and in early 1953, the pharmacist and director of a modest pharmaceutical laboratory, G. Feuillet, developed a tin-based preparation as a treatment for furunculosis, a staphylococcus-induced local or general infection of dermal hair follicles. The medicine was produced by a small French contractual pharmaceuticals processing firm (*façonnier*), Février, Decoisy and Champion (FDC). In June 1953, after toxicological and clinical evaluation, Feuillet submitted his preparation, Stalinon ("Sta" for tin and "lino" for linoleic acid identical to vitamin F), to the French regulatory authorities for premarketing approval.

The *Visa* was granted on August 10, 1953, after certification by the Technical Committee for Specialties (CTS) on June 24, and commercial distribution started in December 1953. In February 1954, the LNCM notified that nonconformity had been observed in the Stalinon preparations that they had compared with the officially authorized formula. The notification indicated that significant parts of the Stalinon capsules examined were under dosed, and this was simply dismissed by the pharmacist Feuillet without any further consequences. From May to July 1954, individual accidents were declared by practicing physicians to the producer Feuillet, who denied any evidence incriminating Stalinon as a possible cause of the deaths.

On June 28, 1954, the occurrence of three similar cases in a hospital in Niort raised a physician's suspicion, and he refused to sign the death certificates. This refusal of a significant medical act initiated an administrative procedure, leading to informing the prefect of the *département* that same evening, and followed by a phone conversation with the Central Service for Pharmacy (SCP) at the Ministry of Public Health and the decision the next morning to suspend the distribution of Stalinon immediately. Furthermore, the accusation of possible manslaughter set off a legal investigation on the possible causes of the deaths. By July 7, a connection between the drug and

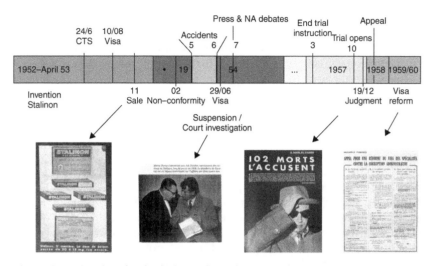

Figure 10.1   Timeline for the Stalinon drug affair in France, 1952–1960

the sudden deaths seemed probable, and the Ministry officially revoked the *Visa* for Stalinon F and prohibited its distribution permanently. The judge in charge of preparing the case (the *juge d'instruction*) publicly mentioned four preliminary causal hypotheses: (1) a decomposition of the active principle under certain conservation conditions; (2) a mysterious idiosyncratic reaction of the victims (specific sensitivity); (3) the combination with vitamin F modified the intestinal absorption of the tin salts; (4) a production error.

Scientific expert inquiries and the preparation of the trial (the *instruction*) had lasted for nearly three years when the Stalinon case came to its conclusion on March 1, 1957. At the same time and after two years of discussions and negotiations, a member of parliament, Maroselli, introduced, on July 24, 1957, a bill to reform drug legislation in order to improve the *Visa* procedure and guarantees for the quality of drugs, to take old products that had been commercialized prior to the *Visa* legislation off the market, and to reduce the number of specialties (over 20,000 products) on the market overall.

The trial of the pharmacist Feuillet who produced Stalinon F opened on October 29, 1957, at the *16e Chambre correctionnelle de la Seine*.[1] Two months later, on December 19, 1957, the Paris court sentenced the pharmacist to the maximum fine possible, two years of jail, and one million francs in compensation. The verdict held Feuillet responsible for the commercialization of a toxic product due to professional misconduct in the production process of the Stalinon specialty. From April 14 to 29, 1958, the court of appeals examined the Stalinon case and confirmed the initial sentence. The ordinance of February 4, 1959, established the "Visa nouvelle législation," a revision of the premarketing authorization procedure of 1941/46, which tightened control over registration and production (Article 601, CSP) and at the same time reestablished patenting rights for drugs (Article 603, CSP). A further decree (N° 60–507, May 30, 1960) defined the application process of the new special patent for drugs.

Stalinon 1: The 1953 administrative way of regulating

In France, the "disaster scenario," where a major drug scandal leads to subsequent regulatory state intervention, applies only partially. The 1941 law passed by the Vichy government regulating the pharmaceutical industry and requiring drug manufacturers to obtain a premarketing authorization (*Visa*) in order to commercialize their products was not drug disaster induced. Updated in 1946, the French premarketing authorization required pharmaceutical producers to present documentation establishing the safety of the new drug to a state commission – the technical commission for medical specialties (CTS) – before being granted state approval for marketing.

Indeed, the Code of Public Health (*Code de Santé publique*, CSP) integrating the *Décret-loi* of September 11, 1941, and the Law of May 22, 1946, stipulated that:

Following article L 599 of the CSP:

> The manufacturing of pharmaceutical compositions or preparations, the packaging of materials of any kind for sale as drugs, prepackaged or not, and restricted to distribution by pharmacists can only be executed under the direct supervision of a pharmacist.

Article L 601 of the CSP established that "a drug specialty – as well as its advertisement – can be sold or published only after having obtained a *Visa* from the Ministry of Public Health upon a proposal of the Technical Commission for Drug Specialties (CTS)."

Article L 665 of the CSP defined that in the case of a "new specialty the CTS has to report the main novelty of the product, its therapeutic interest, and that it does not present any danger to the physical and moral health of the population. For specialties existing prior to the 1941 decree it is sufficient for the CTS to certify the last condition (absence of danger)."

Manufacturing supervision by a trained pharmacist and *Visa* certification by the ministry based on the evaluation of novelty (innovation), therapeutic interest, and nontoxicity were thus basic principles required by law. Following the example of the Food and Drug Administration (FDA) (Marks, 1997; Carpenter, 2010), the creation of the *Visa* in 1941 and its validation in 1946 were intended to increase the protection of the general public but were not prompted by a drug scandal. After 1946, French politicians lobbying for the institution of a European Health Community, the so-called white pool project in reference to the European Coal and Steel Community (Bonah, 2009) considered their regulatory system as the strictest in Europe. At the time, no similar legislation existed in Luxembourg or in the Netherlands, Belgium required some form of authorization, and only Italy had a system of premarketing authorization vaguely resembling the French regulation. The German situation remained based on the National Socialist 1943 *Stopverordnung*, which was an authorization by the Health Ministry but having no specific requirements attached to it, with the different *Länder* applying their own regulations and practices. In the early 1950s, European state control over drug production and production sites existed only in Italy and in France.

In practice, nevertheless, the administrative way of regulating therapeutic agents in France was cast, as indicated by the Stalinon case, in a more ambiguous light. Imagining an organic tin-derivative-based specialty to treat furunculosis, the pharmacist Feuillet contracted with Syntha, a corporation specialized in chemical syntheses, for the production of its active principle, diiododiethyl tin. The raw material was contracted to the pharmaceutical company FDC, acting as a contractual processing firm *(façonnier)* packaging the material in the form of capsules containing 100 mg of vitamin F for 50 mg of diiododiethyl tin. A few grams of the raw material were sent to the LNCM performing toxicology studies of the active principle on a contractual basis. Besides its official role as the analysis laboratory for drug control by the Ministry of Public Health, the LNCM was authorized by the December 31, 1925, clause to conduct remunerated analyses and scientific work for associations and individual clients. It was commissioned under these circumstances by Feuillet to test on mice the toxicity of diiododiethyl tin, "a product to be administered to human beings orally at the maximum dosage of 30 cg a day" (in fact prescription indications finally recommended two capsules containing 50 mg three times a day, that is, only 15 cg). At the same time, Feuillet requested from the SCP at the Ministry of Public Health the transfer of using rights for an old, authorized but abandoned medical specialty based on a tin derivative, Stanolex or Stanomaltine. The acquisition of the rights of the abandoned product enabled Feuillet to present his new specialty as an extension of an already authorized product, thus falling under the restricted authorization procedure of Paragraph 665 of the SCP stipulating that "for specialties existing prior to the 1941 decree it is sufficient for the CTS to certify to the last condition (absence of danger)." On May 1, 1953, Feuillet gave to a friend, Professor Mougenot, a physician at the military hospital Bégin, 260 capsules of Stalinon F in order to fulfill the requirement of clinical testing of the substance. Mougenot treated eight patients for less than a week with six capsules a day without noticing any complications or adverse events.

Based on the LNCM toxicology tests and the open and uncontrolled eight-patient clinical trial, the Stalinon file was submitted on June 24, 1953, to the CTS, the advisory board to the *Visa*-granting SCP. The CTS was made up of Professors René Hazard, Justin Besançon, Raymond Turpin, and Georges Brouet from the Paris Medical School, Professor Guillaume

Valette from the Paris Pharmacy School, Professor Albert Lespagnol from the Pharmacy School in Lille, as well as Mr. Jean Choay, Mr. Chivot, and Mr. Tavernier as representatives of pharmacists and of the pharmaceutical industry, and, finally, the administrator Charles Vaille, head of the SCP and the official secretary of the commission. As indicated in the minutes of the meeting, a favorable vote was given to the modification solicited for the formulation of Stanolex, which then became the authorized Stalinon F.

By the end of July 1954, the Stalinon affair had entered the political arena. The Commission of the French National Assembly on Family, Population and Public Health, an intermediary body between the executive power of the government – especially the Ministry of Population and Public Health – and the legislative assembly auditioned the Minister of Public Health, Louis-Paul Aujoulat, on what had become public debates on the Stalinon affair. Members of parliament questioned the minister on the attitude of his ministry in the Stalinon case and on the measures taken by the minister to avoid the reoccurrence of a similar case. Furthermore, MP Frugier requested information on the conditions under which the *Visa* was granted to specialties and whether they were "of a nature to appease physicians' and patients' legitimate concerns over them following the recent disaster" (CARAN, Procès verbaux de la CFPSP, C 15606, Séance du 29/07/1954, p. 7.) The Minister of Public Health presented to the commission, to its satisfaction and in detail, his ministry's management of the Niort warning and the swift and efficient actions immediately taken by it and the SCP under his responsibility, thus avoiding further comments on the exact functioning of CTS accreditation and the ministry's *Visa*-granting procedures. Anticipating public opinion reactions, the members of parliament considered that:

> The Stalinon affair calls for energetic action and we need to consider without delay a tightening of the inspection of pharmaceutical industries and the establishment of an institute for drug control that can guarantee entirely satisfactory control. (CARAN, Procès verbaux de la CFPSP, C 15606, Séance du 29/07/1954, p. 22)

Refusing to jump to conclusions, the minister insisted that for the time being his services had done what they needed to do and that it was urgent to refrain from further action and let the courts fulfill their role in establishing causation of the harm produced and individual responsibilities. It was the court, as superior regulation body, that should establish whether professional, industrial, and administrative ways of regulating had failed and whether individuals' responsibilities were engaged in terms of illegal or criminal neglect, or misconduct.

Once the accidents had happened, Stalinon victims organized as an association –*Association pour la Défense des Victimes du Stalinon* – that pointed to the malfunctioning of state administration. The criticism was not referring to shortcomings in the forms of evidence mobilized or required, and the aims and values pursued. Rather, the victims' criticism focused on the administrative practices of a "corrupt system" (Association des victimes du Stalinon, undated [1958]) where hundreds of *Visas* were delivered by a 13-member committee meeting eight times a year for three hours, where deliberations were held behind closed doors and remained secret, and where individual experts reporting on specific *Visa* requests were covered by anonymity and thus could not be held responsible individually. The victims' association claimed that state officers could even deliver a *Visa* without consulting the technical committee and insinuated that this had been the case for Stalinon. The association argued that no contradictory hearings had been organized, that members of the technical committee were not truly independent from the pharmaceutical industries, that decisions were never motivated in writing, and that no appeal could be formulated against authorizations that had been granted. All these grievances amounted to the fact that regulatory practices, instead of protecting citizens, produced false illusions of security and control. While deficits in administrative regulatory practices were clearly not substance-related, they were, nevertheless, for victims, politicians, and the press essential causes of the Stalinon disaster. And they were serious enough to consider reforming the 1953 administrative way of regulating drugs in France.

From the perspective of the 1953 Stalinon case, it becomes clear that the French administrative way of regulating therapeutic agents of the time, despite French politicians' pride in considering it the strictest in Europe, is a far cry from the ideal typical model of the administrative way of regulating portrayed by Gaudillière and Hess. While public health and accessibility – two aims of the administrative way of regulating according to table summarizing ways of regulating (WORs) in the introduction – were theoretically included in the Code of Public Health as a legal basis for administrative control, they were not mentioned explicitly as aims in the related practical procedures. Efficacy, the third aim/value of the administrative way of regulating, was theoretically required by law. In practice, it was sidelined by the presentation of Stalinon as a modification of an already existing specialty. Repurchase practice was tolerated by the existing laws. Regarding the forms of evidence, neither statistical evidence nor controlled clinical trials were required in France in 1953 as part of administrative regulatory practices. This underscores a general argument set forth by H. Marks about the slow and very progressive adoption of these evaluation standards in the Western world after 1945 (Marks, 2009). On the other hand, procedures required by the CTS for medical specialties representing the social world of the 1953 French way of granting premarketing *Visas* indicate that administrative regulatory practices were aligned on tacit professional and industrial standards of regulating. Stalinon was tested on eight patients for a week by a sympathetic observer, and toxicology studies were provided for a fee by a laboratory belonging to the regulatory body in charge of market approval. Consequently, it was less the theoretical concept of administrative regulation than its practical implementation that led to converging administrative, professional, and industrial standards for drug safety and efficacy evaluation in the Stalinon case. Whether this expressed successful deterrence by state regulation leading to industrial alignment (there were after all toxicology and patient studies) or whether they should be read as the state regulatory body's lacking independent expertise and tacitly adopting the industrial standards that it was supposed to control needs to be examined beyond the Stalinon case. It needs to be examined from the perspective of industrial and professional ways of regulating and from that of how the three regimes interfered with one another. This is what the following section turns to.

Stalinon 2: Professional regulation and industrial alignments

During the preparation of the Stalinon trial, causation issues around the tortfeasor focused on two major themes. On the one hand, the pharmacist Feuillet was challenged for his personal shortcomings and imprudence in having insufficiently controlled the quality of the product that he marketed. On the other hand, Feuillet was also held responsible for having commercialized an evidently toxic remedy by choosing diiododiethyl tin salt as a therapeutic principal, a compound that major bibliographies qualified as essentially unstable and dangerous. In legal terms, the first implied proving egregious misconduct as required by criminal liability and criminal law. The second related rather to tort liability and questions of legal fairness, balancing multiple causations, liability compensation, and proportional evaluation of the seriousness of the risks generated (Robinson, 1982.) According to the typology of ways of regulating, the first accusation can be compared with the industrial way of regulating, considering quality control as an industrial regulatory tool. In contrast, the grievance of substance choice and the defective understanding of the consequences of mixing diiododiethyl tin and linoleic acids can be interpreted as belonging to the aims and values of professional regulatory ways in terms of competence in preparation. Both grievances were intimately linked to the two professional and industrial realms, and their respective distinctive regulatory characteristics overlap to some extent. Furthermore, during the trial, they were considered as required by law, referring to the administrative way of regulating. Since the legal terms of administrative regulation were inevitably extremely general in their wording, their precise meaning depended intrinsically on the interpretation of what the CTS had adopted from professional and industrial practices as rules to be imposed and the jurisprudence concerning them.

The legal commentaries of Article L 599 of the CSP, "the manufacturing of pharmaceutical compositions...can only be executed under the direct supervision of a pharmacist," insisted that manufacturing control had to cover all the production steps in order to be complete, including an analysis of primary materials, and every moment of the production process, as well as end-product control. They further stipulated that the legal purviews applied to the official producer of the preparation as well as to any contractual processing firm (*façonnier*) (Tribunal correctionnel de la Seine, December 19, 1957).

An explanation by the Minister of Public Health to the commission of the National Assembly on the actual functioning of the CTS disclosed that the SCP often received unofficial inquiries on files to be submitted for drug registration. In this case, the pharmacists working for the SCP, having the official title of "inspectors of pharmacy," advised specialty producers to supplement, if needed, their submission for registration according to the rules imposed by the law or by its interpretation in jurisprudence of CTS decisions (CARAN, Procès verbaux de la CFPSP, C 15606, Séance du 29/07/1954, pp. 25–26). Evaluations covered chemical and clinical tests as well as the physical, microbiological, pharmacological, or therapeutic qualities of the drug involved (Tribunal correctionnel de la Seine, December 19, 1957).

What the Stalinon case highlights, first, is that in the specific French setting of drug production and regulation the two ways of regulating, professional and industrial, were more intricate than in the United States, for example. Second, the Stalinon case indicates that in France industrial regulation by the state emerged more slowly and later than in the United States. This is due to the fact that on the one hand, the pharmaceutical industry in France had for the most part originated in pharmacies (Chauveau, 1999), so its dispersion was enormous: in 1956, the country had more than 1,800 registered pharmaceutical companies and half of the market shares were produced by as many as 43 companies (whereas in Switzerland, for instance, 4 major companies accounted for 50% of market shares in the 1950s) (Chauveau, 1999). While there existed two distinct representative bodies, the Order of Pharmacists (*Ordre des pharmaciens*, 1941) and the Federation of Pharmaceutical Industries (*Chambre syndicale des produits pharmaceutiques*, FPP 1944, and since 1970 the *Syndicat National de l'Industrie Pharmaceutique*, SNIP) the Order of Pharmacists had nevertheless a branch specifically devoted to industry pharmacists. Administrative regulations and law corroborated these specificities in the sense that the public-health code imposed that pharmaceutical companies needed to be majority-owned and run by trained pharmacists. The pharmaceutical industry was dominated by the pharmacists' profession. The social spheres and actors behind the two – professional and industrial – ways of regulating were thus less distinct in France than in many other countries.

The company that produced Stalinon, described as a modest laboratory (less than 10 employees), was thus typical of the 1950s French pharmaceutical scene. Resembling a pharmacy more than a modern industrial plant, Feuillet's aims and values for his company were, nevertheless, described during the preparation of the trial as productivity- and even more as profit-oriented. For instance, it was the high price of diiododiethyl tin that decided the pharmacist to reduce the dosage contained in every capsule from 50 to 15 mg. Similar aims and values are ascribed by the typology to industrial rather than professional ways of regulating. The evidence mobilized for the registration procedure was toxicology-oriented, clinical trials were minimal, and the product followed professional rules for dosage and indications. Animal studies were nonexistent. Given the total absence of statistics and controlled clinical trials, the evidence produced and accepted by the CTS was closer to professional than to industrial ways of regulating. The central issue for political, judicial, and public grievances during the lawsuit against Feuillet and his laboratory crystallized not around poor evidence leading to authorization, but rather around quality control in the production process of the drug. Based on the typology characterizing it as a tool of the industrial way of regulating, the absence of product quality control throughout the entire production process, from the raw materials to the end-product, was interpreted in the courtroom as a substantial causal factor leading to injuries and deaths, implying personal fault of compliance and competency by the individual pharmacist. Although this concerned industrial practices, it was interpreted by the court as serious personal negligence to the administrative regulations required by the 1941/46 *Visa* legislation. Accordingly, this nonobservance of law was the basis for legal sanction. Hearings of the SCP employees established that pharmacy inspectors were far too few to verify in practice the regular observance of law-required quality control. In other words, legal action represented here a mechanism of regulation enforcement, thus acting as control over administrative regulation.

A second fundamental question during the trial, nevertheless, concerned the initial process of granting a *Visa* to Stalinon F in the first place. Were the risk-assessment strategies and practices of the SCP and the CTS efficient? Should Stalinon have received a *Visa* in the first place? Questioning the premarketing authorization procedures and practices of the administration led to a dead end. During the trial hearings, Feuillet refuted the accusation that he was also responsible for having commercialized an evidently toxic remedy. Answering the grievance that he had chosen an active therapeutic principle that major bibliographies qualified as potentially dangerous, that he had presented his product as the reformulation of the abandoned Stanomaltine product whereas it differed significantly from it, and that he had submitted a *Visa* application of questionable experimental evidence, the defendant declared that all these observations were overruled by the fact that the SCP had granted his product a marketing authorization in June 1953. On the one hand, the question of whether toxicological and clinical-test requirements were compliant with the norms and standards of scientific evaluation of the times was a matter of scientific expertise, beyond the scope of a courtroom (Anonymous, 1958). On the other hand, questions on the responsibility of the administration and its members lay outside the field of competence of a criminal court. Negligence in quality control was the proven transgression. Professional or industrial regulatory standards as the foundation for liability in a tortuous act did not need to be detailed, as neglect of administrative regulations was a sufficient basis for the conviction of a professional, who had thus become a criminal.

Finally, the Stalinon case raised a third question. The affair and its negative publicity affected the image of the pharmaceutical industry in general as witnessed by corporate press. By the end of the court proceedings the federation of pharmaceutical industries, *Chambre syndicale des produits pharmaceutiques* (FPP), joined the victims as plaintiff against Feuillet. This astonishing alignment between victims and industrial players of the pharmaceutical industry as plaintiffs in the Stalinon case indicates a strategic distancing of the would-be reputable pharmaceutical industry from what they defined as low-profile, small laboratories run by individual pharmacists that were pharmaceutical corporations only in name and were producing drugs "in a laundry room" (Arnsperger, 1958). While reputation damage was shared by pharmacists and the industry, big or small, serious or dubious, the principal cause designated by the trial was insufficient quality control – a regulatory tool of the industrial way of regulating. Responsibilities were attributed to a dubious individual pharmacist. By aligning the industry with the victims, the individual shortcomings of a tortfeasor were used to dissociate a faulty individual from the profession of pharmacists and, furthermore, pharmacists from the pharmaceutical industry. By transforming potential industrial perpetrators into victims, the alignment was supposed to save the industry's reputation and to designate a clearly identified culprit, suggesting at the same time that only large industry had the capacity to fulfill and monitor correctly the shared administrative–industrial regulatory interface of quality control (Anonymous, *Le Monde*, March 5, 1957). The regulatory tool of quality control was thus in the Stalinon case not simply an external state-imposed requirement, it became at the same time a tool shared by politicians and fractions of the pharmaceutical industry to reorganize what was perceived by both players as a excessively fragmented and overcrowded market (Anonymous, 1954). Associated on the one hand with the victims in the courtroom to fustigate disregard of administrative and professional rules, the industry was also aligned with politicians and the state to demonstrate that everything to ensure quality and safety was being undertaken by an alliance between state administrators, politicians, and "serious" industry. Jean-Paul Gaudillière argues in this volume that the French Colbertism discussed by historians – a unique alliance between private industrial firms and state administration (Minard, 1998. F. Caron, 1997) – does not hold for the pharmaceutical industry during the interwar period. He concludes, rather, that the interwar period was one during which regulation relied "on the expertise and power of medical and pharmaceutical professions" (Gaudillière, this volume). The Stalinon affair, with its outcome of twofold legislative reform (1959/60) – increasing state control over industrial quality-control guidelines and practices (1959) and in return concession of formerly nonexisting patent rights to the pharmaceutical industry (1959/60) – can be interpreted beyond its obvious reinforcement of the administrative way of regulating, as a trade-off between the state administration, politicians, and "progressive" pharmaceutical industrialists. The trade-off consisted in granting patent rights to the industry against its acceptance of increased state control over industrially established quality control in drug-production practices and sites. Thus the 1959/60 reform has to be considered as a defining moment advancing Colbertism in the pharmaceutical domain from the time that the professional representation of the pharmaceutical industry sealed an alliance with state administrators. From the perspective of ways of regulating, this return to Colbertism convincingly illustrates the interplay of administrative and industrial ways of regulating that need to be understood as interacting complementary layers rather than as a chronological succession.

Suffering from the Stalinon crisis and instrumentalizing it at the same time, the pharmaceutical industry used the trial to advance its own professional interests. This led to a significant moment, where in France the pharmaceutical and industrial ways of regulating were redefined. The crisis and the trial served fractions of the industrial corporation in setting standards that were their own, but that integrated the 1959 reformed legislation on drug approval. State administrative regulation was collaboration with and control of industrial regulatory standards.

My preceding remarks converge on the point that the analysis of the different ways of regulating should not be separated from the analysis of the constitution of the entities, processes, and activities that are entangled with them. If this is so, however, then the challenge for historians is to find a way of producing historical narratives that "far from relying on a common-sense understanding of a small set of 'usual suspects,' each clearly assigned to a pre-established, watertight domain (state, industry, science), and of the categories (political, economic, cultural, technical) within which their activities allegedly fall" should instead focus on the shifting composition and modalities of actions of the collectives "that produce and perform the network of conventions and inventions that has come to characterize twentieth-century biomedicine" (Cambrosio, 2010).

Stalinon in court: Toward a juridical way of regulating

Courts, their proceedings, their decisions, and juridical proceedings in general predate the other four ways of regulating presented in the introduction.

Since the 1789 Declaration of Human Rights, individual citizens have had the fundamental right to seek justice in a court – transforming subjective justice into socially recognized justice – and as a corollary have renounced to taking justice into their own hands. The right to seek justice through the courts offsets the state monopoly of legitimate power in modern democracies. In the case of medical and pharmaceutical responsibility, as more generally in the case of industrial or political liability, the idea of being responsible for a given accident is connected to rules that define under what conditions someone who has caused harm to someone else has to repair the consequences. Throughout a long nineteenth century, civil and penal law applied in France to medical and pharmaceutical responsibility in the form of a general set of rules (Bonah, 2007). Beyond professional regulation, whereby the state delegated to pharmacists the exclusive right to deliver medicines under their responsibility of monitoring product safety in their preparation of products (1803 law on pharmacy in France) (Bonah & Rasmussen, 2005), the courts served as remedial bodies of litigation in cases of severe accidents. As a form of regulation, law suits endorsed a role as bodies of ultimate enforcement of professional and administrative regulation. Furthermore, the courts performed a reparatory function –compensation of prejudice of an individual victim establishing causation and the fixation of a compensatory settlement – and in the case of action before a penal court sanctioning social disturbances, in the name of society as a whole. Medical responsibility was tied to the therapeutic act and physicians' activity (in the case of medication, a prescription). Therapeutic products remained in the realm of transactions of goods and merchandises with formal liability for their manufacturer (for the FDA distinction between product qualities and prescription practices, see the contribution by H. Marks in this volume). Since its origins in 1810, French criminal law – the *Code pénal* – prosecuted not only voluntary physical injuries and homicide, but involuntary ones due to misconduct or negligence as well. Prosecutions were individual in nature, as they related to misconduct in respect to social morality and order.

Based on scientific inquiries by experts nominated by the court, the trial preparation for the Stalinon case, between July 1954 and March 1957, investigated proximate and actual causation of evident tort by the Stalinon product, thus establishing a legal basis for judgment. Furthermore, the criminal court needed to establish personal negligence and law transgression in order to convict and sentence an individual thus held responsible for the damage caused. Beyond lengthy debates over "what could be considered as the causes that explained how a *Visa*-certified product could unfurl its toxic action," the experts and the investigating judge preparing the Stalinon case for trial established a long list of shortcomings that could be counted as having contributed to proximate or actual causation of the adverse events and deaths (Robinson, 1982). "Fault" could be imputed to the defendant when his/her conduct disrespected the Code of Public Health, which as administrative guideline had legal character. Moments and places of production where a spoliation could have occurred ranged from the first synthesis of the raw material diiododiethyl tin on February 5, 1953, which went with no control by the pharmacist, to a report by the LNCM in October 1953 providing new physiological testing after a new synthesis of raw material showing reduced toxicity due to low product concentration, and on February 15, 1954, the LNCM detection of the nonconformity of Stalinon capsules with their description in the *Visa* application pointed to a series of different forms of negligence. Their detailed analysis is beyond the scope of this chapter and of limited interest to its central argument. Places and moments where Stalinon could have become toxic and where control should have detected it were divided by the court between those who were legally reprehensible as required by existing public-health law and those who may have been morally reprehensible, but not so in legal terms. The former constituted the long list of negligence drawn up by the prosecution accusing the Stalinon-producing pharmacist, Georges Feuillet.

It is at this point that what had been loosely called by politicians and the press the "Stalinon affair" or the "Stalinon trial" became the criminal court trial *Feuillet et Genet vs. Ministère Publique (Stalinon)* opening on October 29, 1957. Without going here into all the details necessary to justify the accusation, multiple failures to control the product during production steps between March 1953 and June 29, 1954, were the major grievance addressed to a pharmacist described by the court as not very meticulous and too commercially oriented. More especially, the fact that Feuillet had presented Stalinon not as a new product to be certified, but had rather bought an old abandoned tin-containing brand and its name in order to be able to submit Stalinon as the reformulation of the old Stanomaltine – thus avoiding part of the requirements for new drug applications – hinted at dubious maneuvering rather than "ethical" drug production. The prosecution further accused the pharmacist of not having included *Visa* information in advertisements between November 12, 1953, and June 29, 1954, and of having caused – by mistake, inattention, negligence, or nonobservance of the law – injuries to 117 persons and the death of 102.

Accusations were not prioritized according to their importance in the genesis and development of the disaster but instead according to what was legally reprehensible, regardless of whether this put the insignificant omission to print the *Visa* information in an advertisement in a journal alongside having ignored a warning by the SCP that Stalinon capsules were not in conformity with the initial description in the *Visa* application. Once the court had been convinced of the pharmacist's responsibility for the disaster, a single documented negligence considered as a substantial factor leading to injury and death was enough to convict the defendant. When on December 19, 1957, the court sentenced Feuillet to a two-year prison term, a culprit had been designated and justice had condemned his transgressions of the 1941/46 drug regulation (*Visa*) law.

In 1955, victims and their families gathered to set up the Association for the Defense of Stalinon Victims. When the Stalinon case reached the stage where its trial was opened on October 29, 1957, victims and their representatives entered the courtroom proceedings with three clearly identified expectations. They demanded that the responsible pharmacist acknowledge his fault publicly; they requested urgent compensation to pay for the necessary treatment of disabled victims; and, third, they hoped that the court decision would lay the foundations for procedures ensuring that a similar disaster could "never happen again." Adi Zurcher, president of the victims' association, related after the trial that he had been present for the entire hearings and that the pharmacist Feuillet had not, at any point, manifested even the slightest regret or sympathy for the victims. Infuriated by the defendant's attitude, Zurcher, according to his personal testimony, finally approached him personally, and on the last day of the hearings Feuillet, with a broken voice, asked the victims for their forgiveness (Zurcher, 1999).

The compensation issue was even thornier and heralded considerable potential for transfer into the public sphere. Since the accidents in 1954, more than 100 victims with severe and lasting disabilities had received nothing to help them finance long and complicated treatment in rehabilitation clinics. Their seemingly endless waiting, suspended on scientific expertise and court investigations, was first discussed publicly in a radio show called "You are exceptional," hosted by the *Europe N°1* journalist Pierre Bellemare. By presenting the case of a little girl, Marie-Louise N, who lived in the French overseas territories and was not able to finance the trip to Paris to be given care, the journalist lit a fire of public indignation and produced more than 4 million francs in donations. The Stalinon case had once again become a scandal (Boltanski *et al.*, 2007), this time a scandal of the difficulties and delays in compensating the victims rather than of the risk-generating substance itself or its marketing authorization. Major French weeklies followed suit, as, for instance, *Paris Match*, on November 7, 1957, in the midst of the court hearings. Even after the judgment, compensation remained a complicated issue as the insurance company Lloyd, which covered the Feuillet laboratory, further procrastinated payment.

From the preceding analysis, it seems reasonable to consider that the *Feuillet et Genet vs. Ministère Publique* case ushered in a period where court trials were no longer merely a way of regulating the four other ways of regulating, but they began to establish what could be considered as a fifth way of regulating, designated here as juridical. Glen O. Robinson in his 1982 reflections on the diethylstilbestrol (DES) cases in the United States considers that "accident liability rules serve three commonly recognized purposes: compensation of accident victims, fairness, and deterrence of accident-causing behavior" (Robinson, 1982: 736). As indicated by the Stalinon case, compensation issues were at the heart of the procedure. They were raised by plaintiffs well before the trial and they remained after the sentence had been handed down. Since the Stalinon trial, and beginning with the Thalidomide disaster, which fell upon Germany just after the Stalinon case came to a close, victim compensation has become an almost separate issue, in the sense that compensation funds are set up before the cases are judged, which has implied progressively that cases are being referred to the courts but they are more and more frequently settled outside of the courts. The juridical procedure, when interrupted in this way before it ends, has thus been transformed into a way of regulating aiming at compensation, based on the values of responsibility, legal rights, and potential faults. A similar line of events and arguments was illustrated in France in the "HIV-contaminated blood" scandal, where compensation-fund procedures were set up, and the final comment in 1991 by the Minister of Social Affairs, Georgina Dufoix, illustrated the divide when she pleaded that the authorities were "responsible, but not guilty."

Second, the Stalinon case indicates the transition to legal cases where victims are numerous and their legal action becomes collective. The grouping of similar cases in one court, as it happened in the Stalinon case, and the victims' mobilization to set up collective representative associations inaugurated adversarial debates during the second half of the twentieth century, even though they differed significantly from one legal system to another. Criminal- and civil-liability cases became exemplary cases that triggered specific jurisprudence and legal commentaries on "pharmaceutical responsibility." This led not only to multiple causation issues and the "alternative liability" rule for many tortfeasors envisioning an apportionment of damage compensation corresponding to their market share (Robinson, 1982: 743), but also to a development that Sheila Jasanoff has described as the creation of statistical victims with hundreds of anonymous plaintiffs (Jasanoff, 2002). Focused on compensation, disentangling traditional features such as individual plaintiffs and tortfeasors in courtroom procedures, and introducing probability, risk, and apportionment as arguments, the courts have progressively become since the *Feuillet et Genet vs Ministère Publique* case a fifth, juridical way of regulating.

One last element in this transition, deterrence of accident-causing behavior, was present in the Stalinon case. The court's analysis reduced the issue to the individual producer and his professional shortcomings. Adversarial debates with the Stalinon victims' association did not, however. This is the remaining point we are turning to now.

The Stalinon trial never happened:
Victims and the public way of regulating

Even if at the end of the trial Feuillet finally acknowledged his fault of inconsistent production control, the defendant refuted the accusation that he was also responsible for having commercialized a remedy that, given

Le procès du STALINON n'a pas eu lieu !

(Voir pages 4 et 5)

Société de Publicité
et d'Édition
AUBENAS (Ardèche)

ASSOCIATION

POUR LA

DÉFENSE DES VICTIMES DU STALINON

Président-Fondateur de l'Association des Victimes : Pasteur A. ZURCHER, 1 rue Japan, MONTPELLIER (Hérault)

AU NOM DE :

PLUS de 100 MORTS INNOCENTS,

de 150 condamnés à la MUTILATION A VIE,

de 250 familles atteintes dans leur affection, toutes dans le besoin, la plupart dans la Misère, et SANS SECOURS AUCUN depuis 4 ans...

DEVANT :

l'INDIFFÉRENCE ABSOLUE des Services Publics de la Pharmacie :
Ordre National des Pharmaciens, Conseil National et Conseils Centraux de l'Ordre,
Conseil Supérieur de la Pharmacie,
Chambre Syndicale Nationale des Fabricants de Produits Pharmaceutiques,
Académie de Pharmacie,
Ordre de la Santé Publique,
Pharmaciens décorés de l'Ordre National de la Légion d'Honneur,

INDIGNÉES :

de l'attitude des Compagnies d'Assurances refusant les secours qu'ils doivent aux Victimes,

les « VICTIMES du *Stalinon* » ...

(Voir mille page 3)

Figure 10.2   Pamphlet by the Association for the Defense of Stalinon Victims , n.d.
[1958]

the established toxicity potential of tin, represented unreasonable risk.
The defendant declared that these observations were overruled by the fact
that the SCP had granted his product a marketing authorization (*Visa*) in
June 1953. Because of this argument, and the refusal of the penal court,

for procedural reasons of competence, to consider the state administration's omissions, the plaintiffs turned to an administrative court in Paris to sue the state administration and its directors. In the end, this amounted to invoking the responsibility of the French state for having caused damage that was imputable to an administrative decision. The administrative court exonerated the state administration, considering that the *Visa* procedure was intended to strengthen the protection of the public and that it did not alter the responsibilities of the producer. The role of *Visa* control consisted in subordinating authorization for the commercialization of a specialty to proof that a certain number of conditions had been met by the pharmaceutical company, and this did not to free pharmacists of their obligations, which were codified by professional rules and legal prescriptions (Trib. Adm. Paris, April 30: 1957; Trib. Adm. Rennes, 1958). Although it seemed evident that there had been regulatory deficiencies, these were not considered as coming under the jurisdiction of the criminal court and were dismissed by the administrative courts (Trib. Adm. Rennes, 1958). For France of the 1950s, these would have to be dealt with in another arena than the courtroom, as they were essentially of a political nature. And yet, the subsequent reform of pharmaceutical legislation in 1959/60 carefully avoided reforming the state administration's granting of a premarketing authorization. The reform was instead based on the criminal court's interpretation that insufficient quality control by the producer was a principal shortcoming to be corrected through enhanced state control. Meanwhile, the penal court sanctioned a social disturbance caused by a negligent pharmacist. In the name of society as a whole and with prior definition of specific sanctions for specific breaches to the CSP, the pharmacist Feuillet was declared responsible and guilty.

When asked during the trial if Stalinon, if submitted to the CTS in 1957, would still be granted a *Visa*, the secretary of the SCP replied: "Yes, the commission would still grant the substance a *Visa*" (N., Z., 1958). This statement confirmed the victims' misgivings regarding the capacity of the CTS to prevent accident-causing behavior: if dangerous products like Stalinon were not ejected via the administrative way of regulating and the main representative of its government agency declared that a dangerous product would be granted state authorization again five years after the catastrophe, then how could state regulation be depended upon?

In the spring of 1958, after the court of appeals validated the initial judgment and public and media interest faded, the victims and their association engaged in one last act of mobilization. Zurcher and his association published a ten-page pamphlet in the name of "100 innocent deaths, 150 victims condemned to mutilation, 250 families touched in their affection, all of them in a situation of financial distress or even extreme poverty, and with no assistance for over four years now," who requested changes so that a similar disaster could "never happen again." Considering that charges against the state administration responsible for premarketing authorizations were dismissed by the court victims claimed that the "Stalinon trial never happened," at least not the right one. The association presented a catalog of administrative failures and how they had been taken into consideration neither in court, nor in the pharmacy-legislation reform project under debate.

Using a term of the French Revolution, "salubrité publique," to call for a reform of the *Visa* system for specialties, the public appeal focused on what victims saw as administrative corruption. The argument was cast in three columns: (1) corruption of the current system; (2) an insufficient reform project; and (3) victims' proposals for a reform in view of transparency and limitation of corruption. Following their analysis of the corruption of the current system (see section Stalinon 1, p. 252), the victims criticized the reform Bill N°5642 for pharmacy legislation. Examination of the requests was given no specific time frame. Expertise was neither contradictory in form nor independent in status. Most important in the victims' eyes, the reporting experts remained anonymous and were not accountable for their recommendations. The victims greeted the introduction of an appeal procedure but considered again that conflict of interests was not resolved as the appeal remained in the hands of the *Visa* granting CTS. Bitingly, the column dedicated to the reform concluded that obscurantism and arbitrariness remained: what had changed was that there had been 13 irresponsible experts till now, and there would be 50 in the future (referring to an increase in CTS members in the reform project). From the victims' perspective, a solution to corruption and to malfunctioning required the experts' complete independence from the laboratories requesting *Visas*, and the experts should be publicly known and accountable. Decisions should be motivated and made public, and the appeal commission should be independent from the CTS experts. The evidence mobilized should focus on pharmacological properties because the therapeutic value was more complex to define. At the core of the victims' criticism stood not the administrative way of regulating as such, with its forms of evidence and regulatory tools, but its practice and impact. Failings were identified with organizational deficiencies in the structure of the French agency, the CTS, and with "corruption" or negligence in administration.

The association's pamphlet opened a second scandal front but one that was much less attractive for the mass media than the suffering of infant victims. France's general opinion was not yet ready to serve as lobbyist for establishing a public way of regulating. Mass-media attraction to the affair had its limits and beyond the evident scandal potential of Stalinon, more technical aspects related to drug-regulation reform with a detailed analysis of administrative procedures did not serve the media's own interests. The victims' association was run more like a self-help group than like an association of public activists. The trial had convicted an immoral professional playing, to some extent, the role of the scapegoat. The mass media were more attracted by the spectacle of a trial than by a detailed criticism of the practices of the administrative ways of regulating. Drug scandals became standard material for sensational reporting, but sensations were short-lived and needed to be replaced by new ones. Although the institution of the Association for the Defense of Stalinon Victims points in the direction of patient-victim mobilization, the same may be said of their influence on establishing a public way of regulating in France in the late 1950s as what they declared about the trial: it never happened. In their interventions on the trial itself, however, the association introduced seminal developments that would become, 20 years later, elements of the public and the juridical ways of regulating.

Conclusion

It was in the first decades of the twentieth century that most Western countries implemented state regulations that fit the restrictive definition of regulation as the promotion of minimal standards for drugs through state intervention in a specific domain of national economies (Daemmrich, 2004; Cambrosio 2010). This definition can be associated with the more general definition used by Boris Hauray, who characterizes regulation as "permanent and focused control enforced by a public authority on activities of specific value to society" (Hauray, 2006).

Following the conviction of the Central Service for Pharmacy considering that the Stalinon disaster was due mainly to inadequate control of the manufacturing of Stalinon, the beginning of the third legislature of the Fourth Republic integrated the courtroom conclusions into administrative regulations as reforms of pharmaceutical legislation. In 1959, reforms established a new *Visa* legislation requiring wider-ranging reports on the safety, therapeutic utility, and chemical composition of drugs to be submitted before marketing permission was granted, coupled with the patent law allowing patents to be taken out for drugs for the first time in more than 100 years (Ordonnance of February 4, 1959 "relative à la réforme du régime de la fabrication des produits pharmaceutiques et à diverses modifications du code de la santé publique," N° 59–250). Here, considerations on the standards for chemical, biological, and clinical tests as risk-evaluation assets that would be mandatory prerequisites for marketing authorization appeared for the first time in the French context. At the same time, state administration reinforced its own role of production control in laboratories and in the industry, and increased positions for pharmacy inspectors and members of the CTS. While requirements imposed on drug manufacturers were heightened, in return the state offered to accelerate the procedure and granted patenting rights (Brevet spécial de médicine, Décret N° 60–507, May 30, 1960), a transaction that has been described here as a trade-off between state administration, politicians, and "progressive" pharmaceutical industrialists, in short a return to French Colbertism.

What the preceding analysis of the Stalinon case has argued with respect to the theme of regulation is, first, that terminological and procedural standard-setting rules for invention and production of drugs, which, as scientific and industrial activities, had started well before World War I, were conditions for political and regulatory rules codified through law. The latter in their interpretation, however, always remained dependent on scientific and industrial activities. Second, regulation only acquired its full significance when it was applied and enforced with the introduction of possible legal sanctions in the courtroom. In this sense, the courtroom was not a fifth way of regulating but rather an indispensable complement to the professional, industrial, and administrative ways of regulating. At the same time, and beyond the issue of "judicialization," the courts were the place of invention of a fifth juridical way of regulating. By disentangling traditional features such as compensation, causation, and individual plaintiffs and tortfeasors in the courtroom procedures, and by introducing probability, risk, apportionment, and deterrence of accident-causing behavior as elements, the courts here played a new role, not of judging but of regulating pharmaceutical liability within and outside of the courtroom.

Note

1. To be precise the court trial involved three individuals. Henri Genet, the administrator of the *Laboratoires Février, Decoisy, Champion*, was only accused of breaches of the legislation on pharmacy since his activity with the FDC was purely administrative and not technical. Léon Decoisy was accused of manslaughter, but he died during the period of the court investigations, and prosecution against him was then stopped. The pharmacist Georges Feuillet remained the only individual accused of manslaughter.

Bibliography

J. Abraham (1995) *Science, Politics and the Pharmaceutical Industry: Controversy and Bias in Drug Regulation* (London: UCL Press).

Abraham J., Martin P., Davis C. and A. Kraft (2006) "Understanding the 'productivity crisis' in the pharmaceutical industry: Over-regulation or lack of innovation?" in A. Webster (ed.) *New Technologies in Health Care: Challenge, Change and Innovation* (Basingstoke and New York: Palgrave Macmillan).

Abraham J. and Courtney D. (2009) "The role of standards in understanding the modern drug regulatory period: Contextual case analyses of adverse drug reactions," in C. Bonah *et al.* (eds.) *Harmonizing Drugs. Standards in Twentieth Century Pharmaceutical History* (Paris: Glyphe).

Anonymous (1954a), "L'affaire du Stalinon. La multiplicité des laboratoires et des médicaments paraît être une des causes de la défectuosité des produits," *Le Monde*, July 11.

Anonymous (1954b) "La sûreté nationale est chargée de mener dans toute la France une enquête sur les décès attribués au Stalinon," *Le Monde*, July 7.

Anonymous (1954c) "Un médicament peut toujours se révéler dangereux dans certaines circonstances," *Le Monde*, July 8.

Anonymous (1957) "Le parquet va transmettre à M. Golléty le dossier du Stalinon," *Le Monde*, March 5.

Anonymous (1958) "L'affaire du Stalinon," *Revue de sciences criminelles et de droit pénal comparé*, N°1, 87–92.

L. Arnsperger (1958) "Arzneimittel aus der Waschküche?" *Der Spiegel*, February 5, 40–45.

Assemblée Nationale (1954) "Procès verbaux de la Commission de la famille, de la population et de la santé publique (CFPSP)," Centre d'Accueil et de Recherches des Archives Nationales (CARAN): C 15606, Séance du 29/07/1954.

Assemblée Nationale (1957) "Décret du 20 mai 1955. Annexe 3800 (Organisation du laboratoire national de la santé publique)," *Journal Officiel: Documents parlementaires*, January 17, 1247–53.

Association pour la défense des victimes du Stalinon, n.d. [1958], *Le procès du Stalinon n'a pas eu lieu* (Aubenas : Société de Publicité et d'Edition).

L. Boltanski, E. Claverie, N. Offenstadt and S. Van Damme (eds.) (2007) *Affaires, Scandales et Grandes Causes. De Socrate à Pinochet* (Paris: Stock).

C. Bonah, C. Masutti, A. Rasmussen and S. Jonathan (2009) *Harmonizing Drugs: Standards in Twentieth Century Pharmaceutical History* (Paris: Glyphe).

C. Bonah and A. Rasmussen (2005) *Histoire et médicament au XIXe et XXe siècle. Pour une nouvelle histoire du médicament en France* (Paris: Glyphe).

C. Bonah (2007) *L'expérimentation humaine. Discours et pratiques en France, 1900–1940* (Paris: Les Belles Lettres).

Bonah C. (2009) "The birth of the European economic community and questions of drug exchange and standardization" in C. Bonah *et al.* (2009), *Harmonizing Drugs. Standards in Twentieth Century Pharmaceutical History* (Paris : Glyphe).

M. Callon, P. Lascoumes and Y. Barthe (2001) *Agir dans un monde incertain. Essai sur la démocratie technique* (Paris : Seuil).

Cambrosio A. (2010) "Standardization before biomedicine: On early forms of regulatory objectivity," in C. Gradmann and J. Simon (eds), *Evaluation: Standardizing Therapeutic Agents 1890–1950* (Basingstoke and New York: Palgrave Macmillan).

F. Caron (1997) *Les Deux Révolutions industrielles du XXe siècle* (Paris: Albin Michel).

D. Carpenter (2010) *Reputation and Power: Organizational Image and Pharmaceutical Regulation at the FDA* (Princeton: Princeton University Press).

S. Chauveau (1999) *L'invention pharmaceutique, la pharmacie française entre l'Etat et la société au XXe siècle* (Paris: Le Seuil).

A. Daemmrich (2004) *Pharmapolitics: Drug Regulation in the United States and Germany.* (Chapel Hill: University of North Carolina Press).

A. Daemmrich and J. Radin (2007) *The FDA at 100: Perspectives on Risk and Regulation* (Philadelphia, PA: Chemical Heritage Foundation).

B. Hauray, *L'Europe du médicament. Politique- Expertise-Intérêts privés* (Paris: Les Presses de Sciences Po, 2006).

S. Jasanoff (2002) "Science and the statistical victim: Modernizing knowledge in breast implant litigation", *Social Studies of Science*, 32, 37–69.

H. Marks (1997) *The Progress of Experiment: Science and Therapeutic Reform in the United States, 1900–1990* (New York: Cambridge University Press).

Marks H. (2009) "What does evidence do? Histories of therapeutic research," in C. Bonah *et al.* (eds), *Harmonizing Drugs: Standards in Twentieth Century Pharmaceutical History* (Paris: Glyphe).

P. Minard (1998) *La fortune du colbertisme. Etat et industrie dans la France des Lumières* (Paris: Fayard).

Ordonnance (1959a) n° 59–250 du 4 février 1959 relative à la réforme du régime de la fabrication des produits pharmaceutiques et à diverses modifications du code de la santé publique, *Gazette du Palais*, Sem 1, 239–42.

Ordonnance (1959b) n°59–520 du 4 février 1959 relative à la réforme du régime de la fabrication des produits pharmaceutiques et à diverses modifications du code de la santé publique, "Exposé des motifs," *Journal Officiel*, Février 8.

G. O. Robinson (1982), "Multiple causation in tort law: Reflections on the des cases," *Virginia Law Review*, 68:4, 713–69.

S. Timmermans and M. Berg (2003) *The Gold Standard: The Challenge of Evidence-Based Medicine and Standardization in Health Care* (Philadelphia: Temple University Press).

Thouvenin D. (2007) "Est-il pertinent de parler d'une judiciarisation de la médecine" in J.M. Mouillie *et al.* (eds.) *Médecine et sciences humaines* (Paris: Les Belles Lettres).

Tribunal Administratif de Rennes (1958) "Pharmacie," *Recueil Dalloz de doctrine, de jurisprudence, et de législation*, November 24, 108.

Tribunal correctionnel de la Seine (1958), "Feuillet et Genet C Ministère public 'Stalinon' (Jugement en 1ère instance)," *Recueil Dalloz de doctrine, de jurisprudence, et de législation*, Jurisprudence, 257–63.

Zurcher (1999), *Un handicapé au service de Dieu. L'affaire su Stalinon* (Dammarie-lès-Lys: Editions Vie et Santé).

11
Managing Double Binds in the Pharmaceutical Prescription Market: The Case of Halcion

Toine Pieters and Stephen Snelders

Introduction

The epidemic of serious drug safety problems (e.g., Seroxat/Paxil [paroxetine], Vioxx [rofecoxib], Redux [dexfenfluramide] or Ambien [zolpidem]) in the first decade of the twenty-first century has led to public debate on the integrity of the pharmaceutical industry and the effectiveness of drug regulation (Healy, 2004; Moynihan & Cassels, 2005; Anonymous, 2005; Committee on the assessment of the US drug safety system, 2006; Pray, 2007).[1] To improve drug surveillance practices and governance, the various parties participating in the debates proposed a wide variety of changes for processes of international drug development and drug approval. The proposals included the reduction or extension of the patent protection period, abolition of drug patents, strengthening the independence and transparency of regulatory agencies, and incentives for drug development in high-need, high-risk disease areas. But the majority of these proposals failed to address the historical dynamics underlying drug development and use. Employing a historical perspective is a prerequisite to further our understanding of the process of the societal embedding of drugs and the role played by societal concerns and cultural context. Drug trajectories can serve as analytical tool to study the changing scientific, political, and social economies of the prescription drug markets (Pieters, 2005: 3–5; Gaudilliere, 2005; Levy & Garnier, 2007).

The societal embedding of new drugs involves market success as well as regulation and public acceptance. Both the cultural enthusiasm for benefits of drugs and societal concern about the risks and dangers of pharmaceuticals are important in this respect (Davis, 1997). Tranquillizers and other psychotropic drugs have been focal points of cultural enthusiasm in the public sphere, as well as the locus of public controversy. In mutating, adapting, and responding to social, regulatory, economic, and technological

270

events the trajectories of psychotropic drugs include bifurcations, jumps, improvisations, impasses, and dead ends. These twists and turns may result from problems in clinical trials, new indications or a safety panic, and resulting change in marketing practices, public attitudes, or drug policies. Furthermore, the trajectories of psychotropic drugs demonstrate spiral dynamics (Pieters & Snelders, 2009).

Following the thalidomide drug disaster (1958–62), and the subsequent introduction and enforcement of high drug testing and safety standards, drug regulatory agencies came under attack for delaying the introduction of useful medicines and weakening industrial competition. At the same time, the agencies faced political criticism for approving "me too" drugs or drugs later shown to have detrimental side effects (Stephens & Brynner, 2001). In discussing this tension between safety management and drug innovation the US historian Arthur Daemmrich first used the terminology of a so-called double-bind trade-off phenomenon. He identified the double bind between drug safety and drug innovation as a novel characteristic of the drug-regulation culture following World War II (Daemmrich, 2002: 143). We plan to investigate whether the double-bind phenomenon interferes with the trajectory of the most widely prescribed sleeping pill in the 1980s, Halcion. The Halcion case has figured most prominently in international regulatory debates on both drug safety and drug innovation from the late seventies up to this day.

We will compare the precipitate action of the regulatory authorities in the Netherlands, in 1979, with the "wait and see" policy of regulators in the United Kingdom and the United States in the 1980s. We will examine the subsequent precipitate action of the British regulators in comparison to the continued policy of restraint demonstrated by the regulators of the Food and Drug Administration (FDA) in the United States, in the 1990s (Gabe, 2001). Thus, we will compare three distinct political contexts, in which regulatory authorities chose dissimilar paths to manage the double bind between drug safety and drug innovation.

Another form of double bind for cultures of pharmaceutical marketers occurred in the postthalidomide period. This double bind refers to the short-term promotion and "cashing in" on a product versus a focus on establishing credibility and trust over the long term as a marketing strategy for public relations in the pharmaceutical marketplace. Pharmaceutical marketers have to balance aggressive product promotion with careful establishment of brand and company value. On the one hand, aggressive promotion, with exaggerated claims and unrealistic views of potential applications, can be a means to achieve rapid diffusion and market share. On the other hand, carefully building brand and company value with safe and sound health claims and a moderate view of potential applications safeguards a steady return on investment. The increasing costs of research and development and government approval for new drugs and a growing product failure rate in the postthalidomide period appears to favor a focus on high returns on investment and intensely sales-driven marketing strategies in the pharmaceutical industry (Spriet-Pourra & Auriche, 1988; Pathak *et al.*, 1992; Blackett & Robins, 2001; Greene, 2007; Marshall *et al.*, 2009: 103). We will address the risks involved in adopting and unquestioningly maintaining sales-driven marketing strategies. Among other factors, the growing willingness of consumers to defend their rights by seeking recourse in the legal system dramatically changed the risk of liability when producing and marketing prescription drugs. We will argue that the backlash of public opinion in the Halcion case due to persistent aggressive drug marketing policies helped to set the stage for public debate in the twenty-first century on the integrity of the pharmaceutical industry and the effectiveness of drug regulation.

Regulating Triazolam (Halcion®) in the Dutch context

Triazolam was first synthesized and tested by the US pharmaceutical company, Upjohn, in 1969. The new compound promised to be the lead compound of a new generation of safe and short-acting benzodiazepines (benzos). Since the introduction of Librium in 1960, the medications in this chemical class led the top-selling drug list in Western countries (Baenninger *et al.*, 2004). Most notably, Librium, Valium, and Mogadon helped to expand the medical market of mass treatment for nervous problems in "the age of anxiety" (Tone, 2009). However, over time, reports accumulated on the adverse effects of these minor tranquillizers: tolerance, dependence, drowsiness, reduced alertness, and other reactions leading to traffic accidents (van der Lugt, 1961; Nelemans, 1968; Anonymous, 1967; Dunlop, 1970; Medawar, 1992). In addition, a number of other social and cultural consequences became manifest as part of a lingering controversy over defining medical and nonmedical drug use or abuse. The historian Susan Speaker showed how the criticism of the manufacturing and use of benzos in the United States was linked to the fact that benzos had become a symbolic focal point for articulating and addressing social anxieties and tensions (e.g., addiction, alienation, and hypocrisy), and also public expectations for "The Good Life" (Speaker, 1997: 350).

As the first generation of benzodiazepines gradually fell into relative disrepute with the public, pharmaceutical companies tried to develop minor tranquillizers with improved therapeutic and safety profiles (Pieters & Snelders, 2007). Triazolam was one of the most promising compounds. By 1976 a number of clinical trials had been conducted, and demonstrated that the new drug was an effective and safe sleeping pill with a very short half-life that compared favorably with older benzodiazepine analogues. Triazolam allowed users who took the pill at night to function normally and drive safely the next morning without feeling groggy (Schwartz, Kramer & Roth, 1974; Vogel, Thurmond, Gibbons *et al.*, 1975). Apart from an early warning about the risk of amnesia after Triazolam use from the Sleep Research and Treatment Center at the Pennsylvania State University medical school (played down by most other investigators) no remarkable toxic or side effects were reported (Kales, Kales, Bixler *et al.*, 1976: 406). In 1976 Upjohn sought approval to market the new drug under a shining trade name, Halcion, with a recommended daily dose of 0.25–1.0 mg in various European countries. Together with the Belgians, the Dutch regulatory authorities were among the first to initiate the review process for Halcion.

Pharmacists and physicians as part of a corporative alliance system, derived from medieval guild authority, historically regulated the Dutch and German drug market (Daemmrich, 2002: 140). There was no formal drug approval system in place until the Dutch Medicines Evaluation Board (MEB) was established by law in 1958. However, the law would not be fully executed until 1963 (van den Berg, 1964). The members of the MEB were recruited from both professional and governmental bodies. With the help of the administrative body, the so-called medicines control agency, the MEB decides on behalf of the licensing authority, the Department of Health, on specific aspects of the quality, safety, and efficacy of new drugs seeking marketing approval. As in the case of the FDA in the United States, the preregistration burden of proof regarding safety and efficacy for a new drug is relegated to the drug companies.[2] Like their US counterparts, the Dutch regulatory authorities do not actually carry out in-house safety or efficacy testing of new drugs, but review and assess the results of testing conducted and submitted by the drug industry. Although similar, the Dutch regulatory process was less transparent and open to public scrutiny than MEB's US counterpart, the FDA (van der Giesen, 1996; Daemmrich, 2004; Dukes, 1992).

Shortly after Belgium took action, the Dutch MEB issued a market authorization for Halcion. In November 1977, Halcion was licensed at the daily doses of 1.0 mg, 0.5 mg, and 0.25 mg. In the Netherlands, general practitioners and patients alike enthusiastically embraced the marketing promise of "the key to a good night's rest and a good morning."[3] Within a year, there were more than 50,000 registered Dutch Halcion consumers and sales continued to grow exponentially. However, in the summer of 1979, the "joyous pill peddling" atmosphere of Upjohn's aggressive marketing campaign, encouraging widespread dispersal of free Halcion samples and consequential rapid dissemination of the drug, came to an abrupt end due to a series of dramatic public events (Hoofdredactie, 1979; Dukes, 1989).[4]

Throughout the 1960s and 1970s, in the Netherlands, as in the United Kingdom and the United States, pharmaceutical industry marketing practices came under public fire (Packard, 1963; Abraham, 1970; Silvermann; 1974). This occurred in an atmosphere of growing distrust of scientific medicine and big business, in general, as well as serious concerns about the feelings and rights of patients – psychiatric patients in particular. Within the medical profession itself there was a growing faction of critics calling for reform of medicine.

Cees van der Kroef, who had a private practice in The Hague, in the Netherlands, was one of the critical psychiatrists.[5] Van der Kroef had experience in public campaigning from his work for Amnesty International.[6] Throughout 1978, Van der Kroef noted that his patients were reporting serious symptoms that were difficult to reconcile with the psychiatric conditions for which they were individually receiving treatment. Moreover, his patients had symptoms in common, including acute forms of anxiety, amnesia, paranoia, aggression, and severe suicidal tendencies. In the autumn of 1978 Van der Kroef decided that he had collected enough case material to write an article in the *Nederlands Tijdschrift voor Geneeskunde* (Dutch Journal for Medicine) and to send a letter to the *Lancet*. He submitted both manuscripts early in 1979.

Through informal channels, Upjohn senior executives became aware of Van der Kroef's effort at an early warning about Halcion. Upjohn's European medical affairs director visited Van der Kroef and urged him to share the details of his report. Van der Kroef felt intimidated and refused the request. Out of fear that Upjohn was going to interfere with his publication efforts and cover up his findings, Van der Kroef immediately went public in the belief that this would guarantee him immunity from industry pressure (Groeneweg, 1979).

Media frenzy in the Netherlands

On July 5, 1979, the day before the publication of his article in the *Nederlands Tijdschrift voor Geneeskunde* (Dutch Journal for Medicine), Van der Kroef appeared on Dutch television on the popular program *Televizier Magazine* and dramatically described his findings: "This medicine has turned the lives of my beloved patients into a hell upon earth and stopping the pill was experienced as a liberation" (van der Kroef, 1979; van der Kroef, 1980; van der Kroef, 1982). These words were accompanied by images of desperate patients confirming Van der Kroef's message of a therapeutic nightmare that had to be stopped at all cost.[7] The message was not lost on the public.[8] The TV channel headquarters were inundated with requests from alarmed Halcion users for more information and special hotlines had to be established. In response to the public outcry, a follow-up program was broadcast a week later in which the vice-chairman of the MEB, Graham Dukes, was asked to comment. Dukes did his utmost to defend the authority of the board and to reassure the public that a serious inquiry was under way, but his cautionary arguments were lost in the snowballing of emotionally loaded images of innocent victims of what was increasingly perceived as a toxic drug.[9] Within a week the Halcion case dominated the media and was at the top of the Dutch political agenda.

Questions about Halcion were asked in the Dutch parliament and the MEB was called to account for the reports concerning serious adverse drug reactions with Halcion (Anoniem, 1979a; Anoniem, 1979b; Anoniem, 1979c). The regulatory body reacted swiftly by asking the Dutch Health Inspection Service to send a "dear doctor" letter requesting that they report any data on adverse reactions to Halcion (Meijboom, 1979). Within two weeks, doctors filed 600 reports, a number that exceeded the average for annual reports on all other drugs. Consequently, the MEB warned against possible psychiatric side effects from the drug (Aflevering Halcion geschorst, 1979). The MEB's maneuvering space was dramatically reduced by the dynamics in the public arena as Halcion users shouted "Halcion is murder" in the Houses of Parliament. Great pressure was exerted on the double bind between innovation and safety. Overnight, MEB policy shifted from controlled use of Halcion to the implementation of a general ban on the drug (Anoniem, 1980).

On August 6, 1979, under heavy political pressure, the MEB provisionally suspended Halcion from the Dutch market on the grounds of safety and the refusal of Upjohn to revise the labeling to include serious adverse effects under scrutiny (Meijboom, 1989).[10] By putting the administrative blame on Upjohn the MEB seemed to have found a way to restore public trust and credibility. Upjohn regarded the Dutch suspension of Halcion as an immediate threat to the marketing approval of the drug in other countries: in particular the US home market. Moreover, Upjohn faced a civil suit by a group of Dutch consumers, who charged that Halcion was a "defective drug" and that Upjohn had been negligent about addressing known severe adverse reactions.[11]

UpJohn top management immediately switched over to damage control mode in order to prevent the spread, at all costs, of what they called the "Dutch disease or Dutch hysteria" (Abraham & Sheppard, 1999: 33). Gaining access to the most profitable US market was priority number one for the company; Halcion was still making its way through the FDA-approval process. Upjohn already had its hands full countering a series of critical reports on Triazolam by Anthony Kales' renowned sleep research group at the Pennsylvania State University medical school (Kales *et al.*, 1979). Upjohn chose to leave the Dutch market while they continued to fight the Dutch suspension in court.[12] At the same time they started an unscrupulous national and international campaign to undermine and discredit the validity of the safety challenges and to discredit Van der Kroef as a psychiatrist and researcher (Anoniem, 1982).

Coping with the aggregate of conflicting scientific evidence

Upjohn had powerful professional allies in the Netherlands who questioned the legitimacy and validity of the scientific evidence for Halcion.

The number of cases reported by Van der Kroef and others was thought to be too small. Moreover, it was suggested that the description of the case studies was inadequate and that Van der Kroef had jumped to conclusions with his claim that Halcion was a toxic medicine that could cause serious adverse reactions. Van der Kroef's anecdotal evidence was juxtaposed to the apparently rock-solid safety evidence of large-scale randomized controlled trials performed by reputable medical scientists. Furthermore, by going public before publication, Van der Kroef was alleged to have violated the professional code. He was criticized for not granting his medical peers ample opportunity to assess the quality of his data and he was accused of putting the drug on trial by the media. This action was thought to bias the reporting of adverse effects by doctors, pharmacists, and patients alike and to create unnecessary public unrest. The exponential growth of adverse side effects was considered a psychological artifact resulting from media frenzy rather than a serious safety issue (Breimer, 1979; Zelvelder, 1979, 463; Brouwers & Sitsen, 1979; van der Laan & van Limbeek, 1983). The fact that Van der Kroef, in collaboration with psychologists at the University of Utrecht, had started to do a statistically controlled survey with 400 Halcion users was lost in the turmoil of claims and counterclaims by rival groups of medical experts, Upjohn, regulatory authorities, consumer pressure groups, and lawyers (Gieszen *et al.*, 1981; Gabe, 2001: 1238).[13]

According to the Dutch critics of the suspension of Halcion marketing the crisis should be resolved by duplicating the actions of the British drug regulatory authorities; license only the 0.25 mg and 0.125 mg tablets and restrict the maximum daily dose to 0.25 mg. If Halcion was proven to have adverse side effects these were most likely to occur at a higher dose. Upjohn offered to withdraw the 1.0 mg tablet. The MEB, however, was unyielding and offered only to approve the marketing of the 0.25 mg tablet if there was a revised label referring to the adverse effects at higher doses. As a consequence, the Dutch Halcion ban continued. Subsequently, Upjohn succeeded in discrediting the administrative handling of the Halcion case by the MEB in court although they were not successful in reversing the decision to suspend the marketing of the drug. Upjohn was able to partially place blame on the MEB for the Halcion debacle, and thus made a first step toward restoring the balance of force (Moss, 1997). During the court case, Upjohn organized an international conference with renowned experts in the field of psychotropic drugs. The principal outcome of the conference was consensus on a favorable analysis of the clinical trials data and this was published as a letter to the *Lancet* in response to Van der Kroef's September 1979 early warning *Lancet* letter (van der Kroef, 1979; Ayd *et al.*, 1979). In early 1980, Louis Lasagna, a renowned US pharmacologist, defended Halcion in the *Lancet* and suggested that Van der Kroef might be an "alarmist" provoking media hysteria that had been the main cause of Halcion's suspension (Lasagna, 1980). Moreover, Lasagna's suggestion that the symptoms described by Van der Kroef might have been due "to the licensing of doses up to 1 milligram in the Netherlands" meant that both pro-Halcion letters to *Lancet* implicated the MEB as incompetent in deciding on an appropriate and safe dosage regimen.

Graham Duke's attempt to regain MEB's credibility and authority by presenting a vast amount of clinical evidence demonstrating Halcion's problematic side effects at the first international conference of drug regulatory authorities in Annapolis (Maryland) in the spring of 1980 was to no avail. The Dutch effort to sound the alarm had no persuasive power within the context of US regulatory politics since Upjohn successfully amassed clinical evidence of Halcion's effectiveness in preparation for the drug's FDA review. The FDA review division, despite one of its own medical review officer's concerns about Halcion's safety, seemed to have already made up its mind as a result of close contact with Upjohn's scientific officers. The FDA came to three conclusions. First, the proliferation of Dutch cases had been related to high doses (despite the fact that most of the reported Dutch cases were taking moderate doses). Second, the Dutch regulatory authorities had prematurely yielded to public pressure. Third, the Michigan-based company, Upjohn, had a credible track record for producing high-quality clinical trial data; the evidence of Halcion's effectiveness and safety looked convincing. The FDA did not note a reason to deny a product license for up to 0.5 mg of Halcion in 1982 (Ehrlich, 1988: 66–67; Abraham & Sheppard, 1999).

A further indication of Upjohn's success in bending the aggregate of conflicting evidence to its will was the Dutch high court decision in 1985 to allow a new registration procedure for Halcion. The MEB considered this action as akin to the judgment of Solomon and were publicly humiliated for failing to address a drug safety issue in a legally appropriate manner (Boer, 1990). Although Upjohn was victorious in the Dutch public arena, their simultaneous efforts to manage the dissemination of knowledge and consumer experiences with Halcion in a number of countries proved far more difficult.

As far as Upjohn's marketers were concerned, the focus was on drug promotion for Halcion. The ultimate aim was to maximize short-term profits and achieve the number one sales status in the world for what they coined the "Rolls Royce among sleeping pills" (Hirdes, 1984). In order to achieve this sales goal, Upjohn managers worldwide did their utmost to maintain surveillance of Halcion's safety and efficacy profile and to resist any changes in labeling. They continued to attack unflattering research rather than face the possible implications and tried to influence the climate of medical opinion by giving special attention to opinion leaders. In addition, Upjohn resorted to litigation against high-profile critics and at the same time worked to prevent full public disclosure of data on reported side effects to the dismay of whistle-blowers such as the Dutch psychiatrist Van der Kroef (Ehrlich, 1988; CowleySpringer *et al.*, 1991).

Rehabilitation of the Dutch "Syndrome"

In 1982 Ian Oswald, professor of psychiatry at Edinburgh University, and his colleague, Kevin Morgan, published an article in the *British Medical Journal* in which they put the self-evident positive assessment of a specific pharmacological property, Halcion's short half-life, up for debate. They argued that this property was a distinct disadvantage of Halcion-like drugs since it could be associated with a rebound phenomenon involving anxiety, stress, and fear (Morgan & Oswald, 1982). This article was immediately discredited, with the support of Upjohn, for being unscientific. By 1986, however, the British, French, and Italian regulators were facing increasing public concern about the safety of Halcion as evidenced by a significant increase in the number of Halcion-related adverse reaction reports, among other actions. Gradually, the cautious, "wait and see" strategy of the European regulatory authorities gave way to a more proactive postmarketing surveillance policy. The regulators began to seriously consider Oswald and Morgan's position. Even the FDA, with its close ties to the US drug company Upjohn, felt compelled to reconsider Halcion's safety profile. Ultimately, despite Upjohn's intense lobbying, the FDA required the Halcion package insert to describe the dose-related appearance of side effects including amnesia, bizarre behavior, confusion, hallucinations, and agitation (Abraham & Sheppard, 1999: 40–48). By this action the FDA-mandated US package insert now implicitly confirmed Van der Kroef's earlier clinical findings.

Meanwhile Halcion sales figures continued to rise dramatically. By 1987 Halcion claimed 46% of the US hypnotics market, earning $85 million annually for Upjohn, and was the number one branded pharmaceutical in the hypnotic drug class worldwide with revenues of more than $265 million annually. The Halcion profits helped Upjohn to establish and retain its position as one of the top three US pharmaceutical companies (Sigelman, 1989; Anonymous, 1988).

By the beginning of 1988 French, Italian, and German drug regulatory authorities took a further step and banned the 0.5 mg tablet from the pharmaceutical market. The same year, Upjohn halted production of the 0.5 mg tablet altogether. The company rationalized this step by referring to a "novel" drug development concept they called the "lowest effective dose," a term that was also greeted with enthusiasm by regulators in their efforts to improve their checks and balances (Abraham & Sheppard, 1999: 55–56). But the story of trial, error, and learning did not end here as once again consumer-led actions succeeded in placing the product liability of Halcion on the public agenda.

Ten years after the "Dutch hysteria" the Halcion syndrome hit the headlines in Britain, and this was swiftly followed by a case in the United States. A series of media reports on Halcion's serious safety problems ranging from suicidal behavior to murder fueled the public controversy surrounding the most widely prescribed sleeping pill in the world. By 1990, at least 16 lawsuits had been filed against Upjohn claiming that Halcion had driven otherwise peaceful individuals to murder (Abraham & Sheppard, 1999: 93; BBC, 1991; Ehrlich, 1988; Cowley *et al.*, 1991).

Litigation and public opinion backlash

The 1989 Grundberg versus Upjohn case set the media ball rolling in the United States. Just as more than seven million other Americans, Ilo Grundberg had been taking the prescription drug Halcion to help her sleep. While taking the drug she killed her mother. Ilo Grundberg and her lawyer claimed that she had been acting "under the influence of Halcion" (Cowley *et al.*, 1991: 38). She stated that she had become increasingly agitated and paranoid while taking the drug. Because she had no clear motive for the murder and little memory of it, the experts concluded she hadn't acted voluntarily. Consequently, prosecutors dismissed the case. Ilo Grundberg went free and immediately sued Upjohn – in a $21 million civil suit– for failing to warn her about the severe and sometimes fatal adverse reactions of Halcion. On the eve of the trial Upjohn decided to make an out-of-court settlement without admission of liability. Since this was purported to be a substantial sum, there was the inevitable widespread assumption of Upjohn's culpability. It appeared that the pharmaceutical company feared the negative consequences of publicity if there was a long, public coverage of Halcion's disputed trial, safety, and marketing record, and the massive costs of defending the case. Consequently, Upjohn settled the majority of subsequent cases out of court.

The settlements may have spared the company from emotional battles with unhappy customers; however, it did not quell the safety controversy at all. Halcion stories continued to appear throughout the media, from the popular press to medical journals to radio and TV, and even Philip Roth's (1993) best-selling novel *Operation Shylock*. Upjohn was pilloried, not for distributing a dangerous drug, but rather for aggressively distributing a sleeping drug while deliberately keeping itself in the dark about extent of the risks of swallowing this powerful drug (Bogus, 2003). The outcome was heightened public recognition of the risks involved in taking Halcion and other sleeping pills and more broadly speaking, a decline in public trust in the pharmaceutical industry and regulatory authorities.[14] According to *Newsweek* the lesson for consumers was clear: "Neither Upjohn, nor the FDA nor your doctor can guarantee the pill is right for you."

In 1991 the Committee on the Safety of Medicines (CSM) in the United Kingdom had taken notice of the Grundberg litigation case and attendant revelations in the media about unreliable and deficient trial data Upjohn was providing to obtain market authorization for Halcion. Moreover, the CSM members were aware that the BBC was planning to air a show on Halcion, notably critical of the CSM as well as Upjohn, on a highly rated television program, *Panorama*. Against this background of actual and expected public and political turmoil over Halcion's safety, the British regulatory authorities suspended Halcion's license two weeks before the Panorama program "The Halcion Nightmare" was aired (Gabe, 2001: 1249).

Jonathan Gabe and Michael Bury, both sociologists, have documented that the key feature of this final backlash was the connection between benzo (ab)use, dependence, and an aggregate of broader cultural significance including reduced faith in scientific and medical expertise, and an increased manifestation of critics from *within* the medical profession, who allied themselves with patients and lawyers (Gabe & Bury, 1988; Gabe, 1991; Gabe & Bury, 1996). One might wonder whether this phenomenon first began in the Netherlands, or whether it was occurring with other drugs in Britain and the United States before moving to Halcion. It is worthy of note that the most widely prescribed sleeping pill in the world, Halcion, set a new standard for patient activism for evaluating the safety of psychotropic drugs: consumers who filed lawsuits claiming they or their family were injured by taking this class of drugs.

By 1993 the UK Licensing Authority had revoked all Halcion's licenses (Abraham & Sheppard, 1999; Brahams, 1994). In the same year, Halcion worldwide sales dropped 50% to $131 million. Despite this decline, Halcion did not disappear from the international pharmaceutical market and continued its career as a generic, triazolam. As far as the FDA and the US Institute of Medicine (IOM) were concerned the "final" 1997 assessment of Halcion safety and efficacy data did not "support clearly the existence of a unique profile or syndrome of adverse events associated with Halcion relative to those associated with other drugs of its type." This did not, however, exclude the possibility that when used at higher doses and for a longer duration than recommended by the FDA, the syndromes of adverse effects could become manifest. Unfortunately, according to the IOM and the FDA, "Halcion is prescribed and used in a manner that far exceeds the recommended labelling with respect to dose and duration" (Institute of Medicine, 1997). By this statement, the IOM and FDA indirectly blamed doctors and consumers for the "Dutch Syndrome."

Unfortunately, Upjohn was unable to profit from this official verdict as investors' and stockholders' lack of trust sealed the fate of the company. In 1995 Upjohn was taken over by one of its European competitors, Pharmacia, which in turn was acquired by the US drug company, Pfizer.[15]

Conclusion

The halcyon days of Big Pharma, which for many years were the stars of the world's stock markets with ever-expanding sales and profits, are over.

Now more than ever before, the game of "eat or be eaten" is at play in an industry that is haunted by falling profit margins, empty product pipelines, and increasing competition from generics, and last but not least by a further loss of public trust and credibility. Pill making and pill taking is wide open for public debate. As we have shown, the Halcyon case serves to illustrate the changing dynamics in the pharmaceutical market place at the end of the twentieth century, with reemerging consumer concerns about the safety of prescription drugs.

In the postthalidomide period, efforts to restore public trust in the testing and regulation of medicines included new regulatory barriers. Essentially, the new pharmaceutical testing and evaluation procedures were further professionalized and formalized by relying on professionally endorsed evaluators who used methods based in science and medicine. Later, in the 1970s and 1980s, however, this evaluation framework for the regulation of therapeutic drugs came under increasing attack. Patients wanted a larger voice in choosing the treatment they received. Patient-activist groups, notably the activists for psychiatry and AIDS, began pioneering patient power (Epstein, 1996; Shorter, 2009). These activists sought changes in specific disease treatments, as well as the mechanism for evaluating and marketing drugs. The history of Halcion shows how drug companies and regulators alike were confronted with and tried to manage the consequences of the persuasive public barometer.

The Dutch "trial by the media," a term used by the US pharmacologist Lasagna, reflects the significant role of the mass media in establishing and amplifying the associations between Halcion, consumers, and drug safety and credibility issues. Given the media's impressive capacity to exaggerate the message of peril and fear, we would like to stress that in the case of Halcion, media coverage was a mirror of the expectations, legitimations, and opinions circulating in the public realm. This also applies to the wave of publicity for Halcion in United States in the early 1990s. Questions about medical, regulatory and industrial authority and credibility for Halcion as an effective and safe drug were connected to broader concerns about medical care and treatments. The role of Halcion as symbolic focus of social issues was articulated as a result of newly forged alliances between patients, doctors, and scientists. The media storyline of the "Dutch disease" in the United Kingdom and United States is exemplary of the increasing pressure from consumer groups and individuals, who interfere with drug evaluation trajectories through trial at the bar. However, the specific differences in the phases of media and regulatory trajectories for pharmaceuticals in the Netherlands, the United Kingdom, and the United States require further research. This is also relevant for the extent to which structural social and cultural factors or contingency play a role (e.g., the accidental presence of a politically well-informed Dutch whistle-blower or a US litigation champion).

As far as the double-bind phenomenon is concerned, the Halcion case has novel features. The phenomenon implies that actors in the field, including producers, regulators, marketers, prescribers, and users, have a common interest in the swift introduction of new therapeutic drugs while at the same time are concerned about the licensure of drugs that later show only marginal benefits as well as detrimental side effects. As such the double bind might serve as a critical ingredient of a system of checks and balances to account for a culturally acceptable benefit–harm assessment of therapeutic drugs. However, we have to realize that the double bind works differently depending on the actors involved and on the cultural context. The US regulatory trajectory, for instance, clearly demonstrates that managing the double bind for a "home-made" prescription medicine tends to be different from a "foreign-made" prescription medicine; the former favors permissive regulation and thus benefits the drug manufacturers, rather than the latter, for which precautionary drug regulation is more likely to occur (Abraham & Davis, 2009).[16] In addition, within a more litigious society like the United States the double bind is most likely to stimulate investments and drug innovation in disease areas with a high risk/benefit ratio, as in cancer, at the cost of areas with a lower risk/benefit ratio, such as psychiatry.

In the introduction we referred to a slightly different double bind for the drug company and its marketers: short-term promotion and cashing in versus long-term building of credibility and trust as a means of marketing public relations in the pharmaceutical marketplace. We have shown this approach to be intimately related to the innovation and safety double bind. Upjohn's refusal to incorporate long-term credibility and trust into their marketing polices had a dramatic effect as consumers demanded more control over their health care. Changing market forces together with Halcion consumers, who pioneered a new form of patient power by interfering with the process of drug evaluation through trial at the bar, placed the drug testing, licensing, and postmarketing surveillance under a historically new form of public scrutiny with far-reaching implications. What this litigation-enforced transparency in the pharmaceutical marketplace will yield in terms of drug innovation and safety is still an open question.

Notes

1. http://www.associatedcontent.com/article/559197/heath_ledger_and_the_ambien_panic.html?cat=5 (visited November 9, 2009).
2. After registration the burden of proof regarding drug efficacy and safety is relegated to the regulatory authorities.
3. Ede: Upjohn 1978, Halcion Advertisement, code NL 8502.4.
4. Ironically doctors and their family members were first in receiving the free samples of Halcion tablets and experiencing the therapeutic (side) effects.

5. Most information in this chapter is retrieved from Van der Kroef's extensive personal archive that was handed over to Toine Pieters in 2002. I am very thankful to Cees for this valuable gift.
6. van der Kroef C., personal communication, June 2002.
7. Avro's Televizier Magazine (1979), 10ᵉ jaargang afl. 39, Juli 5.
8. The holiday season with its recurrent low tide of news events may have helped to blow up or inflate the newsvalue of the Halcion case.
9. Avro's Televizier Magazine (1979), 10ᵉ jaargang afl. 40, Juli 12.
10. College ter Beoordeling van Geneesmiddelen, Persbericht, 5-2-1980. Up to this day those involved in the DMEB decision to suspend the Halcion registration insist on not being influenced by political pressure.
11. Stafafdeling externe betrekkingen Ministerie van Volksgezondheid en Milieuhygiëne, Documentatieblad, onderwerp: 1.772.7 doss. 4b. 11-02-80.
12. Stafafdeling externe betrekkingen Ministerie van Volksgezondheid en Milieuhygiëne, Documentatieblad Geneesmiddelen, Halcion, doss. 1.772.7, 4b, 1980.
13. As Gabe has indicated the rise to prominence of lawyers may reflect the growing importance of patient empowerment and individual patient rights.
14. In general pharmaceutical companies prefer to settle safety issues out of court in order to prevent the public disclosure of company data and to prevent the use of evidence in new court cases.
15. In 2003 – as part of a wave of major mergers within the pharmaceutical industry to improve the rather low productivity of research and development, widen product portfolios, and optimize sales and marketing costs– the latter in turn would be taken over by the American drug company Pfizer. See http://www.pfizer.com/about/history/pfizer_pharmacia.jsp (visited November 12, 2009).
16. What also played a role in the case of the FDA was the growing political pressure from the 1980s onward to speed up the drug approval process and put more emphasis on drug innovation and deregulation.

Bibliography

J. Abraham and C. Davis (2009) "Drug evaluation and the permissive principle: Continuities and contradictions between standards and practices in antidepressant regulation." *Social Studies of Science* 39, 569–98.
J. Abraham and J. Sheppard (1999) *The Therapeutic Nightmare.* (London: Earthscan), p. 33.
R. E. Abraham (1970) Analyse van de Geneesmiddelreclame door middel van advertenties in medische vakbladen. (Haarlem: De Erven F. Bohn).
Aflevering Halcion geschorst. *Huisarts en Wetenschap*, 1979, 22, 370.
Anoniem (1979a) "Beklag over slaappil; Psychiater krijgt reeks reacties," *Volkskrant*, July 13.
Anoniem (1979b) "Kamer: Halcion moet uit de handel," *Telegraaf,* July 13.
Anoniem (1979c) "Tweede kamer wil verbod slaapmiddel," *Trouw*, July 13.
Anoniem (1980) "Halcion is een nachtmerriepil," *NRC*, 7-2-1980.
Anoniem (1982) "Halcion; Een Hollands Drama," *De Stem,* June 21.
Anonymous (1967) "Geneesmiddelen en Verkeersongevallen," *Geneesmiddelenbulletin.* 1, 9–12.
Anonymous (1988) "Top US companies in 1987", Scrip., 1306, 2.

Anonymous (2005) "a survey of pharmaceuticals: Prescription for change," *Economist*, June 18.·

Anonymous (2007) "Billion dollar pills." *Economist*, January 27.

F. J. Ayd *et al.* (1979) "Behavioural reactions to triazolam." *Lancet* ii, 1018.

A. Baenninger *et al.* (2004) *Good Chemistry: The Life and Legacy of Valium Inventor Leo Sternbach* (New York: McGraw Hill), 65–66, 103–04.

BBC (1991) "The Halcion Nightmare." *Panorama* 1991, October 14.

C. van den Berg (1964) "Herinneringen aan de wordingsgeschiedenis van de wet op de geneesmiddelvoorziening," *Tijdschrift voor Sociale Geneeskunde*, 42, 12–15.

T. Blackett and R. Robins (2001) *Brand Medicine: The Role of Branding in the Pharmaceutical Industry* (Houndmills: Palgrave Macmillan).

E. J. Boer (1990) "Bijsluiter moet Halcion weer acceptabel maken." *NRC Handelsblad*, August 10.

C. T. Bogus (2003) *Why Lawsuits Are Good for America: Disciplined Democracy, Big Business, and the Common Law* (New York: University Press), pp. 106–10.

D. Brahams (1994) "Upjohn (Halcion) libel actions," *Lancet*, 343, 1422.

D. D. Breimer (1979) "Opwinding over Halcion voorkomt objectiviteit." *NRC Handelsblad*, August 1.

J. R. B. J. Brouwers and J. M. A. Sitsen (1979) "Halcion verbod: een hazeslaap van zes maanden?" *Pharmaceutisch Weekblad*, 114, 1272–79.

Committee on the assessment of the US drug safety system (2006) *The Future of Drug Safety: Promoting and Protecting the Health of the Public* (Washington, DC: National Academies Press).

G. Cowley, K. Springer, D. Iarovici and M. Hager (1991) "Sweet dreams or nightmare?" *Newsweek*, August 19, 38–44.

A. A. Daemmrich (2002) "A tale of two experts: Thalidomide and political engagement in the United States and West Germany." *Social History of Medicine*, 15, 137–58, 143.

A. A. Daemmrich (2004) *Pharmacopolitics: Drug Regulation in the United States and Germany.* (Chapel Hill: The University of North Carolina Press).

P. Davis (1997) *Managing Medicines: Public Policy and Therapeutic Drugs* (Buckingham: Open University Press).

M. N. G. Dukes (1989) "Halcion: de achterstallige waarheid." *Nederlands Tijdschrift voor Geneeskunde* 133, 44, 2155–57.

M. N. G. Dukes (1992) *Benzodiazepine Litigation: Medical and Legal Aspects of Adverse Reactions to Halcion® (triazolam)*, unpublished manuscript. January 19.

S. Epstein (1996) *Impure Science; Aids, Activism and the Politics of Knowledge* (Berkeley: University of California Press).

J. Gabe (1991) *Understanding Tranquillizer Use: The Role of the Social Sciences* (London: Tavistock).

J. Gabe (2001) "Benzodiazepines as a social problem: The case of Halcion." *Substance Use & Misuse*. 36, 1233–59, 1238.

J. Gabe and M. Bury (1988) "Tranquillisers as a social problem," *Sociological Review* 36, 320–52.

J. Gabe and M. Bury (1996) "Halcion nights: a sociological account of a medical controversy," *Sociology*, 30, 447–69.

J.-P. Gaudilliere (2005) "Drug trajectories: Historical studies of science, technology and medicine." *Studies in History and Philosophy of the Biological and the Biomedical Sciences*, 36, 602–12.

D. M. Dunlop (1970) "The use and abuse of psychotropic drugs." *Proceedings of the Royal Society of Medicine*, 63, 1279–82.

C. Ehrlich (1988) "Halcion madness: The frightening truth about America's number one sleeping pill." *California.* September, 60–67.

J. Gabe (2001). "Benzodiazepines as a social problem: The case of Halcion." *Substance Use & Misuse,* 36, 1233–59.

W. F. van der Giesen (1996) "Regulering van het geneesmiddelaanbod" in H. Buurma, L. T. W. de Jong van den Berg and H. G. M. Leufkens (eds.) *Het Geneesmiddel.* (Maarssen: Elsevier), pp. 80–102.

J. Gieszen, H. Kennis, M. van Marle, N. Remijn and B. van der Zwan (1981) "Halcion een droom of een nachtmerrie? Research report," Department of psychology, University of Utrecht, July.

J. A. Greene (2007) *Prescribing by Numbers; Drugs and the Definition of Disease.* (Baltimore: The Johns Hopkins University Press).

C. Groeneweg (1979) "Het slaapmiddel Halcion: 'Een hel voor duizenden'." *Nieuwsnet,* Juli 28, V en M 317–19.

D. Healy (2004) *Let Them Eat Prozac: The Unhealthy Relationship between the Pharmaceutical Industry and Depression* (New York: New York University Press).

R. Hirdes (1984) "Upjohn Nederland over Halcion: Een droom van een slaapmiddel ... " *De Medicus,* April 6, 14.

Hoofdredactie (1979) Halcion – ontmaskering van een mythe. *Nederlands Tijdschrift voor Geneeskunde,* 123, 1653–55.

Institute of Medicine (1997) *Halcion; An Independent Assessment of Safety and Efficacy Data* (Washington, DC: National Academy Press), VIII, 3.

A. Kales, M. B. Scharf, J. D. Kales and C. R. Soldatos (1979) "Rebound insomnia; A potential hazard following withdrawal of certain benzodiazepines." *JAMA,* 241, 1692–95.

J. Kales, A. Kales, E. O. Bixler *et al.* (1976) "Hypnotic efficacy of triazolam: Sleep laboratory evaluation of intermediate-term effectiveness." *Journal of Clinical Pharmacology* 16, 399–406.

C. van der Kroef (1979) "Halcion, een onschuldig slaapmiddel?" *Nederlands Tijdschrift voor Geneeskunde.* 123, 1160–61.

C. van der Kroef (1979) "Reactions to triazolam." *Lancet* ii, 526.

C. van der Kroef (1980) "Het Halcionsyndroom, een iatrogene epidemie." *Modern Medicine,* 23–25.

C. van der Kroef (1982) "Het Halcion-syndroom – een iatrogene epidemie in Nederland." *Tijdschrift voor alcohol, drugs en andere psychotrope stoffen,* 8, 156–62.

J. W. van der Laan and J. van Limbeek (1983) "Heeft triazolam (Halcion®) een aparte plaats tussen de benzodiazepinen?" *Tijdschrift voor alcohol, drugs en andere psychotrope stoffen,* 9, 129–31.

L. Lasagna (1980) "The Halcion story: Trial by media." *Lancet,* April 12, 815.

J. J. Levy and C. Garnier (eds.) (2007) *La chaine des medicaments; perspectives pluridisciplinaires* (Quebec: Presses de l'Universite du Quebec).

A. G. Th. van der Lugt (1961) "Nadelen van Librium," *Nederlands Tijdschrift voor Geneeskunde.* 105, 992.

K. P. Marshall, Z. Georgievskava, I. Georgievsky (2009) "Social reactions to Valium and Prozac: A cultural lag perspective of drug diffusion and adoption," *Research in Social & Administrative Pharmacy,* 5, 94–107.

C. Medawar (1992) *Power and Dependence: Social Audit on the Safety of Medicines* (London: Social Audit), pp. 80–92.

R. H. B Meijboom (1979) "Head of the Dutch office of drug side effects," Dear doctor letter on Halcion side-effects, July 16, code BBG 2175 RM/AvN.

R. H. B. Meijboom (1989) "De 'Halcion-affaire' in 1979, een loos alarm?" *Nederlands Tijdschrift voor Geneeskunde*; 133, 2185–90.

F. Moss (1997) "Halcion: sweet dreams or nightmare? A never-ending story?" *Pharmaceutisch Weekblad*. 132: 15–18.

K. Morgan and I. Oswald (1982) "Anxiety caused by a short-life hypnotic," *British Medical Journal*. 284, 942.

R. Moynihan and A. Cassels (2005) *Selling Sickness* (New York: Nation Books).

F. A. Nelemans (1968) "Psychofarmaca in het verkeer," *Nederlands Tijdschrift voor Geneeskunde*, 112, 1862–68.

V. Packard (1963) *The Hidden Persuaders* (Harmondsworth: Pelican books).

S. Pathak, A. Escovitz and S. Kacukarslan (eds.) (1992) *Promotion of Pharmaceuticals: Issues, Trends, Options* (New York: Haworth Press).

L. Pray (2007) *Challenges for the fda: The Future of Drug Safety* (Washington, DC: National Academies Press).

T. Pieters (2005) *Interferon: The Science and Selling of a Miracle Drug* (London: Routledge), pp. 3–5.

T. Pieters and S. Snelders (2007) "From King Kong pill's to mother's little helpers: Career cycles of two families of psychotropic drugs: the barbiturates and benzodiazepines," *Canadian Bulletin of Medical History*, 24, 93–112.

T. Pieters and S. Snelders (2009) "Psychotropic drug use: Between healing and enhancing the mind," *Neuroethics*, 2, 63–73.

J. L. Schwartz, M. A. Milton Kramer and T. Roth (1974) "Triazolam: A new benzodiazepine hypnotic and its effect on mood," *Current Therapeutic Research*, 16, 964–70.

S. Shorter (2009) *Before Prozac: The Troubled History of Mood Disorders in Psychiatry* (New York: Oxford University Press).

D. W. Sigelman (1989) "Halcion: Waking up to the dangers of a sleeping pill," *Trial*, November, 38–43.

M. M. Silvermann (1974) *Pills, Profits and Politics* (Berkeley: University of California).

S. L. Speaker (1997) "From 'happiness pills' to 'national nightmare': Changing cultural assessment of minor tranquillizers in America, 1955–1980". *Journal of the history of Medicine*, 52, 338–376.

C. Spriet-Pourra and M. Auriche (1988) "Drug withdrawal from sale: An analysis of the phenomenon and its implications." *Scrip*, August.

T. Stephens and R. Brynner (2001) *Dark Remedy: The Impact of Thalidomide and Its Revival as a Vital Medicine* (Cambridge, MA: Perseus Publishing), 101–11.

A. Tone (2009) *The Age of Anxiety: a History of America's Turbulent Affair with Tranquilizers* (New York: Basic Books).

F. J. Venema (1990) "Halcion® weer verkrijgbaar," *Pharmaceutisch Weekblad*, 125, 870–71.

G. Vogel, A. Thurmond, P. Gibbons *et al.* (1975) "The effect of triazolam on the sleep of insomniac," *Psychopharmacologia*, 41, 65–69.

W. G. Zelvelder (1979) "Halcion®: Gefundenes fressen." *TGO tijdschrift voor geneesmiddelonderzoek*.

12

Pharmaceutical Patent Law In-the-Making: Opposition and Legal Action by States, Citizens, and Generics Laboratories in Brazil and India

Maurice Cassier

After the first patent laws on inventions were enacted in the late eighteenth century, parliaments, governments, institutions, and medical professions endeavored to limit or suspend the extension of monopolies on remedies, in the interests of public health. In 1844 the French parliament, against the government's advice, prohibited patents on medicines for a century. In this respect it followed the line of the Académie de Médecine, which at the time was fiercely opposed to monopolies on medicinal remedies. After 1944, when pharmaceutical processes could again be patented in France, the government instituted a "special license" justified in the name of public health. This measure granted the Ministry of Health the authority to resort to such a license if it deemed that medicines were not sufficiently accessible in terms of price, quality, or quantity (Cassier, 2000). Despite the early internationalization of patent rights, via the Paris Convention of 1883, many other states also excluded medicines from patenting. This was the case for instance in West Germany, until 1968, and Japan, until 1975. The developing countries of interest to us here, Brazil and India, likewise opted for the nonpatentability of pharmaceutical products, respectively, in 1945 and 1970. In 1994 a new phase of globalization of intellectual property (IP) was, however, initiated with WTO trade regulations that extended 20-year patents on medicines to all member countries (May, 2000). Two years later the Brazilian parliament passed a law recognizing pharmaceutical patents, even though the WTO international standard did not come into force until a decade later. India, on the other hand, stalled until March 2005.

In the late 1990s, the globalization of medicine patents triggered an upsurge of actions by governments and civil society calling for the regulation of the scope of these patents. They endeavored to use every form of flexibility in the WTO agreements and national laws in order to strike a balance between IP and access to new treatments. In response to the new hegemony of 20-year medicine patents,[1] a countermovement developed, with the slogan "patients' rights against patent rights."[2] The initiators and promoters were certain countries of the global South, like Brazil and India, NGOs engaged in campaigns for access to treatment, especially in the context of the AIDS epidemic and the arrival of tritherapies from 1996, and public and private laboratories producing generic medicines, who thus defended their possibility to copy new medicines that fell under the 20-year patent law. The emblematic event in this new period of conflict was the Pretoria medicine trial, from 1998 to 2001, in which an international coalition of 39 pharmaceutical laboratories and the national pharmaceutical producers' union sued the South African Ministry of Health and the NGO Treatment Access Campaign (founded in 1998) supporting it. The plaintiffs challenged the validity of two articles of the South African Medicines and Related Substances Control Amendment Act, which granted the Ministry of Health the authority, "in certain circumstances," to decide on compulsory licenses and parallel importation of medicines. During the same period, in 1999, the French NGO Médecins Sans Frontières (MSF – Doctors without Borders) launched its campaign for access to treatment. During that period MSF initiated cooperation with the Indian laboratory Cipla, to obtain inexpensive tritherapies. Cipla approached the South African government at the beginning of 2001 with an offer of generic medicines in the event of a compulsory license. The pharmaceutical industry's *Sainte Alliance* ("holy alliance") withdrew its complaint, leaving the South African law intact, but negotiated an agreement to ensure that it would not be affected by compulsory licenses. This legal tug-of-war was also emblematic of a new activism regarding medical therapies, bringing together states – especially their Ministries of Health – international NGOs and patient organizations, and generics laboratories in the global South (Biehl 2009; Loyola 2009). The Pretoria trial furthermore marked the globalization of NGOs' struggles over medicine patents (Cassier, 2002).

These struggles through the law (petitioning patent offices, lawsuits) and over the law (public campaigns) have proliferated since the early 2000s, especially in Brazil and India. In Brazil, struggles to obtain compulsory licenses have intensified since 2001, under the impulse of the HIV/AIDS program, MSF Brazil, AIDS patient organizations, and public and private generics manufacturers engaged in reverse engineering of patented antiretrovirals (ARVs). Legal opposition to patents provided for by Brazilian patent law and Indian law has multiplied since 2006 and led to several refusals of patents on ARVs and on a particularly expensive cancer medicine, Glivec, owned by Novartis. This opposition often brings together NGOs and generics laboratories, to defend the accessibility of medicines and the local production of generics. Major court cases have taken place and have contributed to jurisprudence in this domain. In the Merck laboratories versus FarManguinhos

(Brazilian public pharmaceutical laboratory) trial in 2006, the court ruled that a local laboratory had the right to reverse engineer a patented medicine and then to register it as a generic. In the same year, Novartis sued the Indian government for refusing its patent on Glivec, and for an article in the Indian patent law that prohibited the patenting of new formulae for known molecules. The court ruled in 2007 that a state had the right to define the bounds of patentability, in the public interest.

The study of these actions and regulatory measures, and of the special mission to examine pharmaceutical patents that was entrusted to the Brazilian medicine agency in 2001, enables us to reconstruct the process of pharmaceutical patent law in-the-making. We see how legal battles and the intervention of the Brazilian medicines agency in the patent-granting process resulted in stricter patentability criteria and in several patents falling into the public domain. Public campaigns for compulsory licenses, and the Brazilian state president's May 2007 decision to approve a compulsory license on a patented ARV, effectively exploited all the flexibilities in patent law, against all external and internal pressure.[3] These struggles, measures, and decisions have defined a new balance of power between the owners of therapeutic inventions and the actors of public health, and between the monopolies of the laboratories of the global North and the Brazilian and Indian generics producers' right to copy.

This chapter considers several measures and actions to regulate medicine patents in Brazil and India. The first section examines the procedure of prior consent by the Brazilian medicine agency, for the granting of pharmaceutical patent rights. The second section analyses the Merck/FarManguinhos trial in which the court authorized reverse engineering on patented molecules. The third section studies a third form of flexibility, that is, the use of compulsory licenses to authorize the importing or local production of a patented medicine, Efavirenz, used extensively to combat the AIDS epidemic. In the fourth section I describe the legal battles in which NGOs and generics producers in Brazil and India demanded the prohibition of patents on another commonly used ARV, Tenofovir. The fifth section describes the opposition and the lawsuit over Glivec in India. Finally, the sixth section considers the process of legal acculturation of NGOs in the IP field, and the diffusion of legal counter expertise through these new actions.

Prior consent in Brazil: When the sanitary security agency intervenes in the process of granting pharmaceutical patents (2001–)

In 2001, as the Brazilian program for the free and universal distribution of tritherapies for HIV/AIDS was becoming a model for treating the epidemic, Brazilian Minister of Health and development economist José Serra[4] decided to institute a new procedure for examining and granting pharmaceutical patents. The procedure, which involved the Brazilian national sanitary security agency, Agência Nacional de Vigilância Sanitária (ANVISA), in addition to the national industrial property institute, INPI, was incorporated into the 2001 patent law: "The granting of patents on pharmaceutical products or processes shall be dependent on prior consent from the National Sanitary Surveillance Agency – ANVISA" (Article 229C). In terms of this article, the sanitary security agency had the power to refuse a patent granted by the INPI. This dual authority was fiercely challenged by the INPI: "When we send out a patent for analysis by ANVISA it is because INPI has already given its approval. So every time they refuse to grant this patent we have different reports" (the President of INPI at a public hearing at the Chamber of Deputies, November 11, 2009). The coordinator of IP at ANVISA justified his agency's intervention in the patent-granting process in terms of the special status of medicines: "Luiz Lima said that medicines are so important that the entity which is a national authority in public health should be added to the review process" (public hearing at the Chamber of Deputies, November 11, 2009). ANVISA's head of IP argued that patents could not be granted on the basis of industrial interests only, promoted by the INPI; it also had to take public health interests into account. The agency's involvement in the process of examining patent applications and granting patents was consistent with its mission of controlling the quality, safety, usefulness, and accessibility of health products. It had to ensure that pharmaceutical patent rights were not contrary to public health interests and especially the Health Ministry's programs: "Therefore," he said, "there is a need for a careful review because they are giving a monopoly, and must take into account that this implies a lack of competition and, consequently, the final price of the product to the public and for Health Ministry programmes" (Luis Carlos Lima, ANVISA at a hearing on second-use medicine patents, 2008). The coordinator of IP advocated a conception of property that included its "social function,"[5] and ANVISA was the guarantor of that social function of IP rights on medicines. Leaving it entirely up to the INPI, as certain members of parliament proposed, would undermine the public interest: "If approved," he added, "it could cause a serious drop in the quality of the examination of pharmaceuticals patents, generating worrisome economic and social consequences to society" (Lima, 2008). The lawyers of patient organizations involved in discussions on IP in Brazil saw prior consent as a measure to protect patients' interests: "Prior consent by ANVISA is not, therefore, simple interference in the patent-granting procedure. It is a measure to protect patients, by preventing medicine patents from being awarded when they are undeserved" (Chaves *et al.*, 2008). The NGOs involved in battles over IP rights supported ANVISA's coordination of IP rights.

The implementation of prior consent by ANVISA required the creation within the agency of a group of patent examiners. This unit, called the Coordination of Intellectual Property, was located in Rio de Janeiro very close to the INPI, to facilitate interaction between the two. In 2001, ANVISA recruited 16 professionals (chemists, pharmacists, biologists) who had received training in IP. Four teams were created, each supervised by four examiners who had worked at the INPI for two years. The first patents examined were each discussed and evaluated by a team.[6] Today, after that initial learning period, the examiners carry out an individual examination that is then submitted to the Technical Support Group, consisting of four chemical engineers, a doctor, and a lawyer, and which supervises and discusses each file. The Coordenação de Propriedade Intelectual (COOPI) thus brings together patent examiners specialized in the field of pharmaceutical chemistry and biomedicine, and lawyers who have trained their colleagues in patent law. During the period from 2001 to 2009, when over 1,000 files were examined, this combination of scientific and legal competencies and teamwork created a learning dynamic. The legal expertise and examination practices of ANVISA's IP department are very similar to those of the INPI examiners. It even seems that the ANVISA examiners have more time to examine each file than that allocated to their colleagues at the INPI.[7] Moreover, the ANVISA examiners study a patent application after the INPI examiners have done so, and have access to their examination reports. Since 2001, the ANVISA team of examiners has gradually been consolidated and has developed a corpus of rules and knowledge in the process of examining over 1,000 patent applications. It can therefore be said that ANVISA now hosts a sort of counter expertise to that of the INPI as regards the patentability of medicines. It is precisely this counter expertise that was criticized in the Chamber of Deputies in November 2009 when the INPI president challenged the duplication and confusion of roles between the two agencies. He also criticized the quality of ANVISA's expertise: "It's a small group, yet it is much broader and better prepared than the group of only 18 researchers that Anvisa has" (President of INPI, November 2009).

The reality of this counter expertise is evidenced in the existence of a set of rules and interpretations that partially differ from those of the INPI. The ANVISA COOPI evaluates patent applications on the basis of set patentability criteria: novelty, inventiveness, and industrial application. It emphasizes the technical and legal nature of its examination work, thus denying any "ideological" basis, of which it is sometimes accused.[8] By taking Brazilian patent law and the WTO TRIPS agreements as its references, it anchors its work in national and international law. At the same time, the COOPI also develops its own corpus of interpretations and guidelines that differs from that of the INPI in several respects. The most noteworthy disagreements concern the patentability of the second therapeutic use of a known molecule and of polymorphic molecules. Whereas ANVISA refuses these two types of patent, INPI accepts them. ANVISA considers that patents on a second therapeutic application of a known molecule are "detrimental to public health and to the country's scientific and technological development, and could impede access to medicines" (Guimaraes, 2008). The COOPI's argument is grounded on the principle of accessibility of medicines, and on the hypothesis that the granting of new patents on a known molecule would hinder research for the development of new applications. It contends that the preservation of the public domain is conducive to the free development of investigations. In 2008 the interministerial group on intellectual property (GIPI) followed ANVISA's guidelines and decided to limit the granting of patents on the second therapeutic use and polymorphs. In May 2009 the parliamentary commission on social security also passed a bill prohibiting the patentability of the second use and polymorphs.[9] The controversy spread to the public sphere: a pharmacists' union, the FENAFAR, endorsed ANVISA's guidelines and demonstrated against patents on the second therapeutic application.

These differences of interpretation concerning the rules of patentability, especially the fact that ANVISA adopts higher standards of patentability in order to safeguard the public domain, explain many refusals of patents initially approved by the INPI. The statistics produced by the COOPI show a 5% refusal rate on patents approved by the INPI. ANVISA has emphasized the technical nature of these refusals: "simply for strictly legal reasons such as the lack of novelty or inventive activity." The INPI has retaliated by consistently refusing to publish these decisions, so that the medicines in question have never been formally put into the public domain (Chaves *et al.*, 2008).

The intervention of the national medicine agency in the patent domain has generated considerable opposition and controversy: "The INPI argues that the role of Anvisa should be limited, as set out in Bill 3709/08, introduced by Deputy Rafael Guerra (Partido da Social Democracia Brasileira (PSDB)-MG). Anvisa, on the other hand, has criticized the proposal, arguing that it does not serve the public interest" (ANVISA and INPI disagree on Bill that changes Patent Law, November 27, 2009). Pharmaceutical laboratories affected by ANVISA's refusal of their patents have sued the agency on the basis of what they deem to be the illegitimate intervention of the COOPI. For instance, Aventis, to which ANVISA refused to grant a patent on Taxotere, accused the agency of overstepping its prerogatives, which, according to the pharmaceutical firm, were strictly limited to sanitary affairs (Federal Court of Rio, July 15, 2008). This was also the position of private consultants in industrial property, who maintained that ANVISA's prior consent should apply only to sanitary criteria and not to patentability: "We can see that with prior consent, ANVISA cannot reassess the requirements of patentability itself. ANVISA should be limited to its skills and evaluate the chances that new medicines or new uses for known medicines could – even if only potentially – cause harm to the population's health" (Roner *et al.*, October 16, 2009). The INPI, Aventis, and certain legal firms have thus confined ANVISA's and the Ministry of Health's province to sanitary issues, whereas in their opinion industrial property is the prerogative of the INPI alone. ANVISA's COOPI, however, defends a completely different point of view that justifies its "social function" and its duty to strike a balance between industrial and public health interests. The struggle has been intensified by the fact that ANVISA has developed real expertise on medicine patents, which has proved to be an alternative to that of the INPI in certain respects. Pharmacists' unions, AIDS patient organizations, networks of NGOs working on IP (the GTPI of the Rede Brasileira Pela Integração dos Povos (REBRIP) the Brazilian network for the Integration of Peoples), the parliamentary commission on social security, and the deputies of the Partido dos Trabalhadores (PT[10]) all support the COOPI's action.

Obtaining the right to do reverse engineering on a patented medicine, in the name of the public interest: The Merck/FM trial (2004–06)

In 1997 the Brazilian Ministry of Health launched a program for the reverse engineering of ARVs, with a view to developing local production of generic medicines to supply its free and universal distribution of tritherapies (Cassier & Correa 2003). The copying of the first ARVs was licit as Brazil had excluded pharmaceutical products from patenting from 1945. Paradoxically, when the Ministry of Health launched its copying program, parliament adopted the new WTO standard instituting 20-year pharmaceutical patents. The evolution of tritherapies led the Ministry increasingly to use medicines that, under the new standard, were patented and therefore not legally reproducible. From the early 2000s, the Ministry of Health threatened patent-owning companies with compulsory licenses that would authorize the production of molecules patented by third parties in Brazil, without the owners' authorization. To make the threat credible, the Ministry of Health encouraged the federal laboratory FarManguinhos to develop the reverse engineering of patented molecules. One of the most frequently used molecules in tritherapies in Brazil was Efavirenz, patented by Merck. In 2001 Merck wrote to FarManguinhos, asking it to cease its work on a molecule that it, Merck, owned. The federal laboratory replied that its research concerned the raw material of a patented medicine and was licit in so far as it was not for commercial purposes. This research on reverse engineering was used by the Ministry of Health to secure discount prices on a medicine that absorbed 10% of its AIDS program budget. In case of deadlock, the Ministry could opt for a compulsory license to produce Efavirenz locally: "If they agree to a lower price we will not launch local production, but if they don't, then we're ready" (E. Pinheiro, Director of FarManguinhos, March 2001). Merck eventually agreed to a discount price on Efavirenz and the government withdrew its threat of a compulsory license. Reverse engineering work continued in 2002, primarily to produce a standard of the molecule under an R&D program funded by the sanitary security agency.

In parallel with the reverse engineering work, the federal laboratory engaged in negotiations with Merck to obtain a voluntary license to produce

Efavirenz locally.[11] In March 2004 Merck decided to freeze negotiations after several Brazilian laboratories, both public (Lafepe) and private (Cristalia, Labogen and Globe), registered several Efavirenz generics with ANVISA. In September 2004, FarManguinhos put out an international call for tenders for 200 kg of raw material for Efavirenz, with a view to developing the technology and producing the batches required by ANVISA to register the generic. The objective was twofold: to pressurize Merck into resuming negotiations on a voluntary license and to acquire the technology so that it could register the generic and prepare a compulsory license, if necessary. Merck immediately demanded the withdrawal of the call for tenders, claiming that it was the only firm authorized to produce and commercialize Efavirenz. It denounced the call as "an illegal and inconceivable compulsory licence on a patented product."[12] FarManguinhos then turned to Merck to supply it with the 200 kg of raw material for the purposes of "research and technological development" (letter dated September 29, 2004). The call for tenders was suspended and negotiations on a voluntary license were resumed. A Merck team visited the FarManguinhos laboratories in October 2004, but neither the purchase of the raw material nor the voluntary license materialized. FarManguinhos refused Merck's price offer: whereas its reference was generics prices, Merck's was the patented raw material. Moreover, Merck claimed that it was unable to supply the requested raw material: "Given the increasing number of patients and the consequent increase in global demand for Stocrin" (letter dated February 17, 2005). FarManguinhos then addressed its call for tenders for the active principle of Efavirenz to the Indian generics producer Aurobindo. On February 18, 2005, Merck laid charges against FarManguinhos and demanded the cancellation of this call for tenders[13] on the following grounds: the fact that FarManguinhos had broken off negotiations, whereas it (Merck) had shown its goodwill to cooperate with the AIDS program; that the Indian generics laboratory chosen had not presented all the technical guarantees required in such a sophisticated medicine; that a patented molecule could not be open to competition in a call for tenders; that the transactions of the call for tenders were of a commercial nature even though FarManguinhos was a research institution; and, finally, that the aim of this call for tenders was not R&D but the commercialization of a patented product.

FarManguihnos based its position on two exemptions in Brazilian patent law and in the WTO TRIPS agreements: an exemption on research to acquire knowledge and technology and the Bolar exemption that authorized research on a patented medicine with a view to registering its generic. The federal laboratory justified the acquisition of the technology and the registration of the generic on the grounds of the urgency of making treatment available to patients: "Those who are suffering from the unjustified delays in research on HIV are the people carrying the virus, who have once again been put second after private financial interests."

Initially the Rio de Janeiro Federal Court ruled in favor of Merck and ordered the suspension of the imports. It justified this decision on the grounds that reverse engineering with a view to registering generic medicines was a long-term process – until the expiry of the patent in 2012 – and did not correspond to an emergency. FarManguinhos appealed, however, and the Federal Court's ruling was quashed in August 2005.[14] On appeal, the court reversed the decision and ruled in favor of FarManguinhos, in the interests of public health and the necessity to reverse engineer the molecule: "the delay in developing the above-mentioned research will lead to delays in acquisition of the technology and the production of generic medicines"; "the delay or banning of production will be detrimental to public health due to the lack of generic medicines on the market."[15] The new ruling was thus made on the grounds of public health interests and the urgency to acquire the appropriate technology. The priority was no longer the duration of patent rights but the development of generic medicines in the interests of the population's health. This order of priority was clarified in the October 6, 2005, ruling: "In view of the conflict of interests in this case, we have to emphasize that the economic interests of Merck, the holder of the patent on the medicine Efavirenz, does not take precedence over the joint interests of protecting both the economic order and public health" (Federal Court, October 6, 2005). The protection of the economic order refers here to the interests of the public economy of pharmaceutical R&D undertaken at Fiocruz to produce generics that would suffer from any delay in acquiring the pharmaceutical technology. This decision embedded the Bolar exemption in Brazilian law. In December 2006, when the Federal Regional Court confirmed the ruling, the Director General of FarManguihnos encouraged the government to opt for a compulsory license on Efavirenz. He claimed that his laboratory was able to produce the medicine at half the cost of the price proposed by Merck. The lawsuit filed by Merck in February 2005 had, however, brought research on Efavirenz to a standstill for two years.[16]

Policy of universal access to HIV/AIDS medicines (1996) and decision on compulsory licenses in Brazil (May 2007)

The compulsory license is the most emblematic and controversial measure in patent law. It does not consist in "quashing the patent," as the Brazilians say, but in revoking the exclusivity of appropriation. It is, in a sense, a form of public expropriation, except that the owner loses only the monopoly and not the patent itself. The 1994 WTO agreements on IP do not cover the compulsory license but do contain an article that defines and codifies it as follows: the authorization to use a patent without the owner's authorization, especially for reasons pertaining to public health, national emergency, and noncommercial use by the government (Article 31). Debate on the compulsory license revolves around the fact that it challenges the idea underlying the whole patent system: the monopoly. Yet many countries have included compulsory licenses for public health reasons in their patent laws. The United Kingdom and Canada did so in the early twentieth century, and France in the 1950s. Economist F. Scherer has studied its use in the United States for pharmaceutical inventions, especially in the 1960s and 1970s (Scherer, 2000). Activists for access to medicines point out that in September 2001, in the context of a biological anthrax threat, the US Secretary of State for Health threatened Bayer with a compulsory license if the firm did not lower the price of its antibiotic.

In the past the United Kingdom, the United States, Canada, and Germany applied compulsory licenses to undo monopolies that seemed detrimental to public health interests, or for the purpose of systematically producing generic medicines, as in Canada in the 1970s and 1980s. The difference today, of particular interest to us here, is the fact that it is now countries of the global South that have decided on a series of compulsory licenses to import or locally produce medicines made accessible to their populations. Since the early 2000s there has been a real movement in this respect in Southeast Asia (Indonesia, Malaysia, Thailand), Africa (Zimbabwe, Mozambique, Ghana), and Latin America (Brazil, Ecuador). This geopolitics of compulsory licenses seems to be a direct response to the globalization of pharmaceutical patents in countries where medicines were excluded from patenting before the WTO agreements. It is also a movement related to the emergence of new demands in terms of rights to treatment or simply to life, especially in the context of the AIDS epidemic. The vast majority of compulsory licenses concern the importation or local production of ARVs. These compulsory licenses are based on the existence of a pharmaceutical industry of generic medicines that has developed over the past three decades in India (Lanjouwe 1997; Sahu 1998; Scherer and Watal 2001), Brazil (Cohen 2001; Cassier and Correa 2003), and Thailand, where they contribute in turn to developing a South–South generics market. Brazil initially imported Efavirenz from India, under a compulsory license, before producing it locally.

In Brazil the stakes involved in a compulsory license decision were high. First, the policy of free and universal access to HIV/AIDS treatment, including the most recent tritherapies, generated tensions around the new patent law that came into force in 1997. Whereas the local production of a generic version of the first ARVs had limited the Health Ministry's expenditures, by the mid-2000s the purchase of new patented ARVs from proprietary laboratories was absorbing four-fifths of its budget. Second, this extension of the patented pharmacopoeia reduced the possibilities for copying and producing generic medicines to virtually nil. The technological learning dynamic of Brazilian laboratories for duplicating such sophisticated medicines as ARVs was thus being ruined by increasingly broad property rights. Consequently, the use of compulsory licenses seemed to be the most viable solution both to maintain the policy of universal access to HIV/AIDS treatment and to reopen the space for local laboratories to carry on copying. The compulsory license decision therefore seemed to be justified and legitimate on two counts: the Brazilian government's health policy regarding the AIDS epidemic, which was held up as a model of public health management (Teixeira *et al.* 2003) and the development of local pharmaceutical production to supply the domestic demand.

The first struggles for a compulsory license in Brazil erupted in the summer of 2001, in a case involving Roche, and recurred in 2003 and 2005. The Ministry of Health used them as a threat in its negotiations with the proprietary laboratories, to obtain price cuts on patented ARVs. While the clashes of 2001 and 2003 were saluted as victories when the multinational firms agreed to substantially lower their prices, the battle with Abott in 2005, over Kaletra, was seen as a defeat. In this case the Ministry of Health had opted for a commercial agreement rather than settling the matter in court. The battle was waged by 50 national and international NGOs, along with the Director of the HIV/AIDS program, Pedro Chequer, who had campaigned for a compulsory license: "Chequer used his speaking engagements at the conference to emphasize his opinion that the Brazilian government should move forward with compulsory licensing to ensure the best care for its citizens" (Iavi Report, July 9, 2005). The NGOs criticized the "paper tiger." Those Brazilian generics laboratories that had embarked on the reverse engineering of patented molecules believed that the commercial pressure of the United States was too great and that the threat of a compulsory license was nothing but "bravado" (Director of R&D at Cristalia, 2004).[17] Yet in May 2007 the President of Brazil signed a decree on a compulsory license to import or locally produce Efavirenz in the public interest and for noncommercial use. How can we explain a decision for which there was such a strong demand, especially by the Brazilian Ministry of Health and NGOs, and at the same time such strong opposition, primarily from the authorities of the United States and the International Federation of Pharmaceutical Industries and Associations?[18]

A decision of this nature assumed that the public health norm would take precedence over the industrial property norm. That was in any case how the Brazilian president and the heads of the Health Ministry's AIDS program justified the decree on a compulsory license. It also corresponded to the demands of the campaign launched by national and international NGOs. At the beginning of 2003, MSF Brazil acted as a precursor and federator when it urged the Brazilian government to take out a compulsory license. In April 2003, Michel Lostrowska, MSF Brazil's head of the campaign for access to treatment, organized an international forum in Rio de Janeiro to prepare this decision. The meeting was attended by representatives of NGOs such as Oxfam and Consumer Project on Technology,[19] the FarManguinhos Institute (that was producing ARVs for the AIDS program), the Oswaldo Cruz Foundation, the INPI, the Chemicals Industry Union of Brazil, economists from the Federal University of Rio, the heads of Brazil's AIDS program, and lawyers specialized in IP.[20] This forum was emblematic of Brazil's new pharmaceutical governance involving generics laboratories, scientists, NGOs, and jurists. It encompassed several fields of expertise, notably law, science, technology, and public health, to produce an argument in favor of compulsory licenses. In July 2005, 50 national and international NGOs published a joint statement in favor of a compulsory license on Abott's Kaletra: "We, the undersigned Brazilian and international Civil Society Organizations and Networks, urge the Brazilian government immediately to authorize a compulsory license on Lopinavir/Ritonavir and begin local generic production of this important AIDS medicine without delay. This step would be historic, not only for Brazil but for the entire developing world" (Joint Civil Society Statement on Brazilian Compulsory Licensing Dispute, July 2005). The campaign for a compulsory license provided a framework for the integration of Brazilian NGOs, spread across the country, and interaction with international NGOs based both in the global North (e.g., Aides and Act Up Paris and Essential Action in New York) and in the global South (among others, Zimbabwe AIDS Network and Coordinadora Peruana). In August 2005 the National Health Council of Brazil recommended a compulsory license on three ARVs: Kaletra, Efavirenz, and Tenofovir. It justified this in terms of the policy of universal access to ARVs, which was in keeping with the Brazilian constitution; the beneficial impact of this free distribution of medicines on the health of infected persons; the conformity of a compulsory license with Brazil's patent law; and the WTO agreements on IP. The National Health Council recommended the local production of these medicines, an increase of research funding to public pharmaceutical laboratories, and a national debate initiated by the Ministry of Health with a view to amending Brazil's patent law. The idea of a compulsory license became an overriding concern at the Ministry, which surrounded itself with legal expertise on patents.[21] The fate of the compulsory license was playing out at the same time on the legal scene, in the trial between Merck and the FarManguinhos Institute.

As the public health norm started to prevail in the public debate among health experts both at home (National Council of Health) and abroad (the WHO Drug Action Program Essential Drugs, headed by German Velasquez), and among experts on patent law mobilized by Brazil's AIDS program, both nationally (FarManguinhos IP unit) and internationally,[22] the battle was being fought on the technological front in the R&D laboratories of Brazilian generics producers. The compulsory license was not only the product of legal and political work in the public health domain; it was also the fruit of investments in the research of public and private laboratories, to acquire the corresponding technologies and "to be ready" to produce locally. Researchers at the FarManguinhos federal laboratory were asked by the Ministry of Health to prepare the reverse engineering of medicines likely to be placed under a compulsory license. In 2003 the FarManguinhos chemists went to India and China to visit generics laboratories, acquire information on the relevant molecules, and negotiate the acquisition of raw material.[23] Private-sector laboratories were likewise encouraged to develop synthesis technologies with a view to producing active principles in Brazil. The acquisition of know-how served as a credible threat for the government to wield: "Once the national producers have developed reverse engineering, they can talk to multinationals from a strong position, 'if you don't supply us, we have alternatives. We have the knowledge, the know-how'" (Director of R&D at Cristalia, April 2004). In 2004 the private-sector firms complained about the government's backtracking and its inability to recoup its research investments: "The problem with the government, this is the second time that they've said they want to quash the patents. And what happened? They ask us to develop them. I told the government at the beginning of this process: we're going to spend a lot of money to develop the synthesis. We've bought the raw materials, done the research and nothing's happened" (Director of Labogen, Campinas). Having know-how was critical in the process of deciding on a compulsory license.[24] As the government was unsure about the technical capacities of Brazilian generics laboratories to produce patented ARVs industrially, in 2006 one of the main AIDS patient organizations, Associação Brasileira Interdisciplinar de AIDS (ABIA), a partner of MSF, financed an international study on the subject by two chemists.[25] The favorable conclusions of this published study were decisive in the May 2007 decision to finally opt for a compulsory license.

The compulsory license decision in a country as strategic as Brazil regarding medicines – considering its local production and the size of its market – was based on a huge effort to produce knowledge and to shift norms. Technology, public health, and law were closely interlinked in the arguments of the Ministry of Health, the NGOs, and the generics laboratories. Campaigns for a compulsory license in the years from 2001 to 2007 provided the framework for new coalitions between NGOs and the AIDS program, and between the Ministry of Health and private firms. The May 2007 decision in favor of a compulsory license spawned the emergence today of a new organization of industry, with a pharmaceutical consortium consisting of two public and three private laboratories, formed to produce Efavirenz locally. Thus, battles for the compulsory license contributed to reconfiguring Brazil's health policy and its pharmaceutical economy.

Opposition to patents on Tenofovir in Brazil and India: The engagement of NGOs and generics laboratories

Legal opposition to patents enables third parties – citizens, NGOs, ministries, firms – to petition the patent office for the refusal or cancellation of a patent. It is a regulatory procedure embedded in patent law of many countries, as in the European patent law, and which takes place within the industrial property institutions. In Brazil the country's patent law author-izes a pre-grant opposition procedure, while a patent application is being examined, whereas in India both pre-grant and post-grant opposition is recognized. Generally, opposition procedures are used extensively by firms in an attempt to limit their rivals' patent claims. In the field of health, the fact that the procedure is open to third parties enables a far wider range of actors to challenge a patent.[26] In Europe, civil society organizations, patient organizations, scientific societies, medical institutions, political parties, and Ministries of Health all challenge patents. The same applies to Brazil and India, where the number of oppositions to medicine patents has multiplied since 2006. These procedures afford a framework for the structuring of civil society in the field of health and play an important role in regulating IP. Several major patents have been refused in this way in Brazil and India since 2006, thus contributing to expanding the public domain and the space in which generic medicines can be produced. The first successful opposition, which has served as jurisprudence for other oppositions in India, concerned an extremely expensive cancer medicine Glivec, owned by Novartis. Much like the campaigns for compulsory licenses, the oppositions mobilized patient organizations demanding access to treatment, as well as generics laboratories defending the opening of their market in the face of multi-nationals' patents. In this section I analyze the oppositions filed both in India and in Brazil in May 2006 and June 2008 against the US firm Gilead's patents on Tenofovir, an antiretroviral that is one of the WHO's recom-mended medicines and is distributed to over 30,000 patients in Brazil. These oppositions were aimed at: (i) safeguarding the local production of generics in India, where Cipla had been commercializing a Tenofovir generic since 2005; (ii) authorizing the establishment of a local production of generic Tenofovoir in Brazil; and (iii) allowing for the acquisition of Indian generics by the Brazilian AIDS program, in which case Gilead's patent would have to be refused in both countries.

In 2006 the FarManguinhos Institute, which produced 40% of Brazilian ARV generics, filed two acts of opposition to two patents on ARVs: Abbott's Kaletra and Gilead's Tenofovir. These were the first oppositions to medi-cine patents filed in Brazil. The argument for the acts of opposition was drawn up by Wanise Barroso, the FarManguinhos federal laboratory's IP expert. Barrosso is a chemist who worked for 20 years as a patent examiner at Brazil's INPI, before being recruited in the early 2000s to form a tech-nology watch unit. In her Ph.D. she focused on the creation of a system of integrated databases for patents, registered medicines, and scientific publica-tions. Wanise Barroso and the FarManguinhos Institute were therefore fully equipped to study patent applications, either for the purpose of providing references for the chemists doing reverse engineering or to oppose patent applications at the INPI. Barroso describes herself as an expert in indus-trial property who works for public health.[27] She uses her knowledge to provide counter expertise to that of the proprietary laboratories and the INPI examiners in Brazil, whose job she is thoroughly familiar with. The opposition reports that she has drawn up discuss the patentability criteria applied by the Patent Office on the basis of highly technical arguments, to analyze the molecules to which the patent claim applies. The opposition report that she drew up against Gilead's patents on Tenofovir showed that the molecule to which the US firm was laying claim had been in the public domain for a long time, and that it could not satisfy the criterion of inventive activity. Her opposition was thus based on classical patentability criteria applied by the Patent Office.[28] At the same time, this opposition had the more far-reaching objective of defining stricter patentability criteria, so that patents could not be granted on minor alterations to known molecules. The approach was comparable to the one used by ANVISA's IP unit, that is, to defend the public domain by refusing futile patents or the phenomenon of ever-greening. By filing this act of opposition, the federal laboratory was fulfilling its role as a watchdog with regard to medicine patents, and thus protecting Brazil's space for free copying and generic medicine production. Wanise Barroso thinks that the actors of health in Brazil – laboratories, NGOs, and the Health Ministry – should use the opposition procedure as a weapon to defend the local production of generics and accessibility to treatment. She has set up a technology information system to monitor the filing of medicine patents, which can be used to support the policy of free distribution of treatment for carriers of HIV/AIDS.[29] Wanise Barroso's technical argument circulated among Brazilian NGOs working with AIDS and even reached Indian NGOs via MSF, which acted as an intermediary.[30] This circulation between FarManguinhos and MSF was the fruit of multiple exchanges of information and even research partnerships between the two.[31]

The second opposition was filed against Gilead's patent, by a consortium of six NGOs active in the field of AIDS, IP, and the defense of the right to life.[32] It was emblematic of the expansion of the scope of patient organizations' intervention in the field of IP, and of arguments that closely interlinked the interests of public health with technical considerations on the patentability of medicines. The NGOs' opposition report was drawn up by lawyers employed by them, who combined the technical arguments provided by FarManguinhos with a legal argument on the Brazilian constitution, the country's public health laws, and laws on the participation of third parties in administrative processes. These lawyers and the IP experts at FarManguinhos pooled their complementary expertise: "FarManguinhos wrote the technical argument, ABIA wrote the legal one" (Wanise Barroso, November 2008).

Initially the legal argument put forward by the patient organizations aimed to justify their participation in the process of patent examination, on the grounds of patent law, which afforded the possibility of intervention by the "interested parties," and of the federal constitution of 1988, which guaranteed the right to defense for "interested third parties." The NGOs' lawyers showed that the HIV/AIDS patient organizations that were "highly active in the field of access to medicines" were clearly "interested third parties" as regards the patent in question, on a medicine used to treat AIDS. This opposition was written in highly sophisticated legal terms, so that the organizations' legal expertise would be taken seriously and they would be included in the administrative patent-filing process. It was important for the opponents of the patent to stick to both the terms and the spirit of the law. Once the patient organizations had shown that they did indeed fit the legal category of "interested third parties" and could therefore participate in the patent examination process, they endeavored to show the contradiction between the policy of universal access to treatment for AIDS and the granting of a patent on a medicine like Tenofovir that was used so extensively in tritherapies. The argument highlighted Tenofovir's status as an "essential medicine," to challenge the patent's monopoly. The "essential medicine" category encompasses both pharmacological criteria of sanitary security and therapeutic efficiency and criteria of accessibility of the medicines selected, that is, cost and public health criteria. As Tenofovir had been added to the list of essential medicines in Brazil (the RENAME), its availability had to be guaranteed by the state, in keeping with the Constitution of 1998 in which the universal right to health was enshrined. The opposition then put forward technical arguments from patent law to challenge the patent on Tenofovir, primarily the absence of novelty and inventive activity.

In fact, opposition to the patent was based on several arguments: the democratic reform of the state, the right for third parties to oppose an administrative process; the right to health, guaranteed by the state; therapeutic utility; the cost and accessibility of medicines; and the criteria for defining a valid invention. The NGOs stressed that medicine patents had to come to terms with the public and general interest. Considering the absence of novelty, and the fact that it ran counter to the public interest, the Tenofovir patent could not be approved. This was basically the conclusion of the Brazilian NGOs' act of opposition.

Oppositions to the Tenofovir patent were simultaneously filed in India, in May 2006, by a coalition of two HIV/AIDS patient organizations and the generics producer Cipla, which had started to produce Tenofovir in 2005. The NGOs emphasized patients' rights against patent rights: "For many of us living with HIV/AIDS, newer medicines like Tenofovir offer new hope of continuing treatment. With patents interfering with our lives, we have no choice but to oppose them" (the Delhi Network of Positive People, Third World Network, May 23, 2006). They sought the advice of an association of lawyers, the Alternative Law Forum, to draw up their acts of opposition (two oppositions were filed against two pharmaceutical forms of Tenofovir). The Alternative Law Forum was defined as follows: "ALF was started in March,

2000, by a collective of lawyers with the belief that there was a need for an alternative practice of law. We recognize that a practice of law is inherently political. We are committed to a practice of law which will respond to issues of social and economic injustice." It focused on open source licenses and "the commons." From this point of view, these lawyers' position was fairly close to that of the lawyers of patient organizations in Brazil, except that the former had founded an independent organization. The opposition reports that they drew up were highly technical and precise (25–30 pages), and followed the same order of examination of patentability criteria as a patent examiner would do.[33] The preamble of the oppositions justified the patient organizations' action in terms of both the impact that the patents would have on the accessibility of treatment for patients and gravity of the HIV/AIDS epidemic: "This reality creates a difficult situation between the patent system and the matter of life and death." The argument was similar to the one put forward by the Brazilian oppositions: in this context of tension between patents and public health, it was important to grant patents for real innovations only and not for minor alterations to known substances. Here the lawyers could mobilize a highly controversial article in Indian patent law, Article 3d which stipulates that "a mere discovery of a new form of a known substance which does not result in the enhancement of the known efficacy of that substance" is not a patentable invention. They then showed that, even though it presented a gain in bioavailability compared to a preceding form, the new pharmaceutical form concerned by the patent application did not allow for better therapeutic efficacy. The Indian lawyers were able to refer to a recent case concerning the opposition filed by a patient organization against Novartis' patent on Glivec, which the Patent Office had subsequently refused (January 25, 2006). Those opposed to patents on Tenofovir pointed out that this case afforded the Patent Office with a new opportunity to make the law and to set the patentability standards of medicines: "The opponents contend that this patent office has the ability to set the standard of patentability so as not grant to such obvious patenting for the benefit not only of public health but also genuine inventions" (Opposition Report I-MAK: 14). Oppositions made the law and in so doing were able to organize the medicine market: if Indian patents on Tenofovir were refused, the copying and generics markets would be free and the price of medicines would be reduced considerably (the price of the generic Tenofovir produced by Cipla was ten times lower than that of the patented medicine). In May 2006, lawyers and patient organizations organized a demonstration in front of the Indian houses of parliament as they filed their opposition. Chanting "We want Tenofovir!" and wearing T-shirts blazoned with the words "HIV positive," the New Delhi protesters drew stares from passers-by. "It's a matter of life and death," said Loon Gangte, president of the Delhi Network of Positive People. "At any moment I'll be developing resistance to my existing treatment and will be needing the next line of treatment in the form of Tenofovir" (Indians march on parliament over AIDS drug patent, by Andrew Jack and Jo Johnson in New Delhi, May 10, 2006).

The Indian oppositions directly concerned the global generic medicine market as Indian laboratories were supplying most ARVs consumed in the world. Cipla alone produced 40% of the ARVs used. This globalization of the generics market led to a globalization of solidarity and struggles between opponents. MSF South Africa, for instance, supported the Indian opponents: "We have all been waiting impatiently to get Tenofovir as a generic from India. It's clear that the world desperately needs more sources of Tenofovir. If Gilead is granted the patent in India, our patients will face a potentially deadly delay" (Eric Goemaere, MSF in South Africa, New York Times, May 10, 2006). In June 2008 this globalization of struggles against medicine patents resulted in the Brazilian HIV/AIDS patient organization and the Indian organization SAHARA joining forces to file a new opposition at the Indian Patent Office, against a patent on Tenofovir: "However, this is the first time that a foreign patient group has countered an application in India. The case against Gilead's patent application will be filed formally by ABIA and the Centre for Residential Care & Rehabilitation, an Indian NGO. The patent application has also previously been opposed by groups such as the Indian Network for People Living with HIV/AIDS" ("Brazil & India are increasingly challenging prices by opposing patents," *Healthcare Briefing and Forecasts*, July 2, 2008). Opposition by a Brazilian organization in India was justified by the concern to safeguard the possibility for Brazil's AIDS program to import an Indian generic of Tenofovir. Whereas the Brazilian Patent Office had published a negative examination report on the Gilead patent in April 2008, the examination of the Indian patents was still under way at the time: "Though we are confident that patent will not be granted for Tenofovir in Brazil, we must ensure that the option of importing affordable generic versions from India remains open to our AIDS programme," commented Veriano Terto, ABIA's general coordinator. "This will contribute to the sustainability of our national AIDS programme's universal access policy, on which 180,000 Brazilians depend for their lives" (*Business Standard*, New Delhi, June 27, 2008). This globalization of oppositions bears witness to the existence of a South–South generics market, particularly between India and Brazil, and a solidarity between NGOs to preserve the existence of this market: "We want more options to promote competition in the market and bring down medicine prices," Gabriela Chavez, a pharmacist with ABIA, told *The Hindu* over the phone from Brazil. "If the patent is granted in Brazil but not in India, Brazil has the option to apply for a compulsory licence to buy the medicine at lower cost from Indian companies. If the patent is not granted in Brazil or India, Brazil has the option to import either the key ingredients or the finished medicines from Indian companies," she said (*The Hindu*, June 27, 2008). In so doing, the patient organizations that had recruited pharmaceutical and legal experts devised a sophisticated strategy to organize the generics market.

In the case of Tenofovir, the globalization of the struggles against medicine patents translated into multiple interactions and acts of solidarity between Brazil and India. First, oppositions were filed in parallel in both countries in 2006, and were defended by HIV/AIDS patient organizations and generics producers: the federal FarManguinhos laboratory in Brazil and Cipla in India. Second, Brazilian and Indian NGOs and lawyers traded information. For example, the opposition drawn up by Wanise Barrosso at FarManguinhos was communicated to an Indian NGO via MSF. In 2009 an Indian lawyer from the NGO I-MAK thanked his Brazilian counterpart at ABIA: "Thanks to Francisco Neves da Silva of ABIA for pointing out that the application number of the Tenofovir application refused by INPI was ... " (July 20, 2009). Third, in June 2008, a Brazilian and an Indian NGO filed a joint opposition in India. Fourth, in November 2008 the organization ABIA organized a seminar in Rio de Janeiro on the art and way of drawing up and filing oppositions, to which I-MAK, the NGO of alternative Indian lawyers, contributed. The head of IP at I-MAK titled his talk: "Making the patent system more democratic: the role of public participation." This seminar reviewed pharmaceutical patents and the flexibilities in the law that the NGOs could use to defend the interests of patents and public health. Participants included Eloan Pinheiro, who in 1997 had launched the public program to copy generic ARVs at the federal laboratory FarManguinhos, Carlos Correa, author of a report for the WHO on the flexibilities in pharmaceutical patents,[34] the lawyers of Brazilian and Indian NGOs, and Wanise Barroso from FarManguinhos. Fifth, in 2009 the Brazilian and Indian patent offices both refused Gilead's patents on Tenofovir, on the basis of the lack of novelty and inventive activity of the patent in Brazil, and Article 3d of the Indian law that proscribed the patenting of new pharmaceutical forms of a known substance without new therapeutic utility. These parallel decisions opened the space for copying in Brazil and India, and for the generics market between the two countries. In Brazil, the public and private pharmaceutical laboratories worked together to produce a Brazilian Tenofovir.

Public interest and medicine patents: Jurisprudence in the Gleevec case in India (2006–09)

In January 2006 the Indian Patent Office refused the patent application filed by Novartis for Glivec, a cancer medicine used to treat leukemia. This refusal was the outcome of an opposition filed by the Cancer Patient Aid Association and several generics firms, including Cipla. The opponents' demand for a refusal of the patent, based on Article 3d of the Indian patent law, was validated by the patent office on the grounds that Novartis' patent application was for a new pharmaceutical form of a known substance and that it was therefore not patentable in India.[35] This decision was the first to be taken under the new patent law of March 2005. Its implications were crucial for setting the standards of medicine patentability in a country as strategic as India in the generics economy. Novartis grasped the opportunity to kill two birds with one stone. In May 2006 it appealed against the decision on Glivec, and simultaneously challenged the constitutionality of Article 3d and its conformity with the WTO TRIPS agreements. The outcome of the conflict would have a direct impact on the production and availability of Glivec generics (several Indian laboratories supplied a generic at a price ten times lower than the patented medicine). It would also delimit the room to maneuver that states would have in applying the WTO agreements on IP rights on medicines.[36] In parallel with its lawsuit, Novartis highlighted the free distribution of Gleevec that it had set up in India and that concerned 11,000 patients[37] – to which the opponents replied that India recorded 25,000 cases of Chronic Myelloid Leukemia annually.

The legal tug-of-war took place against a background of international public controversy. Patient organizations and international NGOs like MSF and Oxfam engaged in campaigns for access to treatment emphasized the impact of this trial on the global economy of copying and generic production: "If Novartis wins the Glivec trial and manages to change Indian law, India will have to agree to patents that are as broad and numerous as in the rich countries. This means that Indian generics producers will no longer be able to produce as many generics of patented products as in the other countries for the 20 years of the patent's life, and there will be fewer or no essential medicines at low prices available to poor countries" (MSF, December 20, 2006). That was also the Indian patient organizations' position: "This affair is particularly important, as it is the first trial concerning patents. Novartis is busy challenging the legal validity of all the patents refused, not only that of Gleevec. If Novartis wins this case, the price of several medicines will shoot up, not only that of Gleevec," commented M. Park, the lawyer of a cancer organization.[38] The HIV/AIDS patient organizations joined forces with their cancer patient counterparts, and the Indian Communist Party, a member of the governing coalition at the time, published a communiqué that sounded the alarm on predictable ARV price increases.[39]

In August 2007, to the great satisfaction of generics producers and NGOs, the Chennai court ruled that Article 3d of the Indian patent law was not unconstitutional, and dismissed Novartis' case. The court's justifications for this decision shed light on what I see as two essential issues: the making of medicine patents by the patent office examiners and the public interest that lies in balancing ownership of medicine patents. As regards the first point, the examination of patents, Novartis claimed that Article 3d of the Indian patent law, on the evaluation of the novelty of an invention, was vague and arbitrary. In particular, it argued that Section 3d contained no guidelines to help examiners in deciding whether an invention was patentable. The court replied that the Patent Office examiners were fully competent to judge the novelty and gains in efficacy of the patented substance, on the basis of the patent documentation, and that the applicant had complete latitude to demonstrate the gains in therapeutic efficacy with the new substance. The Chennai court highlighted the concrete interpretative and assessment work of the patent examiners. Concerning the second point, it based the constitutionality of Article 3d of the patent law on the public interest. The judges considered that the objective of this article was to prevent the ever greening of medicine patents and, in so doing, to fulfill the constitutional obligation to provide the country's citizens with good health care: "We have borne in mind the object which the amending Act wanted to achieve, namely, to prevent evergreening; to provide easy access to the citizens of the country to life saving medicines and to discharge its constitutional obligation of providing good health care to its citizens."[40] The public health interest can thus legitimately be mobilized by a state to make and amend medicine patent laws.

In August 2007 the Chennai High Court confirmed the legitimacy of the Indian patent law, and in June 2009 the Intellectual Property Appellate Board (IPAB) upheld the decision to refuse Novartis' patent on Glivec: "The IPAB held that Novartis was not entitled to a patent on imatinib mesylate as its claimed product did not meet the requirement of increased therapeutic efficacy" (Lawyers Collective, August 28, New Delhi). The IPAB extended the reasons for its refusal to the excessively high price of Glivec commercialized by Novartis, which it said was disruptive to public order: "Thus, we also observe that a grant of product patent on this application can create havoc to the lives of poor people and their families affected with the cancer for which this medicine is effective. This will have a disastrous effect on society as well. Considering all the circumstances of the appeals before us, we observe that the Appellant's alleged invention won't be worthy of a reward of any product patent on the basis of its impugned application for not only for not satisfying the requirement of Section 3(d) of the Act, but also for its possible disastrous consequences on such grant as stated above, which also is being attracted by the provisions of Section 3(b) of the Act which prohibits grant of patent on inventions, exploitation of which could create public disorder among other things."[41] This new argument had never been heard before in a patent Appeal Court. The European Patent Office, for example, had always refused to take into consideration the impact of patents on the price and accessibility of treatments: "It is not the EPO's duty to take into consideration the economic effects of issuing patents in certain specific areas."[42] Certain opponents naturally deplored this stance of the European Patent Office (EPO). In August 2009 Novartis decided to lodge an appeal with the Indian Supreme Court against this new refusal.

In the meantime, the Indian patent office had refused a second Novartis patent on another form of Glivec – an alfa rather than a beta crystal – which had also been opposed by three Indian firms. This new decision, in April 2009, was taken on the same grounds as the first one: "If granted, it would have been a clear case of frivolous patenting. This different form of Glivec is in no way superior to the other form for which a patent has not been granted," said an attorney involved in the opposition (essentialdrugs.org, June 3, 2009). The cancer patient organizations and generics laboratories played an essential part in producing this jurisprudence.

New players in the intellectual property field and practical alternatives to patent rights

The growing conflict around medicine patents in Brazil and India in the 2000s was characterized by the opening of the circle of players to organizations in civil society, Ministries of Health, and generics laboratories. A process of democratization of IP rights was thus witnessed, through campaigns for compulsory licenses and the oppositions that were launched from 2005 by patient organizations. We also observed a phenomenon of legal acculturation of these civil society organizations, which incorporated legal experts into their teams or obtained assistance from alternative legal organizations. Alternative practices in patent law thus emerged, to defend the interests of patients and of public health, based on a conception of IP that emphasizes its social function.

Until 2001, HIV/AIDS patient organizations in Brazil focused their action on assisting patients and on prevention. They moved into the IP field during the two battles of 2001: first, the Pretoria trial at the beginning of the year, which was marked by an intense globalization of conflicts over medicine patents; and, second, the complaint filed by the United States at the WTO against an article in Brazilian IP law, on compulsory licenses (Shanker 2001; Varella 2002). In December 2002 a coordinator of one of the largest Brazilian HIV/AIDS patient organizations described this shift: "The first meeting was organized at the same time as the alliance of pharmaceutical industries that were busy suing the South African government. We demonstrated outside the US consulate in Rio and Sao Paulo, outside the embassy in Brasilia, and outside the consulate in Recife. There was a meeting with all the organizations working on AIDS in Recife, last year. This meeting was a way of mobilizing the groups to say that we needed a permanent and regular space for discussions on the subject of intellectual property ... In 2002 we organized three meetings between the organizations working in different areas, to discuss intellectual property" (Carlos Pasarelli, ABIA). The Brazilian Ministry of Health encouraged the AIDS organizations to intervene with regard to IP and access to treatment: "The government complained: it considered that NGOs in Brazil don't work on this subject of intellectual property" (Carlos Pasarelli). In 2001 the Ministry of Health engaged in a struggle with the multinationals over the price of

ARVs, and threatened them with compulsory licenses. It needed the support of civil society. Moreover, Minister of Health José Serra had decided to run for president in the upcoming elections, and part of his campaign was on the policy of access to generic medicines and on clashes over compulsory licenses with the multinationals and with the US government. The Minister of Health thus participated in the organization of civil society. MSF Brazil also contributed strongly to the acculturation of AIDS NGOs on medicine patents: "I've been fighting for a year now because the Brazilian nonprofit organizations aren't used to doing so ... for three years a colleague said to me: Michel, you're boring me with your stories of patents. We're pushing the government to give us medicines, it's up to the government to meet its responsibilities, if there's a need to 'quash patents,' as we say, then it's up to it to do so" (M. Lostrowska, December 2002). MSF endeavored to educate people with regard to compulsory licenses: "I can tell you that, in my budget next year, I've provided for an international seminar in Rio on the compulsory licence, and we're going to make a noise. I'm going to get the world's leading specialists to come over ... We're going to invite the Minister of Health, and the INPI, and we're going to say: here's the recipe for issuing a compulsory licence" (M. Lostrowska, December 2002). Acculturation in IP rights is remarkable in ABIA's trajectory since 2002. In 2006 the organization recruited a lawyer to take over the file, and he drew up the oppositions filed by ABIA in the same year. In November 2008 ABIA organized a course on how to go about opposing patents, with the participation of international experts (Carlos Correa) and an NGO of alternative Indian lawyers, I-MAK. The learning process under way was also evidenced in the organization's publications: until 2001 the journal published by ABIA contained no references to patents. In 2009 its Web site featured reports, publications, and the outcome of the oppositions to the medicine patents that it had engaged. That year the lawyers of the HIV/AIDS organizations published a synthesis article on civil society's action on IP (Chavez, Viera & Reis).[43]

We also find a phenomenon of symmetrical acculturation at the FarManguinhos federal laboratory, which set up an IP and technology transfer unit in the late 1990s. It recruited two experienced patent examiners from the INPI and a young chemist who had training in conflict over compulsory licenses. The new IP unit prepared arguments on compulsory licenses, filed oppositions in 2006, and patented new molecules developed through local research. The Ministry of Health also developed expertise in IP, through the experts at FarManguinhos, reports from independent consultants (e.g., the report on the preparation of compulsory licenses drawn up in 2004 by a patent expert), interaction with MSF's campaign for access to essential medicines, and the seminars on IP law and economics that it had organized regularly since 2002. It was the Ministry that prepared the file for the Efavirenz compulsory license in 2007 (Possas, 2008). There is of course also a circulation of knowledge between the NGOs, FarManguinhos, and the Ministry of Health on the subject of patents, oppositions, compulsory licenses, and local production of generics. Two NGOs, MSF and ABIA, financed expertise by two university chemists on the Brazilian generics laboratories' technological capacities, which was subsequently used in the 2007 decision on a compulsory license for importing and locally producing Efavirenz. From 2000 the NGOs have had a working group on IP on a federal scale, in the REBRIP.

There is a network of lawyers in Brazil today, specialized in patents and based in AIDS NGOs, at MSF, in the FarManguinhos federal laboratory and in the IP unit at ANVISA. They are all working together toward a better balance between patents and the public interest. Above all, they endeavor to use flexibilities in the WTO agreements on IP, to support local production of generics, or to invent ways to support pharmaceutical innovation oriented toward neglected diseases.[44] In India, patient organizations rely on the support of groups of alternative lawyers such as the Lawyers Collective HIV/AIDS Unit, the Alternative Law Forum, and the organization I-MAK (Initiative for medicines access and knowledge), which drafted the oppositions to the Tenofovir patents. These lawyers use their expertise in IP to dissolve monopolies and work toward a policy of access to treatment. They describe themselves as lawyers who practice law in an alternative manner, to destroy monopolies rather than creating them, or as public interest lawyers (Tahir Amin, lawyer with the alternative Law Forum in Bangalore). I-MAK presents itself as follows: "The Initiative for Medicines, Access & Knowledge is a not-for-profit public service organisation consisting of lawyers and scientists working to protect the public domain against undeserved patents. I-MAK works to ensure that patents do not act as a barrier to research and restrict the public's access to affordable medicines." In November 2008 the I-MAK coordinator gave a lecture in Rio de Janeiro entitled: "Making the patent system more democratic: the role of public participation." It sought to define a method for filing oppositions (how to oppose a patent?): collect expert assessments from lawyers, chemists and pharmacists; choose the right medicine to oppose, from the list of essential medicines; identify the right patents and check with a chemist which patent is actually used in the industry; obtain examination reports from the European and US patent offices' databases; and, finally, involve patient groups.

Conclusion

The various legal trials and conflicts that we have examined in this chapter have been instrumental in producing a new regulation of IP with regard to medicines, in which countries of the global South and primarily Brazil and India are major players. The trials, oppositions, specific units for examining medicine patents, and campaigns for compulsory licenses have all aimed to exploit and extend the flexibilities in patent law in a way that facilitates the copying of generic medicines and the accessibility of treatments. These struggles through and over the law have intensified since the early 2000s, especially since March 2005, when India adopted patents on pharmaceutical products. This is clearly a case of law in-the-making: within the group of examiners at the Brazilian medicine agency; in the IP unit at the FarManguinhos federal laboratory where a patent expert works to invalidate the pharmaceutical patent holders' arguments; and in the legal organization Initiative for Medicine Access & Knowledge, which prepares oppositions. It is also a case, in particular, of patent law in-the-making during legal battles such as lawsuits, oppositions filed with patent offices, and the preparation of compulsory license decrees by Health Ministries. Social movements are mobilized, and when necessary they participate in these legal actions, filing oppositions with the patent office or as litigants in a trial. For instance, civil society organizations demonstrated outside the Indian parliament to support opposition against patents on Tenofovir, and participated in public debates on compulsory licenses. In 2005 they intervened alongside the Ministry of Health. Law in-the-making is not isolated from socioeconomic struggles. It extends, codifies, and organizes them. This may provide an answer to the question posed by Latour in *La Fabrique du Droit* (2002): "How can power relations be shifted in the law? Where are the vehicles? What are the channels?" The victorious oppositions of Brazilian and Indian NGOs against the patents of the US firm Gilead, over Tenofovir, redefined the relations of ownership of this ARV and the organization of its market. Brazilian and Indian laboratories were consequently authorized to freely copy and produce inexpensive generics. "Law informs economics," noted Michel Foucault (2004), and this has applied here via the struggles for the right to health, including the patient work of the ANVISA examiners and FarManguinhos patent experts who make patent law by mobilizing it and translating it into singular legal acts: examination reports, opposition reports, compulsory license decisions. The patent examiners' interpretative work has led to court cases when the international pharmaceutical laboratories have rebelled against the refusal of major patents, for example, ANVISA's refusal of Aventis' patent on Taxotere, and the Indian patent office's refusal of Novartis' patent on Glivec. Aventis sued ANVISA, while Novartis sued the Indian government. These clashes revealed the plurality of possible interpretations of patent law when Brazilian and Indian examiners raise the patentability standards to protect the public domain. Here the flexibilities in patent law stem from the interpretation of the rules of novelty and inventive activity, and in particular articles of national laws.

Government or citizen regulation has taken place during the process of examination and granting of property rights, via the oppositions filed by NGOs as "interested third parties," and via the prior consent system involving the Brazilian sanitary security agency in the medicine patenting process. In the latter case, the state, especially the Ministry of Health, altered the patenting procedure by granting the sanitary security agency the power to veto patents. The interests of public health were thus embedded in the IP decision-making process. Citizen regulation, on the other hand, appeared mainly through the opposition procedures filed with the patent offices, when these procedures were opened to the "interested parties." In pre-grant oppositions in Brazil and India, patient organizations or consumer unions moved in to participate in the patent examination process. These struggles through the law have involved a legal formatting of arguments by the specialized lawyers who assist the NGOs. The oppositions and public campaigns have structured the field of the forces of this new pharmaceutical economy: they have triggered the creation of consortiums of NGOs, and have often mobilized patient organizations, generics laboratories, and Health Ministries. This triad of actors – civil society, Health Ministries, generics laboratories – have shaped a new pharmaceutical economy involving trade between countries and laboratories in the global South. At the same time, there has been complementarity between civil society and the state during these different actions to promote a biopolicy of access to medicines, promoted by generics producers in both the public sector (FarManguinhos) and the private sector (Cipla).

These oppositions and regulatory measures have changed the nature of IP law, and in turn the regulation of drug patents, which now has to compromise with the public interest, patients' interests, and a universal right to health. The head of ANVISA's IP unit mobilizes the "social function of property" category enshrined in Brazil's constitution, to balance patent-owner's interests. Likewise, the oppositions filed by Brazilian and Indian patient organizations were based on the "essential medicines" category to limit the extension of patent rights. The notion of essential medicines encompasses both their therapeutic use value and a norm of availability for patients. The Rio de Janeiro Federal Court referred to the public interest of patients when it authorized reverse engineering on patented medicines. The assertion of the public interest in the face of IP rights was also the basis of the Chennai court's decision when Novartis challenged the constitutional nature of Indian patent law. Even the WTO incorporated the notion of the interests of public health into certain articles of the 1994 agreements on IP, particularly the Doha Declaration of November 2001, which was adopted under pressure from Brazil and India. Brazil's Ministry of Health and NGOs advocating access to medicines used these initial successes concerning Efavirenz, Tenofovir, Glivec, and Combivir, to explore more general ways in which accessibility of treatments could be guaranteed. Current reflection is oriented toward the collective management of property in patent pools managed by the United Nations, which would distribute nonexclusive licenses to manufacturers, and toward automatic compulsory license measures encompassing an entire therapeutic category to combat an epidemic.[45] Another option is to use new mechanisms for pharmaceutical innovation, especially in public/private partnerships formed deliberately outside of the patent sphere, like the Fixed-Dose Artesunate Combination Therapy (FACT) international consortium to fight malaria.

Notes

1. Cf. the recommendations of the US Academy of Science in a report dated 1997, to extend this hegemony: "America's vital interest in global health: protecting our people, enhancing our economy and advancing our international interest."
2. This was the slogan of the Treatment Access Campaign in 2001 during the Pretoria trial (Cassier 2002; Beigbeder 2004). It sums up the spirit of counter hegemony fostered by NGOs, activist laboratories such as Cipla in India and Farmanguinhos in Brazil, Health Ministries, and studies on the economy of access to HIV/AIDS medicines, for example the two collective volumes published in English by the Agence Nationale de Recherche sur le Sida (ANRS – the French national agency for AIDS research) in 2003 and 2008.
3. On the external pressure exerted on Brazil, see Flynn, 2009.
4. See the article by Andrea de Loyola, 2009.
5. "Regulação sanitária, propriedade intelectual e política industrial," Luis Carlos Lima, COOPI, Anvisa, May 19–21, 2008). See also the article by Brazilian jurist Maristela Basso, 2006, which shows that the "social function of property" is enshrined in the Brazilian constitution. Basso argues that this is the legal basis for the procedure of prior consent by ANVISA on pharmaceutical patents.
6. Cf. the data collected by E. Guimaraes, 2008, Ph.D. thesis at the UERJ.
7. Visit to the COOPI at ANVISA, Rio de Janeiro, March 2006.
8. Public hearing, Chamber of Deputies, November 2009.
9. "The Committee on Social Security and the Family approved on Wednesday (27), a bill prohibiting the granting of patents on therapeutic indications for pharmaceutical products, and on polymorphic substances (Chamber of Deputies, May 29, 2009).
10. The Labour Party, from which President Lula came.
11. Interview with the director of FarManguinhos, Nubia Boechat, April 2004.
12. Archives of the trial, p. 39.
13. In December 2004 the Brazilian laboratory Cristalia also demanded the cancellation of the call for tenders on the grounds that the Indian generics producer had not supplied all the technical specifications. Cristalia withdrew its complaint in early January 2005.
14. August 17, 2005, ruling, Rio de Janeiro Regional Federal Court.
15. October 6, 2005, ruling, Regional Federal Court of the 2nd Region.
16. Interview with the deputy director of FarManguinhos, Jorge Costa, in May 2009.
17. Interview held in April 2004 in Itapira.
18. See, for example, the file on the Brazilian compulsory license, put together by the CPTech.
19. James Love participated in this seminar in Rio de Janeiro.
20. With my colleague Marilena Correa, we were also invited to this forum as experts, financed by the French national agency for AIDS research (*Agence Française de Recherche sur le Sida* – ANRS).
21. For example, the report by a patent specialist at the INPI and the FarManguinhos Institute, to prepare a compulsory license in 2004.

22. The Joint Civil Society Statement on Brazilian Compulsory License Dispute of 2005 was signed by an Associate Professor of Law from West Virginia University, USA.
23. Interviews held at FarManguinhos in April 2004 with the heads of the mission in India and China.
24. Consulting report for the Ministry of Health, 2004.
25. Antunes O. & Fortunak JM, 2006, "ARV Production in Brazil: An Evaluation," report for the Brazilian Interdisciplinary AIDS Association (ABIA) and MSF Brazil, 8 pages.
26. Cassier M. & Stoppa Lyonnet D, "L'opposition contre les brevets de Myriad Genetics et leur révocation totale ou partielle en Europe" ("Opposition to Myriad Genetics patents and their total or partial revocation in Europe: early conclusions"), *Médecins/Sciences*, n° 6–7, vol 21, June–July 2005, p. 658–62. See also Harhoff D, 2000, "Determinants of opposition against EPO patent grants. The case of biotechnology and pharmaceuticals," *Collection les Cahiers de l'innovation*, CNRS, 27 p.
27. Wanise Barroso's talk at a Brazil–India seminar in November 2008 in Rio was titled "Opposition to pharmaceutical patents: arguments in favour of public health."
28. Opposition report addressed to the Director of the INPI, Brazil, on December 6, 2005.
29. "Relatorio sobre o medicamento Tenofovir"," Wanise Barroso, 2006.
30. Information supplied by Wanise Barroso on June 7, 2006. The author of this article worked with Wanise Barroso on a collaborative research project between the French CNRS and the Brazilian Oswaldo Cruz Foundation, from 2005 to 2007, to study the implementation of these oppositions. The social sciences thus contributed to structuring of the intellectual property field.
31. Interviews with Michel Lostrowska of MSF Brazil, and with Eloan Pinheiro, Director of FarManguinhos, from 2002 to 2004.
32. The six organizations that opposed this patent were large HIV/AIDS patient organizations in Rio, Sao Paulo, and in the state of Rio Grande do Sul (ABIA, Conectas Direitos Humanos, Gurpo Pela Vidda, Gapa, Gapa Rio Grande do Sud, Gestos). Opposition report from ABIA, 2006.
33. The opposition reports are available on the Web site of the legal NGO I-MAK, which has taken over these two files since then.
34. Correa Carlos, 2007, "Guidelines for the examination of pharmaceutical patents: developing a public health perspective," ICTSD, WHO, UNCTAD.
35. Decision of January 25, 2006, V. Rengasamy, Asst. Controller of Patents & Designs.
36. This aspect of the process is similar to that in the Pretoria case concerning the legal validity of two articles of the South African medicine law.
37. On this subject, see the article by Stefan Ecks, 2008, "Global pharmaceutical markets and corporate citizenship: The case of Novartis' anti-cancer medicine Glivec," *Biosocieties* 3(2008), 165–81.
38. "La production indienne de médicaments génériques en danger," Essentialmedicines.access, March 16, 2007.
39. Essentialmedicines.access, op cit.
40. In the High Court of Judicature at Madras, August 6, 2007, Novartis vs Union of India, The Controller General of Patents&designs, Natco Pharma, Cipla, Hetro Medicines, Cancer Aid Association, Ranbaxy, Indian Pharmaceutical Alliance, Sun Pharmaceutical Industries.

41. IPAB, ORDER (No.100/2009).
42. EPO communiqué of January 17 2005, on oppositions against Myriad Genetics' European patents on genes and on genetic tests for breast cancer predisposition.
43. "Access to medicines and intellectual property in Brazil: Reflections and strategies of civil society", *Sur- Revista International de Direito Humanos*, vol. 5, n° 8, Sao Paulo, June 2008. See also "IPR and Access to ARV Medicines. Civil Society Resistance in the Global South," ABIA 2009.
44. From 2002 to 2006 MSF and FarManguinhos participated in the FACT consortium to invent new combinations of molecules to fight malaria. Cf. interview with M. Lostrowska, MSF Rio de Janeiro, December 2002.
45. Cf. the talk delivered by Eloan Pinheiro, former Director of FarManguinhos, at the conference on access to treatment, organized by the Ministry of Health, in May 2009 in Rio de Janeiro.

Bibliography

ABIA (2009) "IPR and access to ARV medicines: Civil society resistance in the Global South," report, 135 pages.

Barroso W (2009) *"Le cas du Tenofovir,"* communication au 5ème Congrès VIH/sida de Casablanca, Mars 2010.

Basso M (2006) "Intervention of health authorities in patent examination: The Brazilian practice of the prior consent," *International Journal of Intellectual Property Management*, 1, 1/2, 54–74.

Beigbeder Yves (2004), *International Public Health: Patients' Rights Vs. Patent Rights*, Ashgate.

Biehl J (2009) "Accès au traitement du sida, marchés des médicaments et citoyenneté dans le Brésil d'aujourd'hui, Sciences Sociales et Santé", *Sciences Sociales et santé*, 3, 27, 13–46.

Cassier M (2000) "Genome patents nowadays and pharmaceutical patents in the 19th: A parallel," International Conference, "Technological Policy and Innovation: Economic and historical perspectives," Paris, November 20–22, 22 pages.

Cassier M (2002) "Propriété intellectuelle et santé publique", Revue Projet, Mai, p. 47–55.

Cassier M, Correa M (2003) "Patents, innovation and public health: Brazilian public-sector laboratories' experience in copying aids medicines," in *Economics of AIDS and Access to HIV/AIDS Care in Developing Countries. Issues and Challenge*, Ed. ANRS, pp. 89–107.

Cassier M, Stoppa Lyonnet D (2005) "L'opposition contre les brevets de Myriad Genetics et leur révocation totale ou partielle en Europe" [Opposition to Myriad Genetics patents and their total or partial revocation in Europe: Early conclusions], *Médecine/Sciences*, 6–7, 21, June–July, 658–62.

Chaves G, Viera M, Reis R (2008) "Access to medicines and intellectual property in Brazil: Reflections and strategies of civil society", *Sur Revista Internatcional de Direitos Humanos*, 5, 8, June.

Cohen J C (2000) "Public choices in the pharmaceutical sector: a case study of brazil". LCHSD Paper Series n° 54, Human development Department, The World Bank.

Commission on Intellectual Property Rights – CIPR (2002) *Integrating Property Rights and Development Policy*, 178 pages.

Correa Carlos (2007) *Guidelines for the examination of pharmaceutical patents: Developing a public health perspective*. ICTSD, WHO, UNCTAD, 2007.

D ' Almeida C, et al. (2008) "New antiretroviral treatments and post-2005 TRIPS constraints: First moves towards IP flexibilization in developing countries" in *Political Economy of HIV/AIDS in Developing Countries*, ed. B. Coriat, Edward Elgar, London, 25–51.

Ecks Stefan (2008) "Global pharmaceutical markets and corporate citizenship: The case of Novartis' anti-cancer medicine Glivec," *Biosocieties* 3, 165–81.

Foucault M (2004) *Naissance de la biopolitique*, Cours au collège de France (1978–1979), Seuil.

Flynn M, 2007, "Brazil's use of compulsory licenses for AIDS medicines," *American Sociological Association*, August 1–4, 1–20.

Flynn M, De Oliviera E A (2009) "Regulatory capitalism in emerging markets: An institutional analysis of Brazil's Health Surveillance Agency (ANVISA)," *American Sociological Association*, August 8–11, San Francisco, 1–20.

Galvao J. (2002) "Access to antiretroviral medicines in Brazil," *Lancet*, 360, 9348, 1862–65.

Guimaraes E (2008) "Le droit à la santé et la propriété intellectuelle des médicaments au Brésil: l'examen préalable à l'Agence de Surveillance Sanitaire" mémoire de Mestrado, Institut de Médecine Sociale, UERJ.

Harhoff D (2000) "Determinants of opposition against EPO patent grants: The case of biotechnology and pharmaceuticals", collection *Les Cahiers de l'innovation*, CNRS, 27 pp.

Lanjouwe J.O. (1997) "The introduction of pharmaceutical products patents in India: Heartless exploitation of the poor and suffering?", Yale University, 54 pages.

Latour B (2002) *La Fabrique du droit*, La Découverte.

Loyola MA. (2009) "Sida, santé publique et politique du médicament au Brésil : autonomie ou dependence," *Sciences Sociales et santé*, 3, 27, pp. 47–75.

National Academy of Science (1997) "America's vital interest in global health: Protecting our people, enhancing our economy and advancing our international interest", Board on International Health, Institute of Medicine, National Academy Press, Washington, DC.

May C (2000) *a Global Political Economy of Intellectual Property Rights: The New Enclosures?*, Routledge.

Possas C (2008) "Compulsory licensing in the real world: The case of arv medicines in Brazil," in *The Political Economy of HIV/AIDS in Developing Countries*, ed. B. Coriat, Edward Elgar, London, 150–66.

Sahu S (1998) "Technology transfer, dependence, and self-reliant development in the third world," Praeger, 250 p.

Shanker D. (2001) "Brazil, Pharmaceutical Industry and WTO," University of Wollongong, New South Wales, Australia.

Scherer F (2000) "Le système des brevets et l'innovation dans le domaine pharmaceutique," *Revue Internationale de Droit Economique*, n°1.

Scherer F and Watal J (2001) "Post trips options for access to patented medicines in developing countries," Commission on Macroeconomics and Health, WHO, 2001, 77 pages.

Shelden K (2009) "The political contradictions of incremental innovation in late development: Lessons from pharmaceutical patent examination in Brazil," 1–29.

Tahir Amin (2008) "Making the patent system more democratic: The role or public participation," *Examination of Pharmaceutical Patents: Arguing from a Pro-Public Health Perspective*, Associacao Brasileira Interdisciplinar de AIDS – ABIA Rio de Janeiro, November 18.

Teixeira P, Vitoria MC, Barcarolo (2003) "The Brazilian experience in providing universal access to antiretroviral therapy," in Moatti J.-P., Coriat B., Souteyrand Y., Barnett T., Dumoulin J., Flori Y.-A., eds, *Economics of AIDS and access to HIV/AIDS: Issues and Challenge*, ed. ANRS.

Varella M (2002) La propriété intellectuelle des produits pharmaceutiques au Brésil, 66 pages.

Velasquez G, Correa C, Thirukumaran Balasubramaniam (2004) "who in the frontlines of the access to medicines battle: The debate on intellectual property rights and public health," Intellectual Property in the context of the WTO TRIPS Agreement, ed. J. Bermudez, MA Oliviera, Fiocruz, 83–97.

Villela Pedro (2009) "Activistes du VIH/SIDA et l'économie politique du médicament," Master's thesis, EHESS, Paris.

Index